A HANDBOOK OF COMPUTATIONAL CHEMISTRY

A HANDBOOK OF COMPUTATIONAL CHEMISTRY

A Practical Guide to Chemical Structure and Energy Calculations

---

TIM CLARK

Friedrich–Alexander–Universität Erlangen–Nürnberg
Erlangen, West Germany

A Wiley-Interscience Publication

JOHN WILEY AND SONS

New York · Chichester · Brisbane · Toronto · Singapore

Library of Congress Cataloging in Publication Data:

Clark, Tim, 1949–
  A handbook of computational chemistry.

  "A Wiley-Interscience publication."
  Includes bibliographies and index.
  1. Molecular structure—Computer programs. I. Title.

QD461.C52 1985   541.2′2   84-27055
ISBN 0-471-88211-9

Printed in the United States of America

10 9 8 7 6 5 4 3 2

FOREWORD

As a science becomes more exact, it inevitably moves toward more precise mathematical descriptions. While this principle was first demonstrated by celestial mechanics and has long been accepted in other branches of physics, many chemists resist rather than embrace the quantitative direction in which their discipline is moving rapidly. One reason for this may be that such an evolution demands a fundamental change of attitude.

Chemistry traditionally has been an experimental science—no molecule could be investigated until it had been synthesized or was found in nature. In contrast, computational chemistry requires no preparations, no separation techniques, and no spectrometers or any physical measurements; it does not even require a chemical laboratory. Through the combined use of ever faster computers and increasingly sophisticated programs, a scientific revolution is taking place. It is now possible, and in fact quite easy, to study unknown molecules, reactive intermediates, reaction transition states, and even species that cannot exist by computational means. Many chemical facts can now be obtained more accurately (and much more readily!) by computer than by experiment. Experience demonstrates that results obtained by adequate calculations can be trusted, and need not always require experimental verification.

The chemist now has incredibly powerful mathematical tools at his disposal. These tools are simple to use and are about the only ones that continue to become less expensive. The future directions of chemical research are already programmed!

Why are these methods not in more general use? This is due partly to unfamiliarity, not only with the approach and the results, but also with the practical aspects of the calculations. Once the chemist has employed these methods, he quickly sees their advantages. Through experience, he learns the systems amenable to study and the limitations of the methods. Tim Clark's book will get you started and take you a long way toward becoming a computational chemist.

PAUL VON RAGUÉ SCHLEYER

Erlangen, West Germany
March 1985

PREFACE

This book attempts to introduce practical chemists, both experimental researchers and students, to chemical structure and energy calculations. The ability to "determine" the structure, stability, and other properties of an unknown molecule without synthesizing it opens up new chemical possibilities and turns fantasy molecules into accessible research objects. Because the book is intended for nontheoreticians, I have avoided detailed theoretical derivations and have instead attempted to give qualitative descriptions of the processes involved. These explanations may be oversimplifications in many cases, but are intended to convey the principles involved to those who are allergic to "Consider the Hamiltonian. . . ." The molecular orbital (MO) section of the book is intended to be read sequentially. The *ab initio* examples are chosen to illustrate points not covered in Chapter 4 on semiempirical systems, which in turn depends heavily on topics covered in Chapter 3. Although no training in theoretical chemistry is needed, the book does assume an adequate knowledge of symmetry operations and point groups, which are used throughout without further explanation. In both the molecular mechanics and MO sections the emphasis is on the practical aspects of performing useful calculations and understanding the output.

I have placed rather more emphasis on semiempirical methods than *ab initio* purists may find desirable. This is not because of any particular leaning toward MNDO or MINDO/3, but rather because these two methods are of more use to practical chemists than the more expensive and complicated *ab initio* methods. Many problems are simply not amenable to *ab initio* calculation because of their size—even when they are reduced to a minimum.

One of the most important requirements of the chemist performing structure and energy calculations is the ability to assess their reliability, or lack of same. The review literature on molecular mechanics and *ab initio* calculations is adequate, but the situation for the semiempirical methods is less satisfactory. Chapter 4 therefore includes reviews of the strengths and weaknesses of MINDO/3 and MNDO, along with a subject index that refers to over 500 relevant citations.

Much of this book is based on a lecture course that I have given in Erlangen for several years, and I am grateful to Peter Hofmann and Jayaraman Chandrasekhar, who were actively involved in the early versions of this course. Many will recognize Paul Schleyer's influence throughout. There is probably no adequate way to thank him for introducing me to computational chemistry, for the interest he has taken in my development as a chemist, and for the infectious excitement he brings to every problem. Many colleagues have commented on parts of the manuscript or have helped with the technical details. Alwyn Davies and Brian Roberts provided a suitably academic background for the writing of Chapter 3. Discussions with John Pople were, as always, very fruitful. Andrzej Sawaryn converted GAUSSIAN82 for use on our CDC computer and offered useful criticism of the MO sections. Peter Budzelaar read the entire manuscript and helped eliminate many small errors. Guenther Spitznagel and Elmar Kaufmann also provided valuable comments, especially on Chapter 4. Finally the hundreds of students who made mistakes, asked questions, discovered new problems, and generally approached the task of calculating structures and energies with such enthusiasm contributed much to the approach I have used.

TIM CLARK

Erlangen, West Germany
March 1985

CONTENTS

A HANDBOOK OF COMPUTATIONAL CHEMISTRY

CHAPTER 1

CHEMICAL CALCULATIONS

1.1. INTRODUCTION

Chemical calculations that can predict the structures, energies and other properties of known or unknown molecules have often been heralded as important new tools in chemical research. The analogy with nuclear magnetic resonance (nmr) spectroscopy is often used. There is, however, a fundamental and very exciting difference between calculations, whether quantum-mechanical or force field, and all experimental techniques: calculations can just as easily be performed for compounds that have never been made, or even cannot exist under real conditions, as for those that can be stored in bottles on a shelf. Tetrahedrane's properties are no more difficult to calculate than those of butane (in fact they are somewhat easier), and reaction intermediates with lifetimes of nanoseconds or less present no more problems than do the stable products of the same reaction. The majority of calculations reported so far have, however, been limited to the type of work in which extra information about a known, stable species is obtained by calculation. Even in this respect the information provided by a single calculation is remarkable in comparison with what chemists expect from experimental techniques. What other method can give the molecular structure, heat of formation, dipole moment, ionization potentials, charge densities, bond orders, spin densities, and so on in one experiment? The obvious objection is that the results of the calculations may not be reliable, but the strengths and weaknesses of the common methods are well known, so realistic estimates of their probable accuracies can be made. In some cases, at present rare but becoming more common, the results of calculations may

1

be more reliable than those of experiments. Measuring the heat of formation of a polycyclic alkane, for instance, is a long and difficult task that requires the utmost experimental precision and sample purity. Even with the most careful experimental work there is no guarantee that the value obtained is correct. So many factors can introduce errors large enough to alter the result significantly that only truly independent measurements on different samples can be taken to indicate experimental accuracy. For a very large range of such alkanes, however, the most modern molecular mechanics methods allow the calculation of heats of formation that differ from the "true" values by at most 2 kcal mol^{-1}, and probably by a lot less. And the calculation delivers an accurate molecular structure as a by-product! The total cost of such a calculation is a few seconds of computer time and at most half an hour for the preparation of the data.

This example is still exceptional, and few calculations achieve this level of accuracy. In many cases, however, given a choice between two structures or between two possible reaction intermediates, the commonly used calculational methods can give an answer that is close to definitive. Even molecular orbital (MO) calculations, which are usually less accurate than molecular mechanics methods are now capable of rivaling or bettering experimental accuracy for small molecules. To achieve this sort of performance, however, a huge investment in computer time is needed. The methods involved in such calculations will probably be commonplace in 10 years, but this book is concerned with more practical calculations using published program packages. These are today's standard methods for chemical purposes. They are certainly not perfect, but they have achieved a reliability that makes them a valuable addition to the range of tools available in chemical research. More importantly, they have extended the range of possible research subjects by removing the practical limit that only compounds that can be made can be studied. This feature alone should provide new impetus for experimental research by opening up new areas that would probably remain unexplored without calculations.

The costs and accuracies of the various types of calculations are naturally of interest. Table 1.1.1 summarizes the computer times needed for and the results obtained from the different types of calculation treated here. The example chosen is propane. All programs (MMP2, MOPAC, and GAUSSIAN82) were given the same starting geometry (1.54 Å CC bond lengths, 1.09 Å CH bond lengths, and 109.5° bond angles) and were then allowed to optimize the geometry. C_{2v} symmetry was used for the MO jobs, but not for MMP2. The starting geometry places MINDO/3 at a disadvantage as can be seen from the fact that it optimizes to a considerably different structure. Normally it would be expected to be about 1.5 times faster than MNDO. One important point to note is that propane is a member of the parametrization sets for MM2, MINDO/3, and MNDO, so their general agreement on the heat of formation should not be surprising. The computer time required varies between 0.83 and 4700 seconds, a 5600-fold difference

TABLE 1.1.1
Computer Times and Results for Calculating the Properties of Propane Using Various Calculational Methods

	MM2	MINDO/3	MNDO	3-21G	6-31G*	Exp.
Computer time (sec. CYBER 845)	0.83	9.75	10.32	550	4702	—
r_{CC} (Å)	1.534	1.495	1.530	1.541	1.528	1.526
$\angle_{CCC}$ (°)	111.7	121.5	115.4	111.6	112.7	112.4
$\Delta H°_f$ (kcal mol^{-1})	−24.8	−26.5	−24.9	—	—	−25.0

between the MM2 and 6-31G* optimizations. Each type of calculation has its advantages and applications for which it is particularly well suited. The choice of the method, which this book should make easier, is perhaps the secret of a successful calculation. It is unfortunately not possible to give an equivalent cost in the experimental column. The man hours, equipment, and chemicals needed for a structure and heat of formation determination would, however, certainly be more expensive than the calculations, even at the most horrendous prices for computer time.

This book is not intended as a text on theoretical chemistry, computer programming, force fields, or any of the other specialized subjects involved in calculation chemistry. Its purpose is to provide a sound practical background for performing chemical structure and energy calculations based on molecular mechanics or MO theory. Many chemists may be reluctant to begin doing calculations of this type. The programs and techniques involved are unfamiliar and the results obtained not immediately understandable. After the first few successful calculations, however, the chemist's usual reaction is one of surprise that results are so easy to obtain. The programs are, in fact, usually so simple to use that it is too easy to obtain results without really understanding the principles involved and what the program is actually doing. This often leads inexperienced users to overlook small errors, such as an incorrectly optimized geometry, which might limit the reliability of the results. One important point is that a mistake that causes the job to crash after a few tenths of a second computer time is infinitely preferable to one that allows a large job to run, but renders the results useless. Only a limited amount of computer time is likely to be available, so its optimum use is of paramount importance. There are many warnings in this book that may seem trivial or obvious. They are, however, all related to errors that inexperienced users have made, or to questions that they have asked. I hope that the strength of this work will be in anticipating, and thus avoiding, most of the problems and mistakes that a newcomer to calculations may encounter. However, experience is important, especially in recognizing hidden errors. There is a dangerous phase that most novices go through, in which the results come so quickly and easily that they cannot be thoroughly checked and interpreted. The way to avoid problems is never to allow the computer to control your pace, but to work so that you have time to evaluate the results before you rush off and waste valuable computer time.

Many problems arise from the "black box" approach to calculations and programs. It is difficult in a book of this type to avoid treating at least parts of the calculations as "black boxes." I have tried to find a compromise that does not involve long excursions into quantum chemistry or computer science by giving qualitative descriptions of what the programs do whenever possible. This sort of description is not enough to provide a detailed understanding of the workings of the program, but should at least allow a rough diagnosis. This book cannot stand alone as a theoretical textbook or as a program manual. It is intended to be used in conjunction with such texts, which are more often used for reference than for learning how to use the programs. The program manuals contain a great deal of detailed information about a given program, but usually in a form more suited to experienced users. Similarly, theoretical textbooks usually cover the mathematics of the calculations, but give little indication of how the programs work.

One of the main problems encountered by experimentalists who turn to calculations is that they pose the wrong questions to the programs. The experimental system that may have revealed the effect to be investigated is seldom the best one for calculations. Experiments are usually performed on molecules that are synthetically convenient, not on those that show the effects most clearly. The trick is to eliminate complicating effects in order to leave a system in which the phenomenon of interest can be clearly and uniquely identified. This brings advantages not only in computer time (by making the calculations shorter), but also in the evaluation of the data. One of the problems in the interpretation of results is that the calculations are too much like an experiment, by which I mean that their behavior and the reasons for it may be as complicated as those in the experimental system. The reduction of the calculated species to the bare essentials is therefore often very necessary.

A second common problem is that of chemical selectivity. It is often overlooked that the two alternative reaction paths of a reaction that gives useful synthetic regio- or stereoselectivity may differ by only 2–3 kcal mol^{-1} in their free energies of activation. Many of the methods discussed herein are capable of giving reliable predictions in such cases. The comparison is between similar species and so the errors tend to cancel. The calculations apply, however, to hypothetical motionless molecules *in vacuo*. The difference between this state and the actual reaction medium involves effects, such as entropy, the population of vibrational energy levels, solvation, and aggregation, that alone are enough to determine the selectivity of the reaction. Calculational studies on chemical reactivity have given good results in many cases, but in least as many the experimental trends could not be reproduced. This is not necessarily evident from the literature, as failures are seldom published. The situation is, however, not as bleak as it seems. If calculations that are normally reliable predict behavior other than that observed experimentally, it may be an indication either that entropy or solvation effects are controlling the reaction or that another mechanism is operative. In any case, it is often necessary to know the properties of the unperturbed system in order to assess the effects of factors such as solvation.

Another problem that is not often appreciated is that ionic reactions are essentially condensed phase phenomena. Consider the two possible dissociations shown below, one homolytic and one heterolytic:

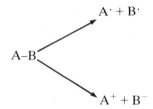

The two processes may be competitive in solution because of solvation of the ions, A^+ and B^-, but any separation of charge in the gas phase leads to a loss of electrostatic attraction, and hence a large increase, typically several hundred kilocalories, in energy. Calculations therefore cannot be used to decide between polar and nonpolar processes that may compete in solution. Similarly zwitterionic intermediates may not be treated well by gas-phase calculations, although the results often show that structures that conventional bonding theories describe as zwitterionic are in fact remarkably nonpolar in the gas phase. In general there is a strong tendency for formal π charges to be compensated by an opposite charge shift in the sigma framework, so that charge separation may be smaller than that observed in solution. A striking example is the 1-aza-3-bora-allene (**1**), which is written classically as a zwitterion, but is calculated to have very little charge separation:

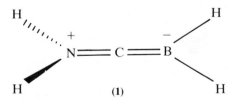

(1)

Another problem, which arises from the calculational methods normally used rather than from differences between gas phase and solution, is that of calculating activation energies for homolytic bond fission reactions. These processes cannot be calculated realistically at simple levels of theory because they involve splitting a pair of electrons. Restricted Hartree–Fock (RHF) theory, which is usually used for closed-shell molecules, limits the electrons to a pairwise occupation of the molecular orbitals, so that only a heterolytic bond fission—one in which one fragment takes both electrons from the bond—is possible. This is, in fact, what happens if, for instance, the bond length in H_2 is steadily increased in RHF calculations. This leads to a dissociation that is more like an ionic one than like the desired homolytic cleavage, and therefore to an energy curve that deviates considerably from that required. The solution to the problem lies in mixing the two electronic configurations to obtain a good description of the bond breaking process. This can be achieved either by considering more than one configuration from the start, as in generalized valence bond or multiconfiguration self consistent field (MCSCF) calculations, or by mixing excited states into the single-determinant wave function, as in configuration interaction (CI) calculations. These techniques will be discussed in the chapter on *ab initio* calculations. Figure 1.1.1 shows the situation for H_2.

The last important problem in all calculational work lies in assessing the likely accuracy of the method used. This is often possible by comparison with literature data for related compounds, if not, it may be necessary to carry out

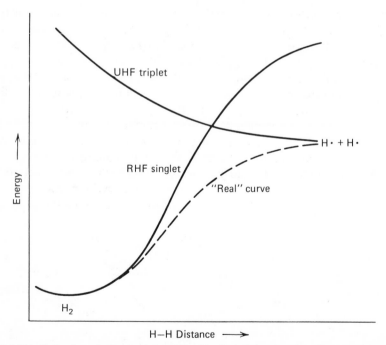

FIG 1.1.1. Behavior of single determinant calculations for the dissociation of H_2. Restricted Hartree–Fock theory fails for the dissociated atoms and unrestricted Hartree–Fock calculations on the triplet give excited H_2 at small HH distances.

extra calculations on similar systems for which experimental data are available. For work on unknown systems the only alternative is to use better and better levels of calculation and hope that the results converge. This technique is used in *ab initio* studies, but is often limited by the computer time and disk space needed for the calculations.

The published literature is extremely important in calculational studies, probably more so than in experimental work. Several reviews and a recent book on molecular mechanics calculations have been published, so it is fairly easy to find relevant work. Similarly, there are a number of very comprehensive bibliographies of published *ab initio* calculations. The *Carnegie–Mellon Quantum Chemistry Archive* is a particularly useful source. It is a compilation of the results of all the calculations run on the VAX 11/780 at Carnegie–Mellon University. Each job is automatically added to the archive as it is successfully completed. The *Archive* is available directly from Carnegie–Mellon. It contains complete optimized geometries, total energies and a variety of other information for several thousand molecules, much of it unpublished elsewhere. The *Archive*, the bibliographies, and the relevant reviews are cited in full in the chapters on the individual calculational methods.

The situation for semiempirical calculations is less satisfactory. There are no comprehensive bibliographies and the few reviews are limited in scope. I have therefore included a subject index of published MINDO/3 and MNDO calculations in Chapter 4. This compilation is by no means complete as it used the *Chemical Abstracts* entries under "MINDO/3" and "MNDO" as the principal source. There are, however, some 350 MINDO/3 references and over 150 for MNDO. In each case the index covers literature cited until the end of *Chemical Abstracts* volume 98.

1.2 THE PROGRAM

Chemical programs are unfortunately very large and sometimes very machine dependent (i.e. are adapted to the special features of one type of computer). They are written almost exclusively in FORTRAN, a programming language long outdated for other purposes, but still the language of choice for scientific programs. The reason for this is largely historical. FORTRAN was used in the early 1970s for the first versions of the programs, which have since been constantly extended and improved, so that translation to another language would now be a horrendous task. Large program packages consist of up to 130,000 lines of FORTRAN, although many of the earlier versions were only about 6000 lines long. It is not possible to load program systems of this size into the computer all at once, so many programs are divided into sections that may be loaded consecutively into the memory. The conventional way to do this is by means of overlays. An overlaid program is one in which a control section with some very commonly used subroutines remains in the memory continuously and controls the loading and removal of the other sections, or overlays, as they are needed. Data are transfered from overlay to overlay *via* common blocks or files written to a magnetic storage device. Newer programs, such as GAUSSIAN82, use a different technique in which separate programs are run by the control program in place of the overlays. These programs run consecutively and access a set of files in which the data are stored. Some machines, such as the VAX, use a technique known as virtual memory, in which the operating system loads only those parts of the program that are actually needed and swaps the program sections from disk to memory as required. This is a very attractive technique from the programmer's point of view, as there are essentially no limits to the size of the program. The disadvantage is that a large program, which has to undergo more swapping than a smaller one, runs more slowly because the swapping process also takes time. The disadvantage for users of nonvirtual memory machines is that a program written for virtual memory may have to be reduced in size or overlaid to make it fit.

The precision with which a given computer works is governed by the length of the word used to store each number. Molecular orbital programs need to calculate more precisely than is possible using a 32-bit word, which is standard on many computers, such as the VAX or those manufactured by IBM. In this case two words must be used to store each number. This is known as double precision, and is achieved simply in FORTRAN programs by declaring the variables as double-precision species, to which two words are assigned, and using the appropriate double-precision functions to process them. This results in a larger, slower program than the equivalent in single precision, but is very necessary if the program is to run properly. Some programs for IBM machines are not written explicitly in double precision, but are intended to be converted by the extended FORTRAN "H" compiler (a program that translates the FORTRAN source code into a machine-executable version). These programs should not be run in single precision on IBM computers. Other computers, such as those from CDC, use word lengths long enough to give the necessary precision, so that running programs in double precision on them would be an unnecessary waste of time and space. Conversion from double to single precision is easy, even for large programs, but the reverse process needs more effort if it is to be done properly. Single- and double-precision versions of most programs are generally available, so you should be able to find one suitable for your machine. Molecular mechanics programs usually run in single precision, even on 32-bit machines.

One of the major problems in converting *ab initio* programs from one type of computer to another is the packing and storage of integrals. Moderately sized *ab initio* calculations calculate and store hundreds of thousands of integrals with indices to identify them. The integrals and their indices are usually packed into a convenient form and written onto disk files. On rereading by another part of the program the integrals and indices must be unpacked (i.e. converted back to their original form). This packing and unpacking of integrals and their writing and reading are often performed by subroutines written in assembler language, a far more direct method of communicating with the computer than

FORTRAN. Assembler is not transferable from one machine to another, and so the conversion of such routines may be one of the major tasks in converting a program. Generally the conversion of molecular mechanics or semiempirical programs for use on another machine is a fairly simple task, given an adequate knowledge of FORTRAN. Conversion of *ab initio* programs, however, is far more difficult and time consuming. Luckily there are several different versions of GAUSSIAN80 available for different computers and several new versions of GAUSSIAN82 should soon be available. Table 1.2.1 lists the currently available MM2, MINDO/3, MNDO, and GAUSSIAN programs.

The easiest way to obtain a program that runs on your computer is from a colleague who is using the program on the same machine. Usually, however, the program must be bought, either from the Quantum Chemistry Program Exchange (QCPE) or *via* a software agreement. QCPE, which is based in Indiana University, is a nonprofit organization for the distribution of a wide range of programs for chemistry. Authors submit their programs to QCPE where they are tested and made available to members. Membership currently costs less than $20 per year and includes the *QCPE Bulletin*, in which new programs are announced. QCPE prices for programs cover essentially only the costs of magnetic tapes, computer time to make copies, administration and mail. Programs are supplied as the source (usually FORTRAN) version and no restrictions are placed on their use. From the user's point of view this is an ideal situation, but it means that writers and developers of programs derive no benefit from their work and have no control over modifications to the programs. There is therefore a move toward marketing programs under copyright. In many ways this is regrettable, as QCPE has been of great service to chemistry in general for many years. On the other hand, an effective program package is a very valuable commodity and improperly modified versions may wreak havoc if they become generally available. At present two of the programs discussed here, GAUSSIAN82 and MMP2, are available on a commercial basis.

GAUSSIAN82 is available as a VAX and CRAY1 source (FORTRAN) version directly from Professor J.A. Pople at a cost of $250 for academic and $500 for industrial users. For this price one obtains the program and the manual but no further support, such as help with installation. The program may be run on one computer only, and it or any version derived from it may not be passed on to other users. A number of other versions of GAUSSIAN82 (IBM and CDC) should soon be available.

The MM2 and MMP2 programs in their latest versions are available from Molecular Design Ltd. A binary version of the program is delivered with documentation, usually as part of a more extensive program system. The fact that a binary version of the program (that produced by the compiler) is supplied rules out program modifications by the user and limits use to machines for which Molecular Design supplies suitable programs. Once again, the program may be used only on one computer and may not be passed on to other users. Older versions of MM2 (but not MMP2) are available as source versions from QCPE.

TABLE 1.2.1

MM2, MINDO/3, MNDO and GAUSSIAN Programs Available from QCPE or Commercially

Program	Source	Computer	Comments
MM2 MM2 MM2	QCPE395 QCPE423 QCPE448	IBM CDC Perkin-Elmer 3220	Earlier MM2 version limited to nonconjugated molecules.
MMP2	Molecular Design Ltd. 1122 B Street Hayward, California 94541	Various	Full version including SCF calculations for conjugated molecules and up-to-date parameter set. Available only as compiled version.
MINDO/3 MINDO/3 MINDO/3	QCPE279 QCPE308 QCPE309	CDC IBM (H) IBM	Original MINDO/3 program. No frequency calculations, etc.; only half-electron (HE) calculations for open-shell molecules.
MOPN	QCPE383	IBM (H)	Unrestricted Hartree–Fock (UHF) version of MINDO/3. Program is more comfortable to use than the earlier versions and includes a useful symmetry package. Can do only UHF calculations (not HE) for open-shell molecules and RHF for closed shell.
RPI/MINDO/3	QCPE431	IBM	Extended MINDO/3 version with a convergence-forcing procedure.
MNDO MNDO	QCPE353 QCPE379	IBM CDC	Original MNDO program. Very similar to the original MINDO/3 versions.
MNDO	QCPE428	VAX	Improved MNDO program with pseudodiagonalization to speed up the SCF and McIver–Komornicki transition-state search. Can do either UHF or HE calculations for open-shell molecules.

Program	QCPE No.	Computer	Description
MOPAC MOPAC	QCPE455 QCPE464	VAX IBM	Combined MINDO/3 and MNDO program package. Similar to QCPE428, but with additional normal vibration and thermodynamics calculations and several options to improve the SCF convergence.
GAUSSIAN70	QCPE236	IBM	First GAUSSIAN program package. No d-orbitals or automatic optimization and no post-SCF procedures.
GAUSSIAN76 GAUSSIAN76	QCPE368 QCPE391	CDC IBM	Similar to GAUSSIAN70, but with d-orbitals.
GAUSSIAN79	QCPE421	DEC-10	A modified version of GAUSSIAN76 with an added overlay to calculate one-electron properties.
GAUSSIAN80 GAUSSIAN80-UCSF	QCPE437 QCPE446	IBM (H) VAX	QCPE437 is a conversion of the original VAX GAUSSIAN80 program, which is no longer available. Automatic optimization, MP2, MP3, and CID are available. QCPE446 additionally contains a one-electron properties package, Morokuma component analysis, and a variety of electrostatic potential programs.
GAUSSIAN82	Prof. J. A. Pople Dept. of Chemistry 4400 Fifth Avenue Carnegie–Mellon University Pittsburgh, Pennsylvania 15213	VAX CRAY1	Improved and more precise version of GAUSSIAN80. Post-SCF calculations to MP4SDTQ. Transition-state search. One-electron properties package. Normal vibration calculations. Soon to be available for CDC and IBM. Supplied as source (FORTRAN) code.

CHAPTER 2

MOLECULAR MECHANICS

2.1. INTRODUCTION

Molecular mechanics, or force field, calculations are based on a simple classical-mechanical model of molecular structure. The model nature of the calculations should be emphasized. There is very little physical significance in the parameters and energies in molecular mechanics calculations, as was shown by the early force fields, which often gave very similar results, but for completely different reasons. Nevertheless, molecular mechanics can already be considered to have achieved chemical accuracy, at least for certain types of molecule. The molecular mechanics method and its applications have recently been very comprehensively reviewed by Burkert and Allinger,[1] so this chapter will deal mainly with the background and principles of molecular mechanics calculations and with Allinger's MMP2 program, the latest in a series of programs that are probably the most popular with users.

Molecular mechanics treats the molecule as an array of atoms governed by a set of classical-mechanical potential functions. The principle is best illustrated by considering the bond-stretching term in a molecule. The potential at any given interatomic distance r is described by the well-known Morse curve, as shown in Fig. 2.1.1. The energy minimum occurs at the equilibrium bond length, r_0. The expression for a Morse curve is, however, complicated and would require too much computer time, much more than other types of potential function. This is not a critical problem, as the vast majority of molecules have bond lengths within a limited range, symbolized by the shaded portion of Fig. 2.1.1. Within this part of the curve one of the

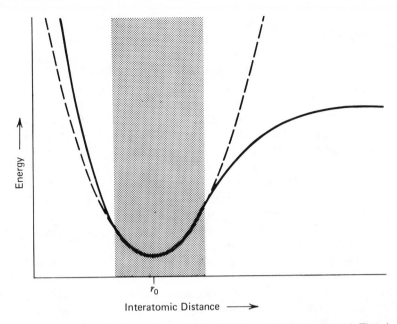

FIG. 2.1.1. The potential-energy curve for stretching a chemical bond. The dashed curve represents a simple Hooke's law potential-function.

simplest types of potential function, Hooke's law, gives a good fit to the more realistic energy profile, as shown by the dashed curve in Fig. 2.1.1. The Hooke's law expression

$$V = k(r - r_0)^2/2 \,,$$

where V is the potential energy and k is a constant, is particularly simple to calculate and gives very fast execution of the program, so it is used in many force fields. The only problems that may arise are for molecules with very long bonds, and are usually caused by steric effects. These molecules may have bond lengths that lie outside the shaded area in Fig. 2.1.1, so that the simple Hooke's law expression is no longer appropriate. This situation can be improved by adding a term proportional to $(r - r_0)^3$ to the expression given above. This improved type of function also has its limits, but does yield considerable improvement for sterically crowded molecules. One problem that may occur with such cubic terms is that they become more attractive at very long bond lengths, so that molecules may fly apart if a very poor starting geometry is used. The MMP2 program includes the cubic terms at a relatively late stage in the optimization when the molecule has almost reached its final structure.

The second type of function is the angle-bending potential. This is in principle exactly the same as the bond-stretching function. The potential energy rises as a given bond angle is deformed away from its optimum value, θ_0. Once again, a simple potential function proportional to $(\theta - \theta_0)^2$ is often used, although higher-order terms may also be included. The forerunners of today's force fields often included only bond-stretching and angle-bending functions, and were used on estimated structures without geometry optimization to calculate strain energies.

Modern force fields contain many more types of potential function designed to give a good fit to the experimental data. The most important of these additional functions are the torsional potentials. The rotation barrier of ethane, for example, cannot be reproduced by a reasonable force field that does not include an inherent threefold potential function for the carbon–carbon bond. Steric interactions alone do not give good results. Threefold torsional potential functions were therefore included in most force fields fairly early in their developments. A further important advance was made with the inclusion of one- and twofold torsional functions in the MM2 force field. This gave a dramatic improvement in the results for some molecules. The combination of bond-stretching, angle-bending, and torsional potential functions is often known as a *valence force field* because it accounts for the properties normally attributed to chemical bonds.

Valence force fields are, however, still not adequate for high-quality quantitative calculations. Force fields intended for such applications include the so-called van der Waals functions used to account for steric interactions. Van der Waals potentials often take the form of a "6/12" function; that is, they have two terms, one proportional to the 6th power of the interatomic distance and one dependent on the 12th power. The MM2 force field, however, uses an "exponential minus 6" expression, in which the 12th-power term is replaced by an exponential. The form of such functions is shown in Fig. 2.1.2. Note that there is a shallow minimum at an ideal interatomic distance. The inclusion of steric interactions in molecular mechanics force fields is, however, not without problems. Steric repulsions can never be completely separated from other interactions, and are therefore difficult to define and strongly dependent on the other potential functions used in the force field. One particular problem arises from the fact that steric (van der Waals) repulsions between atoms bonded to a common center (1,3 interactions) are not usually included in the calculations. This is not particularly important in most cases because the 1,3 interactions can be implicitly included in the angle-bending potentials. It makes little difference to the program whether the energy rises because a bond angle is made smaller or because two atoms that are bonded to the same center approach too closely. In the special case of three- and four-membered rings, however, this implicit inclusion of the 1,3 repulsions in the angle-bending functions breaks down. In all rings with more than four atoms there is the same number of 1,3-interactions as there are atoms in the ring. For a four-membered ring, however, there are only two sets of atoms 1,3 to each other, and in a three-membered ring there are no nonbonded 1,3 pairs at all. This means

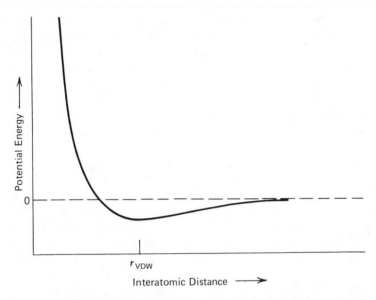

FIG. 2.1.2. The potential-energy curve for the van der Waals interaction between two atoms. The minimum occurs at a distance (r_{VDW}) equal to the sum of the van der Waals radii of the two atoms.

that three- and four-membered rings cannot be treated successfully using the potential functions appropriate to acyclic molecules or larger rings. An obvious solution would be to include 1,3 interactions in the force field. Force fields that consider these interactions (known as Urey–Bradley force fields) do not, however, generally give good results. The accepted solution is to use separate parameters for atoms in three- and four-membered rings, so that the carbons in cyclopropanes and cyclobutanes, for instance, are actually treated as different types of atoms from those in other alkanes.

The omission of 1,3 interactions may also make using more complex types of potential function necessary. Bonds tend to become longer as the angle between them becomes smaller, for instance. This effect can be reproduced by introducing a compound function that depends on both the bond length and the bond angle. Such compound functions, often called *cross terms*, may also include torsional contributions.

Other types of interaction may have to be considered for molecules with polar groups. The charges that build up on these groups and the dipole moments associated with them interact with each other to affect the energy of the molecule. It is therefore not uncommon to use electrostatic terms and dipole–dipole interactions in the force fields. The charges on a given type of group are relatively constant, so simple electrostatic calculations can be used for the charge–charge interactions. The total dipole moment of a molecule may also be represented as the vector sum of the dipoles attributable to each bond. Generally the conformation or isomer in which the total dipole moment is the lowest is the most stable. The bond dipoles are included in many force fields to make possible calculation of the dipole–dipole terms. This has the useful side effect that a dipole moment for the entire molecule is easy to calculate once the geometry has been optimized.

Each of these types of potential function is assumed to be transferable from molecule to molecule. This means that a given type of bond, for instance, is assumed to have the same characteristics in each molecule in which it occurs. That this is often a good approximation is shown by the success of molecular mechanics calculations. In some cases, for example when there are strong interactions between bonds, this approximation may break down. One obvious case is that of conjugated molecules, which are not treated by standard force fields of the type described above. There are many other examples in which electronic effects change the characteristics of the bonds in a molecule and render it unsuitable for molecular mechanics calculations. Nonclassical ions, for instance, could not be handled by a force field parametrized for classical species, but a molecular mechanics calculation would presumably give a good representation of the classical structure. In principle at least this technique could be used to detect and estimate the degree of energetic and structural effects due to nonclassical behavior. The through bond effect in 2,2-para-cyclophane has been demonstrated in this way.[2]

The force field defines the mechanical model used to represent the molecule. The purpose of the molecular mechanics program is to determine the optimum structure and energy based on this mechanical model. The input to the program must therefore define a starting structure for the molecule. This involves giving Cartesian (x,y,z) coordinates for the individual atoms and defining the bonds joining them. In practice often only the heavy (nonhydrogen) atoms are so defined. The program then adds the hydrogens as necessary. The model nature of molecular mechanics calculations requires that bonds be defined in the input. The model corresponds strictly to the classical valence bond picture of chemical bonding. Carbon atoms, for instance, may be either sp^3, sp^2, or sp, and there are three completely different force fields for the three different types. This is in contrast to MO calculations, in which the electronic state determines the bonding pattern, which is therefore not defined in the input.

The first step in the molecular mechanics calculation is determination of the interatomic distances, bond angles, and torsional angles in the starting geometry. The values obtained are then used in the different potential-function expressions to calculate an initial *steric energy*, which is simply the sum of the various potential energies calculated for all the bonds, bond angles, torsional angles, nonbonded pairs of atoms and so forth in the molecule. It is important to note that this steric energy is specific to the force field. It does not correspond to any classical definition of strain energy, although it is related to the heat of formation by a simple expression. Because all other factors remain constant throughout the optimization of a structure it is enough to find a minimum with respect to the steric energy.

Early programs performed geometry optimizations by steepest-descent or pattern search procedures, but these and similar techniques have now been replaced by optimization techniques, such as the Newton–Raphson method, which use analytically evaluated second derivatives of the molecular energy with respect to the geometrical parameters. In the early methods the atomic forces (the gradients or first derivatives) were calculated by the finite-difference method. Fig. 2.1.3 shows a typical potential curve for bond stretching. The finite-difference method calculates the gradient of the curve at a given point by first calculating the energy and then changing the geometry by a small amount, δr in Fig. 2.1.3. The energy is then recalculated and the energy difference between the two points, δE, is used to determine the gradient, $\delta E/\delta r$. It is then assumed that the distance from the minimum is proportional to $\delta E/\delta r$ and the geometry is altered to obtain the next structure, which should be closer to the minimum.

Modern programs determine the first derivatives, dE/dr, and the second derivatives, d^2E/dr^2, analytically. The relatively simple form of the potential functions used in the force fields makes them particularly suitable for this type of treatment, since they differentiate to give expressions that are equally simple, and therefore are also easily evaluated. The second derivatives, or force constants, indicate the curvature of the potential-energy

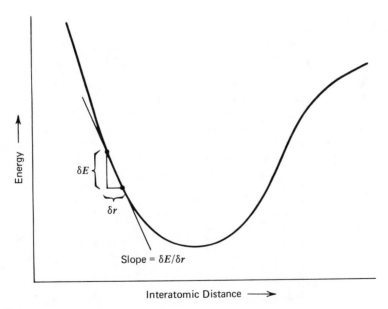

FIG. 2.1.3. Determination of atomic forces by the finite difference method.

curve, and can therefore be used to estimate the position of the minimum. Figure 2.1.3 shows a simple two-dimensional potential curve, but the principles are the same for the multi-dimensional potential surfaces for real molecules. The full Newton–Raphson optimization procedure involves calculating the entire matrix of second derivatives (the force-constant matrix). This matrix consists of $3n \times 3n$ elements for a molecule consisting of n atoms. The $3n$ dimensions correspond to the three degrees of freedom (movement in the x-, y-, and z-directions, for instance) for each atom. The optimization of the geometry, however, requires only $3n - 6$ degrees of freedom, because the three translations and three rotations lead to no change in the energy. The six trivial degrees of freedom are removed from the force-constant matrix, which is then inverted to obtain the predicted minimum-energy geometry. This process can be repeated until the calculation converges on the minimum.

Calculating the full $3n \times 3n$ force constant matrix is, however, not generally necessary in molecular mechanics calculations. Most elements that involve x-, y-, and z-movements of different atoms are small, so one can calculate only the 3×3 matrices attributable to the individual atoms, as shown in Fig. 2.1.4. This technique, known as the *block diagonal method*, requires only $9n$ elements to be calculated, but results in a loss of information about the nature of the potential surface. Optimizations using the block diagonal method are, however, faster and often more reliable. The MMP2 program uses a block diagonal Newton–Raphson optimization.

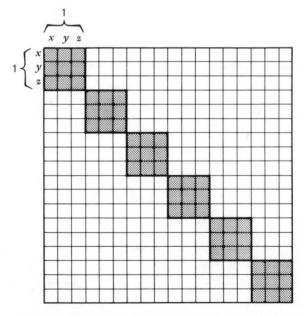

FIG. 2.1.4. The elements of the force constant matrix used in the block diagonal Newton–Raphson method.

Once the optimization has converged (i.e., once the energy and structure remain constant from iteration to iteration and the first derivatives are all close to zero) the program prints the final steric energy and the optimized geometry. This geometry may then be used to calculate such properties as the moment of inertia and the dipole moment (from the vector sum of the bond moments). The heat of formation can be calculated from the steric energy by a group or bond increment method. As with all other approximations in the molecular mechanics model these methods assume transferability of properties, in this case the contribution of a particular group or type of bond to the heat of formation. This is generally a good approximation for the types of molecule that the standard force field method treats well. Consider, for instance, the n-alkanes. The difference between adjacent molecules in the homologous series is one methylene (CH_2) group. If the group increment concept is valid there should therefore be a constant difference between the heats of formation of adjacent members of the series. Figure 2.1.5 shows a plot of the n-alkane heats of formation against the number of carbon atoms, N. There is a good linear correlation, the slope

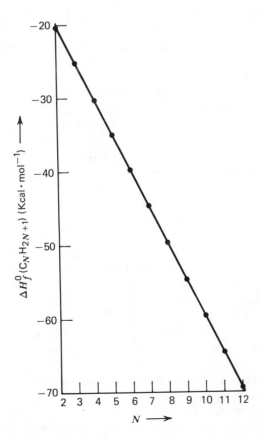

FIG. 2.1.5. The heats of formation of the n-alkanes.

being $-5.13\,\mathrm{kcal\,mol^{-1}}$/carbon. We can therefore assign this energy contribution, or *group increment*, to the CH_2 group. However, because each *n*-alkane consists of $(N-2)$ CH_2 groups and two CH_3 groups, the intercept of the correlation line with the $N = 2$ axis corresponds to the group increments for two methyl groups. Figure 2.1.5 leads to the equation

$$\Delta H^{\circ}_{f(C_N H_{2N+2})} = 2(-10.05) + (N-2)(-5.13) \text{ kcal mol}^{-1},$$

where the group increments for CH_3 and CH_2 are -10.05 and $-5.13\,\mathrm{kcal\,mol^{-1}}$, respectively. These CH_2 and CH_3 group increments can then be used in conjunction with experimental data to determine the increments for CH and C groups, and so on. This approach, which can be based equally well on the numbers of the individual types of bond in the molecule, giving bond increments, has been used very extensively to estimate heats of formation, and can give very accurate results, especially if steric corrections are included.[3] The group and bond increments used in molecular mechanics programs are based on exactly the same principles, except that they are designed to be added to the calculated steric energy to give the heat of formation. This relationship implies that the calculated steric energies are also transferable between nonstrained molecules, that is, that a similar plot to that shown in Fig. 2.1.5 could be drawn for the steric energies calculated for the *n*-alkanes. This assumption is valid because exactly the same mechanical model is used, for instance, for each methylene group so that the steric energies are additive. The group and bond increments used to calculate heats of formation in molecular mechanics calculations are, however, variable parameters like those used in the potential functions, and can be changed to obtain a good fit to the experimental data.

The next step in the calculation, the determination of the *strain energy*, is not specific to molecular mechanics programs, but is also performed using experimental heats of formation. The definition of strain energy is, however, not simple, because the heat of formation of a strain-free reference molecule must be defined. The conventional definition states that the strain energy is the difference in heat of formation between the molecule itself and a totally strain-free molecule of the same constitution (i.e. one that consists of the same numbers of each different type of group). The heat of formation of the strain-free reference compound can be calculated by using strain-free group increments. Several factors must be considered in choosing strain-free group increments. The CH_2 group increments derived above are, at first sight, strain-free. They are derived from the heats of formation of the *n*-alkanes, which must be strain-free. The major difficulty lies in the fact that acyclic alkanes are conformationally flexible, and thus the experimental heats of formation refer to an equilibrium mixture that contains small proportions of conformations that are higher in energy than the most stable, strain-free reference conformation. The proportion of higher-energy conformations can, however, easily be calculated and the experimental heats of formation corrected to the strain-free values. If such heats of formation are then used to derive group increments (*single-conformation group increments*), a more realistic strain-free model is obtained than that arrived at by using the normal experimental data. The molecules used to define the strain-free increments must also be considered carefully. In many branched species, skew interactions are unavoidable, even in the most stable conformation. In such cases the experimental data must also be corrected for the skew interactions, which arise when two alkyl groups that are 1,4 to each other make a dihedral angle of 60° in a staggered conformation.

The total process used by molecular mechanics programs to calculate heats of formation and strain energies is shown diagrammatically in Fig. 2.1.6. The steric energy calculated from the force field is converted to a heat of formation by adding the group and bond increments from the parametrization. These group increments have no physical significance other than that they give a very good fit to experimental heats of formation when used in conjunction with the steric energies given by the force field. From this point in the calculation onward there is no difference between the experimental and calculated heats of formation. The method used to calculate the strain energy is the same. The single-conformation group increments for the molecule are added together to give a strain-free heat of formation, which is then subtracted from the actual heat of formation to give the strain energy.

The calculation can therefore give a structure, some spectroscopic data, a dipole moment, and the heat of formation, from which the strain energy can be evaluated. The structure/energy force fields discussed here are, however, not suitable for the calculation of vibrational frequencies, although other

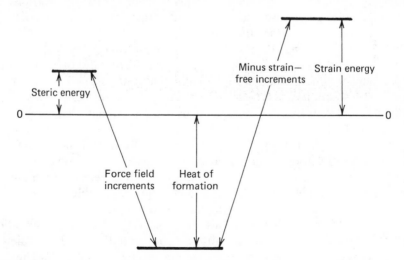

FIG. 2.1.6. Schematic energy diagram for determination of the strain energy in a molecular mechanics calculation.

force fields have been developed for this purpose. It is a general phenomenon that force fields developed for a particular purpose do not give
good results when applied to other problems. This is a reflection of the
model nature of the calculations, which are best regarded as good solutions
of a set of simultaneous equations specific to one or two physical properties.
It cannot be expected that the same equations will reproduce a property
different from that for which they were derived.

One outstanding problem for the force fields outlined above is how to
treat conjugated molecules. Consider 1,4- and 1,3-pentadiene, shown below.
It is obviously desirable to compare the structures and energies of these two
compounds, or of any similar pair of isomers, using molecular mechanics
methods.

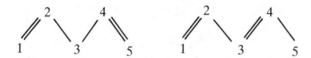

1,4-Pentadiene can easily be handled by conventional force fields. The
C_1–C_2 and C_4–C_5 double bonds are very similar to their counterpart in
propene, for instance, as are the C_2–C_3 and C_3–C_4 single bonds. In this case the
transferability of potential functions and group increments from one molecule
to the other is still a good approximation. 1,3-Pentadiene, however, presents
more problems. Because of the conjugated system the C_1–C_2 and C_3–C_4 bonds
are weaker than the double bond in propene. Even more problems are
encountered with the C_2–C_3 bond. It is stronger and shorter than a normal
single bond and has a completely different torsional-potential function.
Furthermore, there are no single bonds between two sp^2 carbons in
nonconjugated molecules, so that the relevant parameters for this type of bond
are in any case missing from the force field.

One obvious solution to the problem would be to define a new force field for
butadienes. This would involve two new types of atom, the terminal and central
carbons, and two new types of bond, "double" and "single" in butadiene
systems. This solution is, however, unsatisfactory, because it simply postpones
handling the real problem. Conjugated trienes or aromatic molecules, for
instance, would then also need a complete force field of their own. This would
lead not only to an immensely complicated force field with an endless series of
different types of carbon atom, but also to parametrization problems. There
simply are not enough experimental data on 1,3-butadienes or conjugated
trienes to make possible derivation of a force field comparable in accuracy to
those used for alkanes and nonconjugated olefins.

Allinger and Sprague[4] have used a simple, but very effective, combination
of quantum and classical mechanics to treat conjugated molecules. A simple
π-only MO section is built into the molecular mechanics program. The first
step of any calculation is then to perform a MO calculation *on the conjugated
system only* at the input geometry. This calculation gives a set of bond orders

and a total electronic energy. The bond orders are then used to modify the force field for the conjugated system. The equilibrium bond length, the stretching force constant and the rotation barrier are all assumed to depend on the calculated bond orders. The program then uses the normal force field for those parts of the molecule that are not involved in the conjugated system and the modified force field for the delocalized bonds. An ordinary geometry optimization is then performed with this force field. In principle, therefore, only one MO calculation is necessary to treat a conjugated molecule. The available data for the parametrization are also adequate as all conjugated and aromatic molecules can be used. In practice the results of the MO calculation are dependent on the geometry, which changes during the optimization. It is therefore necessary to repeat the calculation of the bond orders when the geometry changes to an extent greater than preset tolerances. The molecule is considered to be optimized when the energy and structure remain constant, even after new MO calculations. The total delocalization energy given by the SCF calculation must be included in the calculation of the heat of formation because it will automatically correct for the energetic effects of conjugation or aromaticity. This simply requires a different set of group increments parametrized to reproduce experimental data. This procedure was first introduced in the MMP1 program, and has been improved for MMP2. It gives remarkably good results for a fraction of the cost of even semiempirical MO calculations.

2.2. APPLICATIONS AND PARAMETRIZATION

The early history of molecular mechanics has been reviewed several times,[1,5,6] and will not be discussed in detail here. The more recent development of the method is, however, an excellent example of the interaction between theory and experiment that eventually led to dramatic improvements in the calculational method. The two force fields most widely used in the early 1970s were Allinger's MM1, introduced in 1973,[7] and the so-called EAS force field, developed by Engler, Andose and Schleyer in 1973.[8] Schleyer et al.[8] also surveyed the predictions and reliabilities of the two force fields. The predictions were usually in good agreement with each other, with some notable exceptions. This general good agreement between the force fields was not, however, based on similarities in the potential functions used, but rather resulted from the fact that they both represented fairly good solutions to the problem of reproducing experimental data.

This was exemplified by the controversy surrounding the cause of strain in *gauche*-butane. In parametrizing alkane force fields one of the most important considerations was that they should be able to calculate the energy difference between *anti*- and *gauche*-butane correctly. The MM1 force field, which reproduced this energy well, did so because the van der Waals repulsions between hydrogens in the *gauche* form were particularly high. The EAS force anti   gauche

field also gave the correct answer, but without significant differences in H–H repulsions between the two conformers. This resulted from hydrogen's being particularly "hard" in MM1 (i.e. it gave rise to strong repulsions at relatively large distances), but "soft" in EAS. To compensate for these differences carbon was "soft" in MM1 and "hard" in EAS. This problem stimulated a discussion[7,9] of the factors responsible for the *gauche/anti* energy difference in butane that is now accepted[1] to have been largely without physical significance. The parametrization procedure produces only a force field that gives the right results, not one that does so for the right reasons. The MM2 force field also reproduces the butane rotational curve well, but in this case one- and twofold torsional potential functions are largely responsible, the hydrogens being "softer" than in MM1.

Schleyer's *Critical Review of Molecular Mechanics*[8] was, however, more than a review. It also contained data for a series of unknown molecules and for others for which few or no experimental data were available. Thus the roles of theory, which usually tries to reproduce experimental data, and experiment were reversed. In one case in particular, adamantane, there were severe experimental problems in measuring the heat of formation. Adamantane and diamantane compounds represented extrapolations beyond the range of molecules used in the parametrization of the 1973 force fields, so that accurate experimental data were needed to test the reliability of the calculations. When these data became available[10] they revealed small systematic errors in the MM1 and EAS heats of formation for diamondoid molecules, but the newer MM2 force field was able to reproduce the experimental data extremely well, with one or two exceptions. The experimental data for these exceptions were, however, not internally consistent with those for related molecules studied, so the MM2 values are probably more reliable.[1,10]

This illustrates one of the major problems in parametrizing force fields. Experimental data are not always correct. In a small set of molecules, one wrong experimental number may bias the parametrization so that the results for a wider range of substances are not as good as they would have been without the rogue number. There is, however, no foolproof way to identify erroneous experimental results, so in the long run only a large set of parametrization molecules can guarantee a reliable force field.

Another serious problem encountered with very accurate calculations is that the concept of molecular structure is by no means well defined. Even ignoring the fact that the calculations use a motionless molecule to represent a vibrating molecule,[1] the "position" of an atom depends on the method used to detect it. Consider, for instance, two structures for a molecule AH, one determined by a diffraction method, either x-ray or electron diffraction, and the other by microwave spectroscopy. The experimental AH bond lengths determined by the two methods would differ by up to 10%. This is not because of experimental errors, but because the two experiments measure two different distances, both of which are called bond lengths. Diffraction methods detect the centers of electron density, which are usually closely associated with the atomic nuclei. This is, however, not the case for hydrogen atoms, which have no nonvalence electrons. The center of electron density due to a hydrogen is often up to 10% closer to the other atom in the bond than is the hydrogen nucleus itself. Microwave spectroscopy, on the other hand, measures the moments of inertia, which depend on the positions of the nuclei. The "bond lengths" to hydrogen determined by microwave spectroscopy are therefore consistently longer than those found in diffraction experiments. The force field can be parametrized to fit either, but not both. The shift of hydrogen electron density into the bond is reflected in the fact that centers used to calculate hydrogen van der Waals energies in MM2 are also shifted toward the other atom in the bond. In this way the position of the nucleus can be adjusted to fit the microwave data while a moderately realistic center is still retained on which to base the van der Waals calculations. The shift of electron density away from the hydrogen nucleus is nicely demonstrated in FOGO calculations, as shown in Section 5.3.

Although the above discussion is based on alkanes, for which the MM2 force field is most highly developed, parameters have been published for a wide range of different organic or organometallic groups. The MMP2 program itself contains parameters, many of them unpublished, for many of these groups. The parameters contained in a program can be listed by using a single blank card as input. Osawa[11] has published a useful bibliography of available force fields, although it should be emphasized that the force field as a whole should be used, not a mixture of, for instance, alkane functions from MM2 with the carbenium ion parameters from EAS. Among the more recent and exotic applications of MM2 are its use for determining structures and energies of free radicals[12] and to investigate the formation of a bond between two radicals.[13] The use of MM2 to calculate the stability of anti-Bredt olefins[14] seems to be an example of an extrapolation beyond the limits of the parametrization set giving useful results.

This introduction has concentrated on "organic" force fields, but a great deal of work has been devoted to the use of force field methods for biochemical problems.[15,16] This area represents a parallel development to that described above, and suitable programs are available from QCPE.[17] The review literature on molecular mechanics is extensive,[1,5,6,8,15,17-19] so that the background to any given problem is usually easy to find.

2.3. THE MMP2 PROGRAM: INPUT AND OUTPUT EXAMPLES

The following examples all use the full version of the MMP2 program including self consistent field (SCF) calculations for conjugated systems, but those not involving SCF calculations can be used with the MM2 program by omitting the second line of input, which is blank in those cases. The timings given at the ends of the ouputs are for a CYBER 845. The program used for these examples was converted for use on the CDC machine from an IBM source supplied by Molecular Design Ltd. The words "line" and "card" are used interchangeably in the descriptions of the inputs, although punched cards are now rarely used.

The first example, cycloheptene in the chair conformation, includes no conjugated system, and the calculation can therefore also be run on the more limited MM2 program. The first line of the input, shown in Fig. 2.3.1, gives the title and defines some options to be used by the program. The full list of options is given in the MM2 or MMP2 manual. In this case the name of the molecule is used as the title, and may appear in columns 1–60. The number of atoms to be read in is given in columns 61–65, but is always right justified. In this example a "7" is typed in column 65 because we intend to read in only the positions of the seven carbon atoms; the program will add the hydrogens. Note that because the number of atoms (and all other integer options) is right justified, a "7" in column 64 would be interpreted as 70 by the program. The "2" in column 67 controls the information printed by the program. For this example we have requested full printing of the data for the initial and final stages of the calculation (the starting and optimized geometries, respectively), but this is needed only for illustration of the details of the calculation. Normally using a zero or a blank, to give minimum printing of the initial stage and full printing of the data for the optimized geometry, is a reasonable choice. The only other option in this card is the time limit, which is typed in columns 76–80. This time

```
!........!........!........!........!........!........!........!........!
CYCLOHEPTENE                                                  7 2        1.0

    1                                       7               1        0
    1     2     4     6     7     5     3    1
   0.67000  0.00000  0.00000    2        -0.67000  0.00000  0.00000    2
   1.49000  1.10000  0.60000    1        -1.49000  1.10000  0.60000    1
   1.29000  2.45000 -0.10000    1        -1.29000  2.45000 -0.10000    1
   0.00000  3.10000  0.00000    1
    4     1     2     3
    4     2     1     4
    2     3     1     5
    2     4     2     6
    2     5     3     7
    2     6     4     7
    2     7     5     6
```

FIG. 2.3.1. One possible MMP2 input for cycloheptene. In this and the following input examples the first and last lines are included to help determine the column numbers. They are not included in the real input for the program.

limit is an internal one used by the program itself to determine whether it has time to perform another optimization cycle or not. If not enough time is left, the program writes the input for a new job that can then continue from the current geometry. It is therefore important that the time limit given in this card (here 1 minute) not be larger than that requested for the job from the computer operating system. Most computers (the VAX is an exception) require that the maximum time to be used be specified in the first line of the job itself, as opposed to the input for the program. The computer operating system will then terminate the job after it has used the given amount of time, regardless of whether the information has been saved or not. It is therefore necessary to warn the program when this is going to happen so that it can save the current geometry, which would otherwise be lost.

The next line of the input would normally contain information on the delocalized π system for MMP2, but is left blank in this case because the isolated double bond in cycloheptene can be treated classically by the force field (this line is simply omitted for the MM2 program). It is, however, an important feature of the MMP2 program that it would give the same answer if the double bond were treated as a π system (i.e. if the force field were determined using SCF calculations for the double bond) as it does using the standard alkene force field.

The third line of the input requests that the heat of formation be calculated (the "1" in column 65) and gives extra information about how the molecule is to be defined. To explain the definition used here, we must define three important expressions used throughout the MM2 and MMP2 manuals.

Connected Atoms

The bonds between atoms may be defined in three ways, the first of that is the *atom connection list*, of which there may be up to 20 in the input. The atom connection list is simply a list of the numbers of the atoms that are bonded to each other in the order in which they are connected. Consider, for instance, cyclohexane as shown below. The atom connection list could be given as:

1   2   3   4   5   6   1

Note that atom 1 appears twice in the atom connection list to indicate that the structure is cyclic (i.e. atom 1 is bound to both atom 2 and atom 6).

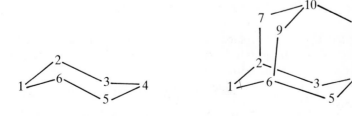

A more complex example is adamantane, which can be defined using three consecutive atom connection lists. One possibility would be to define the lower six-membered ring as before, followed by the bridge joining C_2 and C_4, and finally that joining C_6 and C_{10}:

1	2	3	4	5	6	1
2	7	10	8	4		
6	9	10				

The structure could, however, be defined equally well by starting with an eight-membered ring and then adding two methylene bridges:

1	2	7	10	8	4	5	6	1
2	3	4						
6	9	10						

The numbering of the atoms must agree with that used later in the input when the coordinates are read in, and only atoms whose coordinates are read in can be used in the atom connection lists. One atom connection list occupies one line in the input. The atoms that appear in these lists are termed *connected atoms*.

Attached Atoms

Attached atoms are those that, for the definition of the molecule, are bound only to one other atom. The *attached atom lists* therefore correspond to atom connection lists in which only two atoms appear. If, as in all the examples used here, the program is used to add hydrogens to the carbon atom skeleton a methyl carbon would be considered an attached atom. The MMP2 program works through the input sequentially and actually constructs a molecule composed of only those atoms whose coordinates are read in before modification of or addition to the structure. An attached atom may therefore later be modified, added to and so forth, but can still be defined in the original attached atom list *as long as it is bound to only one other atom whose coordinates are given in the input*. Attached atoms may be converted automatically to connected atoms by the program when it makes a substitution. In contrast to atom connection lists, attached atom lists are written together, up to eight lists per line (16 numbers) and up to a total of 100 attached atoms.

Coordinate Calculations

Coordinate calculation or replacement cards are the third way to define the bonds and atomic positions in the molecule. Atoms defined in this way are not given in the list of atomic coordinates in the input, but are added by the program. Their Cartesian coordinates are calculated from a set of standard bond lengths and angles, and they are added to the atom connection lists and attached atom lists as required. The most common use of the coordinate calculation options is to add hydrogens to the heavy-atom framework, but they may also be used to add other types of atom or to replace an atom by a group such as methyl, methylene, or phenyl. The coordinate calculation options are carried out sequentially, so that an atom that has already been defined *via* a coordinate calculation card may be modified by a later one. This will be illustrated in later examples.

The options given in the third line of Fig. 2.3.1 simply define the number of atom connection lists ("1" in column 5), attached atom lists (zero in column 30), and coordinate calculation cards ("7" in column 40) to be read in. The zero need not be typed explicitly, because a blank is interpreted as a zero.

The fourth card is the atom connection list for cycloheptene, with connections as shown in Fig. 2.3.2. This is exactly analogous to the cyclohexane example given above. There is only one list in this case, as the third card told the program to expect only one atom connection list and no attached atom lists, which would normally come next. Both the atom connection lists and the attached atom lists are typed right justified in columns 5, 10, 15, 20, 25, and so on.

Cards 5 to 8 in Fig. 2.3.1 give the x-, y-, and z-coordinates in Å for the seven carbon atoms in cycloheptene. There are several ways to estimate these

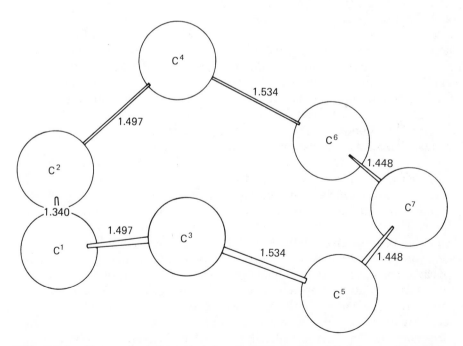

FIG. 2.3.2. The carbon-atom skeleton defined in Fig. 2.3.1 for cycloheptene. Carbons 1 and 2 are the sp^2 atoms.

coordinates, either by way of molecular models or using a suitable program to convert bond lengths and angles into Cartesian coordinates. One very effective and fast method is to make a model of the molecule and place it on a piece of paper marked with a grid the boxes of which correspond to 1 Å on the scale of the model. This grid can be used to obtain x- and y-coordinates, and the distance from the paper gives the z-coordinate. Other more sophisticated (but not necessarily more effective) methods include using the EUCLID program (described in Appendix B) to convert bond lengths, angles, and dihedral angles to Cartesian coordinates; writing a Z-matrix input for MOPAC or GAUSSIAN82 to convert to Cartesian coordinates; or using one of the more sophisticated interactive graphics systems in which a molecular structure can be drawn on the screen in order to obtain a trial set of coordinates.

The coordinates for two atoms are given per line, with code numbers for the atom types. These code numbers are listed in the program manual (in the table of van der Waals parameters), but the relevant ones for the examples given here are "1" for an sp^3 carbon, "2" for an sp^2 carbon, and "5" for a hydrogen. Ten columns are allowed for each coordinate (FORTRAN format F10.5), followed by five columns for the atom type (right justified). The decimal points may be typed to the left of those given in Fig. 2.3.1, but not to the right. The integer numbers (the atom types) must be typed in the correct column. In Fig. 2.3.1 there are four coordinate cards (for seven atoms, as given in line 3). Atoms 1 and 2 are the olefinic carbons and the remainder are the central atoms of the methylene groups, as shown in Fig. 2.3.2.

The definition so far has thus defined the positions and types of the carbon atoms and the bonds between them. The remaining lines of the input are the coordinate calculation cards that add the hydrogens to these carbons. The first number given on each of these cards is the type of substitution or addition to be performed. The options required here are addition of a single hydrogen to an sp^2 carbon atom (option 4) and addition of two hydrogen atoms to an sp^3 carbon atom (option 2). The full list of options is described in the program manual. The second number in each line in this example is the number of the atom to which the additions are to be made. The third and fourth numbers give two other atoms that are also bound to this atom. Thus the first coordinate replacement card (line 9) indicates that a single hydrogen atom (option **4**) will be added to sp^2 carbon number **1**, that is bound also to atoms **2** and **3**. Note that this part of the program does not check that the coordinate calculation requested is consistent with the rest of the input. It will, for instance, happily add hydrogens to an sp^2 carbon with the correct bond lengths and angles for an sp^3 carbon or make a tetracoordinate sp^2 carbon atom. Many of these mistakes will result in a "fatal" error later in the calculation, but some may lead to a very poor starting geometry or even to wrong results. It is therefore always necessary to check that you have added the atoms you want to add with the correct options. This sort of hidden mistake, which frequently occurs when one is changing an input slightly to obtain a related molecule, is often the most difficult to detect in an output. The seven coordinate calculation cards given in

Fig. 2.3.1 simply add the hydrogens to the carbon framework to complete the molecule. A schematic view of the most common type of MMP2 input is shown in Fig. 2.3.3.

The output produced by MMP2 from the above input is shown in Fig. 2.3.4. As noted above, this is the full output for both the initial and the final stages of the calculation. The program first provides a box in which to draw the structure of the molecule, and then summarizes the input data. The atom connection list and the coordinates of the seven carbon atoms are printed, followed by the information that seven coordinate options were used, although these are not printed. After "NEW HEADER CARD" the program prints what would be the input for the molecule if the coordinates for all the atoms, including the hydrogens, were given. The label "(PUNCHED)" does not necessarily mean that the new input is punched on cards: rather, it indicates that the information is written to the output device assigned to channel 7 in the FORTRAN program. Historically, this was the card punch, but it is now more often a disk file. The program writes an input in which there are 19 atoms, one atom connection list (that for the carbon atoms), and 12 attached atoms (the hydrogens). The input coordinates for the seven carbons and those calculated by the program for the 12 hydrogens are then given.

The next section of the output is an analysis of the molecular mechanics energy calculation for the initial geometry. The first table gives the bond lengths, the force field parameters for the bond-stretching energy, and the calculated steric energy for each bond. The expression used by the MMP2

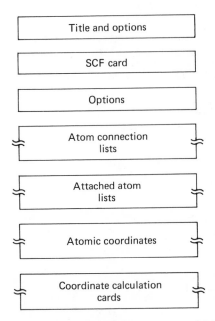

FIG. 2.3.3. Schematic representation of typical MMP2 input.

CYCLOHEPTENE

DATE 84/02/29.

```
          * M M P 2 * * * * * * * * * * 1 9 8 2 *
          *                                      *
          *                                      *
          *                                      *
          *                                      *
          *                                      *
          *                                      *
          *                                      *
          *                                      *
          *                                      *
          *                                      *
          *                                      *
          * * * * * * * * * * * * * * * * * * * *
```

THE COORDINATES OF 7 ATOMS ARE READ IN.

ADDITIONS AND/OR REPLACEMENTS WILL BE MADE TO 7 OF THESE ATOMS.

CONFORMATIONAL ENERGY, PART 1: GEOMETRY AND STERIC ENERGY
                                OF INITIAL CONFORMATION.

CONNECTED ATOMS
    1- 2- 4- 6- 7- 5- 3- 1-

INITIAL ATOMIC COORDINATES

ATOM	X	Y	Z	TYPE
C(1)	.67000	0.00000	0.00000	(2)
C(2)	-.67000	0.00000	0.00000	(2)
C(3)	1.49000	1.10000	.60000	(1)
C(4)	-1.49000	1.10000	.60000	(1)
C(5)	1.29000	2.45000	-.10000	(1)
C(6)	-1.29000	2.45000	-.10000	(1)
C(7)	0.00000	3.10000	0.00000	(1)

----
7 COORDINATE OPTIONS USED

NEW HEADER CARD

CYCLOHEPTENE   19   0   0   (PUNCHED)

    1   0   12   0   (PUNCHED)

NEW CONNECTED ATOM LISTS
    1- 2- 4- 6- 7- 5- 3- 1-   (PUNCHED)

NEW ATTACHED ATOM LISTS
    1- 8, 2- 9, 3-10, 3-11, 4-12, 4-13, 5-14, 5-15,   (PUNCHED)
    6-16, 6-17, 7-18, 7-19,

FIG. 2.3.4. MMP2 output generated using the input shown in Fig. 2.3.1.

NEW ATOMIC COORDINATES

ATOM	X	Y	Z	TYPE	
1	.67000	.00000	.00000	2	(PUNCHED)
2	-.67000	.00000	.00000	2	(PUNCHED)
3	1.49000	1.10000	.60000	1	(PUNCHED)
4	-1.49000	1.10000	.60000	1	(PUNCHED)
5	1.29000	2.45000	-.10000	1	(PUNCHED)
6	-1.29000	2.45000	-.10000	1	(PUNCHED)
7	-.00000	3.10000	.00000	1	(PUNCHED)
8	1.09164	-.89193	-.48651	5	(PUNCHED)
9	-1.09164	-.89193	-.48651	5	(PUNCHED)
10	2.54934	.81199	.53038	5	(PUNCHED)
11	1.20639	1.19788	1.65829	5	(PUNCHED)
12	-1.20639	1.19788	1.65829	5	(PUNCHED)
13	-2.54934	.81199	.53038	5	(PUNCHED)
14	1.47309	2.31687	-1.17645	5	(PUNCHED)
15	2.01681	3.16747	.30863	5	(PUNCHED)
16	-2.01681	3.16747	.30863	5	(PUNCHED)
17	-1.47309	2.31687	-1.17645	5	(PUNCHED)
18	-.00000	3.86462	-.79079	5	(PUNCHED)
19	-.00000	3.59157	.98405	5	(PUNCHED)

BOND LENGTHS AND STRETCHING ENERGY   (19 BONDS)

$$ENERGY = 71.94(KS)(DR)(DR)(1+(CS)(DR))$$
$$DR = R-RO$$
$$CS = -2.000$$

BOND	LENGTH	R(O)	K(S)	ENERGY	E-NEG	R(O)-CORR. ANOM
C(1)- C(2)	1.3400	1.3370	9.6000	.0062		
C(1)- C(3)	1.4975	1.4970	4.4000	.0001		
C(1)- H(8)	1.1000	1.1010	4.6000	.0003		
C(2)- C(4)	1.4975	1.4970	4.4000	.0001		
C(2)- H(9)	1.1000	1.1010	4.6000	.0003		
C(3)- C(5)	1.5338	1.5230	4.4000	.0360		
C(3)- H(10)	1.1000	1.1130	4.6000	.0574		
C(3)- H(11)	1.1000	1.1130	4.6000	.0574		
C(4)- C(6)	1.5338	1.5230	4.4000	.0360		
C(4)- H(12)	1.1000	1.1130	4.6000	.0574		
C(4)- H(13)	1.1000	1.1130	4.6000	.0574		
C(5)- C(7)	1.4480	1.5230	4.4000	2.0497		
C(5)- H(14)	1.1000	1.1130	4.6000	.0574		
C(5)- H(15)	1.1000	1.1130	4.6000	.0574		
C(6)- C(7)	1.4480	1.5230	4.4000	2.0497		
C(6)- H(16)	1.1000	1.1130	4.6000	.0574		
C(6)- H(17)	1.1000	1.1130	4.6000	.0574		
C(7)- H(18)	1.1000	1.1130	4.6000	.0574		
C(7)- H(19)	1.1000	1.1130	4.6000	.0574		

NON-BONDED DISTANCES, VAN DER WAALS ENERGY
   116 VDW INTERACTIONS (1,3 EXCLUDED)

$$ENERGY = KV*(2.90(10**5)EXP(-12.50/P) - 2.25(P**6))$$
$$RV = RVDW(I) + RVDW(K)$$
$$KV = SQRT(EPS(I)*EPS(K))$$
$$P = (RV/R) OR (RV/R\#)$$
$$(IF P.GT.3.311, ENERGY = KV(336.176)(P**2))$$

FIG. 2.3.4. *(continued)*

IN THE VDW CALCULATIONS THE HYDROGEN ATOMS ARE RELOCATED
SO THAT THE ATTACHED HYDROGEN DISTANCE IS REDUCED BY .915

IN THE VDW CALCULATIONS THE DEUTERIUM ATOMS ARE RELOCATED
SO THAT THE ATTACHED DEUTERIUM DISTANCE IS REDUCED BY .915

ATOM PAIR	R	R#	RV	KV	ENERGY	(1,4)
C(1), C(4)	2.4971					
C(1), C(5)	2.5292					
C(1), C(6)	3.1391		3.840	.0440	.1339	*
C(1), C(7)	3.1716		3.840	.0440	.1071	*
C(1), H(9)	2.0336					
C(1), H(10)	2.1148					
C(1), H(11)	2.1148					
C(1), H(12)	2.7759	2.7362	3.440	.0455	.2301	*
C(1), H(13)	3.3623	3.2833	3.440	.0455	-.0485	*
C(1), H(14)	2.7197	2.6866	3.440	.0455	.3084	*
C(1), H(15)	3.4557	3.3729	3.440	.0455	-.0524	*
C(1), H(16)	4.1650	4.0763	3.440	.0455	-.0321	
C(1), H(17)	3.3682	3.3353	3.440	.0455	-.0513	
C(1), H(18)	4.0012	3.9255	3.440	.0455	-.0379	
C(1), H(19)	3.7837	3.7230	3.440	.0455	-.0461	
C(2), C(3)	2.4971					
C(2), C(5)	3.1391		3.840	.0440	.1339	*
C(2), C(6)	2.5292					
C(2), C(7)	3.1716		3.840	.0440	.1071	*
C(2), H(8)	2.0336					
C(2), H(10)	3.3623	3.2833	3.440	.0455	-.0485	*
C(2), H(11)	2.7759	2.7362	3.440	.0455	.2301	*
C(2), H(12)	2.1148					
C(2), H(13)	2.1148					
C(2), H(14)	3.3682	3.3353	3.440	.0455	-.0513	
C(2), H(15)	4.1650	4.0763	3.440	.0455	-.0321	
C(2), H(16)	3.4557	3.3729	3.440	.0455	-.0524	*
C(2), H(17)	2.7197	2.6866	3.440	.0455	.3084	*
C(2), H(18)	4.0012	3.9255	3.440	.0455	-.0379	
C(2), H(19)	3.7837	3.7230	3.440	.0455	-.0461	
C(3), C(4)	2.9800		3.800	.0440	.2801	*
C(3), C(6)	3.1687		3.800	.0440	.0849	*
C(3), C(7)	2.5652					
C(3), H(8)	2.3037					
C(3), H(9)	3.4370	3.3533	3.340	.0460	-.0538	*
C(3), H(12)	2.8983	2.8891	3.340	.0460	.0217	
C(3), H(13)	4.0502	3.9586	3.340	.0460	-.0324	
C(3), H(14)	2.1533					
C(3), H(15)	2.1533					
C(3), H(16)	4.0813	4.0001	3.340	.0460	-.0309	
C(3), H(17)	3.6629	3.6105	3.340	.0460	-.0468	
C(3), H(18)	3.4348	3.3556	3.340	.0460	-.0538	*
C(3), H(19)	2.9284	2.8830	3.340	.0460	.0247	*
C(4), C(5)	3.1687		3.800	.0440	.0849	*
C(4), C(7)	2.5652					
C(4), H(8)	3.4370	3.3533	3.340	.0460	-.0538	*
C(4), H(9)	2.3037					
C(4), H(10)	4.0502	3.9586	3.340	.0460	-.0324	
C(4), H(11)	2.8983	2.8891	3.340	.0460	.0217	
C(4), H(14)	3.6629	3.6105	3.340	.0460	-.0468	

FIG. 2.3.4. (*continued*)

34

C(4), H(15)	4.0813	4.0001	3.340	.0460	-.0309	
C(4), H(16)	2.1533					
C(4), H(17)	2.1533					
C(4), H(18)	3.4348	3.3556	3.340	.0460	-.0538	*
C(4), H(19)	2.9284	2.8830	3.340	.0460	.0247	*
C(5), C(6)	2.5800					
C(5), H(8)	3.3700	3.2926	3.340	.0460	-.0534	*
C(5), H(9)	4.1219	4.0360	3.340	.0460	-.0296	
C(5), H(10)	2.1602					
C(5), H(11)	2.1602					
C(5), H(12)	3.3002	3.2749	3.340	.0460	-.0530	
C(5), H(13)	4.2215	4.1311	3.340	.0460	-.0263	
C(5), H(16)	3.4083	3.3318	3.340	.0460	-.0538	*
C(5), H(17)	2.9684	2.9213	3.340	.0460	.0071	*
C(5), H(18)	2.0353					
C(5), H(19)	2.0353					
C(6), H(8)	4.1219	4.0360	3.340	.0460	-.0296	
C(6), H(9)	3.3700	3.2926	3.340	.0460	-.0534	*
C(6), H(10)	4.2215	4.1311	3.340	.0460	-.0263	
C(6), H(11)	3.3002	3.2749	3.340	.0460	-.0530	
C(6), H(12)	2.1602					
C(6), H(13)	2.1602					
C(6), H(14)	2.9684	2.9213	3.340	.0460	.0071	*
C(6), H(15)	3.4083	3.3318	3.340	.0460	-.0538	*
C(6), H(18)	2.0353					
C(6), H(19)	2.0353					
C(7), H(8)	4.1670	4.0803	3.340	.0460	-.0280	
C(7), H(9)	4.1670	4.0803	3.340	.0460	-.0280	
C(7), H(10)	3.4663	3.3852	3.340	.0460	-.0535	*
C(7), H(11)	2.7970	2.7611	3.340	.0460	.1096	*
C(7), H(12)	2.7970	2.7611	3.340	.0460	.1096	*
C(7), H(13)	3.4663	3.3852	3.340	.0460	-.0535	*
C(7), H(14)	2.0414					
C(7), H(15)	2.0414					
C(7), H(16)	2.0414					
C(7), H(17)	2.0414					
H(8), H(9)	2.1833	2.1116	3.000	.0470	1.1883	*
H(8), H(10)	2.4622	2.3799	3.000	.0470	.2485	*
H(8), H(11)	2.9968	2.8473	3.000	.0470	-.0487	*
H(8), H(12)	3.7747	3.6475	3.000	.0470	-.0293	
H(8), H(13)	4.1466	4.0063	3.000	.0470	-.0179	
H(8), H(14)	3.3042	3.2338	3.000	.0470	-.0482	
H(8), H(15)	4.2387	4.0881	3.000	.0470	-.0160	
H(8), H(16)	5.1743	4.9969	3.000	.0470	-.0049	
H(8), H(17)	4.1654	4.0759	3.000	.0470	-.0162	
H(8), H(18)	4.8897	4.7431	3.000	.0470	-.0067	
H(8), H(19)	4.8431	4.6890	3.000	.0470	-.0072	
H(9), H(10)	4.1466	4.0063	3.000	.0470	-.0179	
H(9), H(11)	3.7747	3.6475	3.000	.0470	-.0293	
H(9), H(12)	2.9968	2.8473	3.000	.0470	-.0487	*
H(9), H(13)	2.4622	2.3799	3.000	.0470	.2485	*
H(9), H(14)	4.1654	4.0759	3.000	.0470	-.0162	
H(9), H(15)	5.1743	4.9969	3.000	.0470	-.0049	
H(9), H(16)	4.2387	4.0881	3.000	.0470	-.0160	
H(9), H(17)	3.3042	3.2338	3.000	.0470	-.0482	
H(9), H(18)	4.8897	4.7431	3.000	.0470	-.0067	
H(9), H(19)	4.8431	4.6890	3.000	.0470	-.0072	

FIG. 2.3.4. (*continued*)

H(10), H(11)	1.7957				
H(10), H(12)	3.9404	3.8476	3.000	.0470	-.0223
H(10), H(13)	5.0987	4.9186	3.000	.0470	-.0054
H(10), H(14)	2.5172	2.4202	3.000	.0470	.1853	*
H(10), H(15)	2.4251	2.3401	3.000	.0470	.3248	*
H(10), H(16)	5.1427	4.9707	3.000	.0470	-.0051
H(10), H(17)	4.6215	4.4939	3.000	.0470	-.0093
H(10), H(18)	4.1908	4.0516	3.000	.0470	-.0168
H(10), H(19)	3.7988	3.6801	3.000	.0470	-.0281
H(11), H(12)	2.4128	2.4610	3.000	.0470	.1329
H(11), H(13)	3.9404	3.8476	3.000	.0470	-.0223
H(11), H(14)	3.0593	2.8962	3.000	.0470	-.0523	*
H(11), H(15)	2.5214	2.4239	3.000	.0470	.1801	*
H(11), H(16)	4.0112	3.9367	3.000	.0470	-.0197
H(11), H(17)	4.0580	3.9449	3.000	.0470	-.0195
H(11), H(18)	3.8164	3.6850	3.000	.0470	-.0279
H(11), H(19)	2.7640	2.7443	3.000	.0470	-.0331
H(12), H(13)	1.7957				
H(12), H(14)	4.0580	3.9449	3.000	.0470	-.0195
H(12), H(15)	4.0112	3.9367	3.000	.0470	-.0197
H(12), H(16)	2.5214	2.4239	3.000	.0470	.1801	*
H(12), H(17)	3.0593	2.8962	3.000	.0470	-.0523	*
H(12), H(18)	3.8164	3.6850	3.000	.0470	-.0279
H(12), H(19)	2.7640	2.7443	3.000	.0470	-.0331
H(13), H(14)	4.6215	4.4939	3.000	.0470	-.0093
H(13), H(15)	5.1427	4.9707	3.000	.0470	-.0051
H(13), H(16)	2.4251	2.3401	3.000	.0470	.3248	*
H(13), H(17)	2.5172	2.4202	3.000	.0470	.1853	*
H(13), H(18)	4.1908	4.0516	3.000	.0470	-.0168
H(13), H(19)	3.7988	3.6801	3.000	.0470	-.0281
H(14), H(15)	1.7957				
H(14), H(16)	3.8870	3.7547	3.000	.0470	-.0253
H(14), H(17)	2.9462	2.9151	3.000	.0470	-.0533
H(14), H(18)	2.1712	2.1024	3.000	.0470	1.2455	*
H(14), H(19)	2.9091	2.7492	3.000	.0470	-.0341	*
H(15), H(16)	4.0336	3.9101	3.000	.0470	-.0204
H(15), H(17)	3.8870	3.7547	3.000	.0470	-.0253
H(15), H(18)	2.4005	2.3016	3.000	.0470	.4137	*
H(15), H(19)	2.1688	2.1003	3.000	.0470	1.2592	*
H(16), H(17)	1.7957				
H(16), H(18)	2.4005	2.3016	3.000	.0470	.4137	*
H(16), H(19)	2.1688	2.1003	3.000	.0470	1.2592	*
H(17), H(18)	2.1712	2.1024	3.000	.0470	1.2455	*
H(17), H(19)	2.9091	2.7492	3.000	.0470	-.0341	*
H(18), H(19)	1.7957				

BOND ANGLES, BENDING AND STRETCH-BEND ENERGIES   (36 ANGLES)

$$EB = 0.021914(KB)(DT)(DT)(1+SF*DT**4)$$
$$DT = THETA-TZERO$$
$$SF = .00700E-5$$

$$ESB(J) = 2.51124(KSB(J))(DT)(DR1+DR2)$$
$$DR(I) = R(I) - RO(I)$$
$$KSB(1) = .120 \quad X-F-Y \quad F = 1ST\ ROW\ ATOM$$
$$KSB(2) = .250 \quad X-S-Y \quad S = 2ND\ ROW\ ATOM$$
$$KSB(3) = .090 \quad X-F-H \quad (DR2 = 0)$$
$$KSB(4) = -.400 \quad X-S-H \quad (DR2 = 0)$$
$$(X,Y = F\ OR\ S)$$

FIG. 2.3.4.   (*continued*)

```
    A T O M S        THETA   TZERO    KB     EB      KSB    ESB
C( 2)- C( 1)- C( 3) 123.202 122.000                  .12    .0013
   IN-PLN  2- 1- 3  123.202 122.000  .550   .0174
   OUT-PL  1- 8- 1   0.000           .050  0.0000

C( 2)- C( 1)- H( 8) 112.539 120.000                  .09   -.0051
   IN-PLN  2- 1- 8  112.539 120.000  .360   .4393
   OUT-PL  1- 3- 1   0.000           .050  0.0000

C( 3)- C( 1)- H( 8) 124.259 118.200                  .09    .0006
   IN-PLN  3- 1- 8  124.259 118.200  .360   .2897
   OUT-PL  1- 2- 1   0.000           .050  0.0000

C( 1)- C( 2)- C( 4) 123.202 122.000                  .12    .0013
   IN-PLN  1- 2- 4  123.202 122.000  .550   .0174
   OUT-PL  2- 9- 2   0.000           .050  0.0000

C( 1)- C( 2)- H( 9) 112.539 120.000                  .09   -.0051
   IN-PLN  1- 2- 9  112.539 120.000  .360   .4393
   OUT-PL  2- 4- 2   0.000           .050  0.0000

C( 4)- C( 2)- H( 9) 124.259 118.200                  .09    .0006
   IN-PLN  4- 2- 9  124.259 118.200  .360   .2897
   OUT-PL  2- 1- 2   0.000           .050  0.0000

C( 1)- C( 3)- C( 5) 113.097 109.500  .450   .1276    .12    .0122
C( 1)- C( 3)- H(10) 108.039 109.410  .360   .0148    .09   -.0001
C( 1)- C( 3)- H(11) 108.039 109.410  .360   .0148    .09   -.0001
C( 5)- C( 3)- H(10) 109.095 109.410  .360   .0008    .09   -.0008
C( 5)- C( 3)- H(11) 109.095 109.410  .360   .0008    .09   -.0008
H(10)- C( 3)- H(11) 109.420 109.400  .320   .0000
C( 2)- C( 4)- C( 6) 113.097 109.500  .450   .1276    .12    .0122
C( 2)- C( 4)- H(12) 108.039 109.410  .360   .0148    .09   -.0001
C( 2)- C( 4)- H(13) 108.039 109.410  .360   .0148    .09   -.0001
C( 6)- C( 4)- H(12) 109.095 109.410  .360   .0008    .09   -.0008
C( 6)- C( 4)- H(13) 109.095 109.410  .360   .0008    .09   -.0008
H(12)- C( 4)- H(13) 109.420 109.400  .320   .0000
C( 3)- C( 5)- C( 7) 118.670 109.500  .450   .8297    .12   -.1775
C( 3)- C( 5)- H(14) 108.566 109.410  .360   .0056    .09   -.0021
C( 3)- C( 5)- H(15) 108.566 109.410  .360   .0056    .09   -.0021
C( 7)- C( 5)- H(14) 105.676 109.410  .360   .1100    .09    .0633
C( 7)- C( 5)- H(15) 105.676 109.410  .360   .1100    .09    .0633
H(14)- C( 5)- H(15) 109.420 109.400  .320   .0000
C( 4)- C( 6)- C( 7) 118.670 109.500  .450   .8297    .12   -.1775
C( 4)- C( 6)- H(16) 108.566 109.410  .360   .0056    .09   -.0021
C( 4)- C( 6)- H(17) 108.566 109.410  .360   .0056    .09   -.0021
C( 7)- C( 6)- H(16) 105.676 109.410  .360   .1100    .09    .0633
C( 7)- C( 6)- H(17) 105.676 109.410  .360   .1100    .09    .0633
H(16)- C( 6)- H(17) 109.420 109.400  .320   .0000
C( 5)- C( 7)- C( 6) 125.975 109.500  .450  2.6903    .12   -.7450
C( 5)- C( 7)- H(18) 105.212 109.410  .360   .1390    .09    .0712
C( 5)- C( 7)- H(19) 105.212 109.410  .360   .1390    .09    .0712
C( 6)- C( 7)- H(18) 105.212 109.410  .360   .1390    .09    .0712
C( 6)- C( 7)- H(19) 105.212 109.410  .360   .1390    .09    .0712
H(18)- C( 7)- H(19) 109.420 109.400  .320   .0000
```

DIHEDRAL ANGLES, TORSIONAL ENERGY (ET)   (52 ANGLES)

$$ET = (V1/2)(1+COS(W))+(V2/2)(1-COS(2W))+(V3/2)(1+COS(3W))$$

FIG. 2.3.4. (*continued*)

SIGN OF ANGLE A-B-C-D : WHEN LOOKING THROUGH B TOWARD C,
IF D IS COUNTERCLOCKWISE FROM A, NEGATIVE.

A T O M S				OMEGA	V1	V2	V3	ET
C(1)	C(2)	C(4)	C(6)	63.352	-.440	.240	.060	-.126
C(1)	C(2)	C(4)	H(12)	-57.507	0.000	0.000	-.240	-.001
C(1)	C(2)	C(4)	H(13)	-175.789	0.000	0.000	-.240	-.003
C(1)	C(3)	C(5)	C(7)	66.746	.170	.270	.093	.349
C(1)	C(3)	C(5)	H(14)	-53.819	0.000	0.000	.500	.013
C(1)	C(3)	C(5)	H(15)	-172.688	0.000	0.000	.500	.018
C(2)	C(1)	C(3)	C(5)	-63.352	-.440	.240	.060	-.126
C(2)	C(1)	C(3)	H(10)	175.789	0.000	0.000	-.240	-.003
C(2)	C(1)	C(3)	H(11)	57.507	0.000	0.000	-.240	-.001
C(2)	C(4)	C(6)	C(7)	-66.746	.170	.270	.093	.349
C(2)	C(4)	C(6)	H(16)	172.688	0.000	0.000	.500	.018
C(2)	C(4)	C(6)	H(17)	53.819	0.000	0.000	.500	.013
C(3)	C(1)	C(2)	C(4)	0.000	-.100	15.000	0.000	-.100
C(3)	C(1)	C(2)	H(9)	-180.000	0.000	15.000	0.000	0.000
C(3)	C(5)	C(7)	C(6)	-41.809	.200	.270	.093	.314
C(3)	C(5)	C(7)	H(18)	-164.043	0.000	0.000	.267	.044
C(3)	C(5)	C(7)	H(19)	80.425	0.000	0.000	.267	.069
C(4)	C(2)	C(1)	H(8)	180.000	0.000	15.000	0.000	0.000
C(4)	C(6)	C(7)	C(5)	41.809	.200	.270	.093	.314
C(4)	C(6)	C(7)	H(18)	164.043	0.000	0.000	.267	.044
C(4)	C(6)	C(7)	H(19)	-80.425	0.000	0.000	.267	.069
C(5)	C(3)	C(1)	H(8)	116.648	0.000	0.000	.010	.010
C(5)	C(7)	C(6)	H(16)	163.839	0.000	0.000	.267	.045
C(5)	C(7)	C(6)	H(17)	-80.221	0.000	0.000	.267	.068
C(6)	C(4)	C(2)	H(9)	-116.648	0.000	0.000	.010	.010
C(6)	C(7)	C(5)	H(14)	80.221	0.000	0.000	.267	.068
C(6)	C(7)	C(5)	H(15)	-163.839	0.000	0.000	.267	.045
C(7)	C(5)	C(3)	H(10)	-172.996	0.000	0.000	.267	.009
C(7)	C(5)	C(3)	H(11)	-53.511	0.000	0.000	.267	.008
C(7)	C(6)	C(4)	H(12)	53.511	0.000	0.000	.267	.008
C(7)	C(6)	C(4)	H(13)	172.996	0.000	0.000	.267	.009
H(8)	C(1)	C(2)	H(9)	0.000	0.000	15.000	0.000	0.000
H(8)	C(1)	C(3)	H(10)	-4.211	0.000	0.000	.520	.514
H(8)	C(1)	C(3)	H(11)	-122.493	0.000	0.000	.520	.518
H(9)	C(2)	C(4)	H(12)	122.493	0.000	0.000	.520	.518
H(9)	C(2)	C(4)	H(13)	4.211	0.000	0.000	.520	.514
H(10)	C(3)	C(5)	H(14)	66.438	0.000	0.000	.237	.007
H(10)	C(3)	C(5)	H(15)	-52.430	0.000	0.000	.237	.009
H(11)	C(3)	C(5)	H(14)	-174.077	0.000	0.000	.237	.006
H(11)	C(3)	C(5)	H(15)	67.054	0.000	0.000	.237	.008
H(12)	C(4)	C(6)	H(16)	-67.054	0.000	0.000	.237	.008
H(12)	C(4)	C(6)	H(17)	174.077	0.000	0.000	.237	.006
H(13)	C(4)	C(6)	H(16)	52.430	0.000	0.000	.237	.009
H(13)	C(4)	C(6)	H(17)	-66.438	0.000	0.000	.237	.007
H(14)	C(5)	C(7)	H(18)	-42.013	0.000	0.000	.237	.049
H(14)	C(5)	C(7)	H(19)	-157.545	0.000	0.000	.237	.073
H(15)	C(5)	C(7)	H(18)	73.927	0.000	0.000	.237	.030
H(15)	C(5)	C(7)	H(19)	-41.605	0.000	0.000	.237	.051
H(16)	C(6)	C(7)	H(18)	-73.927	0.000	0.000	.237	.030
H(16)	C(6)	C(7)	H(19)	41.605	0.000	0.000	.237	.051
H(17)	C(6)	C(7)	H(18)	42.013	0.000	0.000	.237	.049
H(17)	C(6)	C(7)	H(19)	157.545	0.000	0.000	.237	.073

FIG. 2.3.4. (*continued*)

DIPOLE INTERACTION ENERGY (DIELECTRIC CONSTANT = 1.50)

```
   DIPOLE(1)     MU(1)      DIPOLE(2)     MU(2)      R12(A)    E(KCAL*DC)
 + C( 3)- C( 1)   .300    + C( 4)- C( 2)   .300    2.16000     .1113
```

INITIAL STERIC ENERGY IS   24.2941 KCAL.

```
        COMPRESSION        4.7522
        BENDING            7.1786
        STRETCH-BEND       -.5559
        VANDERWAALS
          1,4 ENERGY      10.1852
          OTHER           -1.4385
        TORSIONAL          4.0611
        DIPOLE              .1113
```

DIPOLE MOMENT =   .502 D

    COMPONENTS WITH PRINCIPAL AXES

        X=   .0000   Y=   .4407   Z=   .2404

INITIAL ENERGY REQUIRES   .47 SECONDS.

CYCLOHEPTENE

CONFORMATIONAL ENERGY, PART 2:   ENERGY MINIMIZATION

INITIAL ENERGY CALCD FROM SYMMETRIZED COORDS:

```
TOTAL ENERGY IS      24.2941 KCAL.
  COMPRESS    4.7522    VANDERWAALS                TORSION      4.0611
  BENDING     7.1786    1,4          10.1852
  STR-BEND    -.5559    OTHER        -1.4385       DIPL/CHG      .1113
```


* * * * * * * * * * * *   C Y C L E   1 * * * * * * * * * * * * * *

                    (CH)-MOVEMENT = 1

```
     ITERATION    1    AVG. MOVEMENT =   .03992 A
     ITERATION    2    AVG. MOVEMENT =   .01091 A
     ITERATION    3    AVG. MOVEMENT =   .00730 A
     ITERATION    4    AVG. MOVEMENT =   .00587 A
     ITERATION    5    AVG. MOVEMENT =   .00469 A
```

```
TOTAL ENERGY IS      10.1766 KCAL.
  COMPRESS     .3298    VANDERWAALS                TORSION      3.9667
  BENDING    1.5838    1,4           5.3724
  STR-BEND    .1522    OTHER        -1.3385        DIPL/CHG      .1101
```

FIG. 2.3.4.   (*continued*)

39

```
DELTA(T) =     .63 SEC.        ELAPSED TIME =    1.10 SEC.

        ITERATION    6    AVG. MOVEMENT =  .00380 A
        ITERATION    7    AVG. MOVEMENT =  .00311 A
        ITERATION    8    AVG. MOVEMENT =  .00258 A
        ITERATION    9    AVG. MOVEMENT =  .00216 A
        ITERATION   10    AVG. MOVEMENT =  .00182 A


TOTAL ENERGY IS      9.9808 KCAL.
   COMPRESS    .2958    VANDERWAALS              TORSION      3.9533
   BENDING    1.5411    1,4         5.3095
   STR-BEND    .1448    OTHER      -1.3728       DIPL/CHG      .1091


DELTA(T) =     .62 SEC.        ELAPSED TIME =    1.72 SEC.

        ITERATION   11    AVG. MOVEMENT =  .00155 A
        ITERATION   12    AVG. MOVEMENT =  .00132 A
        ITERATION   13    AVG. MOVEMENT =  .00113 A
        ITERATION   14    AVG. MOVEMENT =  .00097 A
        ITERATION   15    AVG. MOVEMENT =  .00084 A


TOTAL ENERGY IS      9.9440 KCAL.
   COMPRESS    .2820    VANDERWAALS              TORSION      3.9216
   BENDING    1.5862    1,4         5.2876
   STR-BEND    .1428    OTHER      -1.3848       DIPL/CHG      .1086


DELTA(T) =     .62 SEC.        ELAPSED TIME =    2.34 SEC.

        ITERATION   16    AVG. MOVEMENT =  .00072 A
        ITERATION   17    AVG. MOVEMENT =  .00063 A
        ITERATION   18    AVG. MOVEMENT =  .00054 A
        ITERATION   19    AVG. MOVEMENT =  .00047 A
        ITERATION   20    AVG. MOVEMENT =  .00041 A


TOTAL ENERGY IS      9.9356 KCAL.
   COMPRESS    .2759    VANDERWAALS              TORSION      3.9020
   BENDING    1.6176    1,4         5.2789

   STR-BEND    .1420    OTHER      -1.3890       DIPL/CHG      .1083

DELTA(T) =     .62 SEC.        ELAPSED TIME =    2.95 SEC.

        ITERATION   21    AVG. MOVEMENT =  .00036 A
        ITERATION   22    AVG. MOVEMENT =  .00031 A
        ITERATION   23    AVG. MOVEMENT =  .00028 A
        ITERATION   24    AVG. MOVEMENT =  .00024 A
        ITERATION   25    AVG. MOVEMENT =  .00021 A


TOTAL ENERGY IS      9.9335 KCAL.
   COMPRESS    .2729    VANDERWAALS              TORSION      3.8911
   BENDING    1.6357    1,4         5.2747
   STR-BEND    .1417    OTHER      -1.3908       DIPL/CHG      .1081
```

FIG. 2.3.4. (*continued*)

```
DELTA(T) =     .60 SEC.          ELAPSED TIME =    3.56 SEC.


         ITERATION  26      AVG. MOVEMENT =   .00018 A
         ITERATION  27      AVG. MOVEMENT =   .00016 A
         ITERATION  28      AVG. MOVEMENT =   .00014 A
         ITERATION  29      AVG. MOVEMENT =   .00012 A
         ITERATION  30      AVG. MOVEMENT =   .00010 A


TOTAL ENERGY IS        9.9330 KCAL.
   COMPRESS     .2714      VANDERWAALS              TORSION       3.8854
   BENDING     1.6453      1,4          5.2728
   STR-BEND     .1415      OTHER       -1.3915      DIPL/CHG       .1081


DELTA(T) =     .57 SEC.          ELAPSED TIME =    4.13 SEC.



* * * * * * * * * * * *    C Y C L E   2 * * * * * * * * * * * * *

                        (CH)-MOVEMENT = 0


         ITERATION  31     AVG. MOVEMENT =   .00005 A


TOTAL ENERGY IS        9.9330 KCAL.
   COMPRESS     .2711      VANDERWAALS              TORSION       3.8850
   BENDING     1.6461      1,4          5.2729
   STR-BEND     .1414      OTHER       -1.3917      DIPL/CHG       .1081

DELTA(T) =     .13 SEC.          ELAPSED TIME =    4.26 SEC.


* * * * * ENERGY IS MINIMIZED WITHIN   .0015 KCAL * * * * *


      * * * * * ENERGY IS     9.9330 KCAL * * * * *


--------------------------------------------------------------------
   .67042    -.05553     .09876    2       -.66996   -.05639     .10131    2
  1.52380    1.09712     .55526    1      -1.52306   1.09527     .56078    1
  1.29293    2.37001    -.26998    1      -1.29664   2.36834    -.26542    1
   -.00173   3.11741     .08088    1       1.19625   -.95667    -.26289    5
  -1.19600    -.95821    -.25831    5       2.59375    .79243     .46243    5
  1.34436    1.28929    1.63875    5      -1.33992   1.28777    1.64359    5
  -2.59297    .78930     .47172    5       1.31910   2.12981   -1.35953    5
  2.14597    3.06578    -.07822    5      -2.14988   3.06306    -.07077    5
  -1.32628    2.12797   -1.35485    5       -.00332   4.08796    -.47280    5
   .00002    3.37784    1.16610    5

--------------------------------------------------------------------
```

FIG. 2.3.4. (*continued*)

41

CYCLOHEPTENE

C(7) H(12)

```
     * M M P 2 * * * * * * * * * * 1 9 8 2 *          DATE  84/02/29.
     *                                      *
     *                                      *
     *                                      *
     *                                      *
     *                                      *
     *                                      *
     *                                      *
     *                                      *
     *                                      *
     *                                      *
     *                                      *
     * * * * * * * * * * * * * * * * * * * *
```

CONFORMATIONAL ENERGY, PART 3:   GEOMETRY AND STERIC ENERGY
                                 OF FINAL CONFORMATION.

CONNECTED ATOMS
     1- 2- 4- 6- 7- 5- 3- 1-

ATTACHED ATOMS
     1- 8, 2- 9, 3-10, 3-11, 4-12, 4-13, 5-14, 5-15,
     6-16, 6-17, 7-18, 7-19,

FINAL ATOMIC COORDINATES AND BONDED ATOM TABLE

ATOM	X	Y	Z	TYPE	BOUND TO ATOMS			PI #
C(1)	.67042	-.05553	.09876	(2)	2	3	8	
C(2)	-.66996	-.05639	.10131	(2)	1	4	9	
C(3)	1.52380	1.09712	.55526	(1)	1	5	10	11
C(4)	-1.52306	1.09527	.56078	(1)	2	6	12	13
C(5)	1.29293	2.37001	-.26998	(1)	3	7	14	15
C(6)	-1.29664	2.36834	-.26542	(1)	4	7	16	17
C(7)	-.00173	3.11741	.08088	(1)	5	6	18	19
H(8)	1.19625	-.95667	-.26289	(5)	1			
H(9)	-1.19600	-.95821	-.25831	(5)	2			
H(10)	2.59375	.79243	.46243	(5)	3			
H(11)	1.34436	1.28929	1.63875	(5)	3			
H(12)	-1.33992	1.28777	1.64359	(5)	4			
H(13)	-2.59297	.78930	.47172	(5)	4			
H(14)	1.31910	2.12981	-1.35953	(5)	5			
H(15)	2.14597	3.06578	-.07822	(5)	5			
H(16)	-2.14988	3.06306	-.07077	(5)	6			
H(17)	-1.32628	2.12797	-1.35485	(5)	6			
H(18)	-.00332	4.08796	-.47280	(5)	7			
H(19)	.00002	3.37784	1.16610	(5)	7			

AROMATIC TORSIONAL CONSTANTS ARE FACTORED BY   0.0000

BOND LENGTHS AND STRETCHING ENERGY   (19 BONDS)

     ENERGY = 71.94(KS)(DR)(DR)(1+(CS)(DR))
               DR = R-RO
               CS = -2.000

FIG. 2.3.4. (*continued*)

42

BOND	LENGTH	R(O)	K(S)	ENERGY	E-NEG	ANOM
					R(O)-CORR.	
C(1)- C(2)	1.3404	1.3370	9.6000	.0078		
C(1)- C(3)	1.5051	1.4970	4.4000	.0203		
C(1)- H(8)	1.1042	1.1010	4.6000	.0034		
C(2)- C(4)	1.5051	1.4970	4.4000	.0203		
C(2)- H(9)	1.1042	1.1010	4.6000	.0034		
C(3)- C(5)	1.5345	1.5230	4.4000	.0406		
C(3)- H(10)	1.1164	1.1130	4.6000	.0037		
C(3)- H(11)	1.1149	1.1130	4.6000	.0012		
C(4)- C(6)	1.5345	1.5230	4.4000	.0407		
C(4)- H(12)	1.1149	1.1130	4.6000	.0012		
C(4)- H(13)	1.1164	1.1130	4.6000	.0037		
C(5)- C(7)	1.5355	1.5230	4.4000	.0484		
C(5)- H(14)	1.1160	1.1130	4.6000	.0030		
C(5)- H(15)	1.1174	1.1130	4.6000	.0063		
C(6)- C(7)	1.5355	1.5230	4.4000	.0484		
C(6)- H(16)	1.1174	1.1130	4.6000	.0063		
C(6)- H(17)	1.1160	1.1130	4.6000	.0030		
C(7)- H(18)	1.1174	1.1130	4.6000	.0063		
C(7)- H(19)	1.1160	1.1130	4.6000	.0030		

NON-BONDED DISTANCES, VAN DER WAALS ENERGY
   116 VDW INTERACTIONS (1,3 EXCLUDED)

$$\text{ENERGY} = KV*(2.90(10**5)EXP(-12.50/P) - 2.25(P**6))$$
$$RV = RVDW(I) + RVDW(K)$$
$$KV = SQRT(EPS(I)*EPS(K))$$
$$P = (RV/R) \text{ OR } (RV/R\#)$$
$$(\text{IF P.GT.3.311, ENERGY} = KV(336.176)(P**2))$$

IN THE VDW CALCULATIONS THE HYDROGEN ATOMS ARE RELOCATED
SO THAT THE ATTACHED HYDROGEN DISTANCE IS REDUCED BY .915

IN THE VDW CALCULATIONS THE DEUTERIUM ATOMS ARE RELOCATED
SO THAT THE ATTACHED DEUTERIUM DISTANCE IS REDUCED BY   .915

ATOM PAIR	R	R#	RV	KV	ENERGY	(1,4)
C(1), C(4)	2.5198					
C(1), C(5)	2.5312					
C(1), C(6)	3.1428		3.840	.0440	.1307	*
C(1), C(7)	3.2434		3.840	.0440	.0589	*
C(1), H(9)	2.1038					
C(1), H(10)	2.1332					
C(1), H(11)	2.1527					
C(1), H(12)	2.8692	2.8241	3.440	.0455	.1265	*
C(1), H(13)	3.3915	3.3117	3.440	.0455	-.0502	*
C(1), H(14)	2.7061	2.6736	3.440	.0455	.3318	*
C(1), H(15)	3.4570	3.3738	3.440	.0455	-.0525	*
C(1), H(16)	4.2081	4.1165	3.440	.0455	-.0306	
C(1), H(17)	3.2966	3.2690	3.440	.0455	-.0475	
C(1), H(18)	4.2366	4.1498	3.440	.0455	-.0295	
C(1), H(19)	3.6574	3.6107	3.440	.0455	-.0501	
C(2), C(3)	2.5198					
C(2), C(5)	3.1430		3.840	.0440	.1305	*
C(2), C(6)	2.5311					

FIG. 2.3.4. (*continued*)

C(2), C(7)	3.2435		3.840	.0440	.0589	*
C(2), H(8)	2.1038					
C(2), H(10)	3.3916	3.3118	3.440	.0455	-.0502	*
C(2), H(11)	2.8692	2.8241	3.440	.0455	.1266	*
C(2), H(12)	2.1527					
C(2), H(13)	2.1332					
C(2), H(14)	3.2970	3.2694	3.440	.0455	-.0475	
C(2), H(15)	4.2083	4.1167	3.440	.0455	-.0306	
C(2), H(16)	3.4570	3.3738	3.440	.0455	-.0525	*
C(2), H(17)	2.7060	2.6736	3.440	.0455	.3319	*
C(2), H(18)	4.2367	4.1498	3.440	.0455	-.0295	
C(2), H(19)	3.6574	3.6107	3.440	.0455	-.0501	
C(3), C(4)	3.0469		3.800	.0440	.1938	*
C(3), C(6)	3.2007		3.800	.0440	.0642	*
C(3), C(7)	2.5756					
C(3), H(8)	2.2349					
C(3), H(9)	3.5048	3.4182	3.340	.0460	-.0530	*
C(3), H(12)	3.0695	3.0518	3.340	.0460	-.0317	
C(3), H(13)	4.1291	4.0364	3.340	.0460	-.0296	
C(3), H(14)	2.1851					
C(3), H(15)	2.1596					
C(3), H(16)	4.2134	4.1252	3.340	.0460	-.0265	
C(3), H(17)	3.5825	3.5380	3.340	.0460	-.0496	
C(3), H(18)	3.5120	3.4282	3.340	.0460	-.0528	*
C(3), H(19)	2.8101	2.7735	3.340	.0460	.0985	*
C(4), C(5)	3.2008		3.800	.0440	.0642	*
C(4), C(7)	2.5756					
C(4), H(8)	3.5048	3.4182	3.340	.0460	-.0530	*
C(4), H(9)	2.2349					
C(4), H(10)	4.1291	4.0364	3.340	.0460	-.0296	
C(4), H(11)	3.0695	3.0518	3.340	.0460	-.0317	
C(4), H(14)	3.5827	3.5382	3.340	.0460	-.0496	
C(4), H(15)	4.2134	4.1253	3.340	.0460	-.0265	
C(4), H(16)	2.1596					
C(4), H(17)	2.1851					
C(4), H(18)	3.5120	3.4282	3.340	.0460	-.0528	*
C(4), H(19)	2.8100	2.7735	3.340	.0460	.0986	*
C(5), C(6)	2.5896					
C(5), H(8)	3.3281	3.2534	3.340	.0460	-.0524	*
C(5), H(9)	4.1560	4.0680	3.340	.0460	-.0284	
C(5), H(10)	2.1719					
C(5), H(11)	2.1940					
C(5), H(12)	3.4300	3.3969	3.340	.0460	-.0533	
C(5), H(13)	4.2602	4.1690	3.340	.0460	-.0251	
C(5), H(16)	3.5175	3.4343	3.340	.0460	-.0526	*
C(5), H(17)	2.8453	2.8073	3.340	.0460	.0715	*
C(5), H(18)	2.1616					
C(5), H(19)	2.1794					
C(6), H(8)	4.1557	4.0678	3.340	.0460	-.0285	
C(6), H(9)	3.3281	3.2534	3.340	.0460	-.0524	*
C(6), H(10)	4.2601	4.1689	3.340	.0460	-.0251	
C(6), H(11)	3.4300	3.3969	3.340	.0460	-.0533	
C(6), H(12)	2.1940					
C(6), H(13)	2.1719					
C(6), H(14)	2.8454	2.8073	3.340	.0460	.0714	*
C(6), H(15)	3.5175	3.4343	3.340	.0460	-.0526	*
C(6), H(18)	2.1617					

FIG. 2.3.4. (*continued*)

```
C( 6), H(19)    2.1794
C( 7), H( 8)    4.2605    4.1723    3.340    .0460    -.0250
C( 7), H( 9)    4.2605    4.1723    3.340    .0460    -.0250
C( 7), H(10)    3.5054    3.4220    3.340    .0460    -.0529    *
C( 7), H(11)    2.7533    2.7210    3.340    .0460     .1501    *
C( 7), H(12)    2.7533    2.7209    3.340    .0460     .1502    *
C( 7), H(13)    3.5053    3.4220    3.340    .0460    -.0529    *
C( 7), H(14)    2.1897
C( 7), H(15)    2.1542
C( 7), H(16)    2.1542
C( 7), H(17)    2.1897
H( 8), H( 9)    2.3923    2.3028    3.000    .0470     .4107    *
H( 8), H(10)    2.3534    2.2812    3.000    .0470     .4681    *
H( 8), H(11)    2.9466    2.8005    3.000    .0470    -.0432    *
H( 8), H(12)    3.8864    3.7544    3.000    .0470    -.0253
H( 8), H(13)    4.2363    4.0901    3.000    .0470    -.0159
H( 8), H(14)    3.2778    3.2064    3.000    .0470    -.0494
H( 8), H(15)    4.1372    3.9969    3.000    .0470    -.0181
H( 8), H(16)    5.2337    5.0530    3.000    .0470    -.0046
H( 8), H(17)    4.1316    4.0447    3.000    .0470    -.0170
H( 8), H(18)    5.1895    5.0239    3.000    .0470    -.0048
H( 8), H(19)    4.7181    4.5797    3.000    .0470    -.0083
H( 9), H(10)    4.2363    4.0901    3.000    .0470    -.0159
H( 9), H(11)    3.8863    3.7543    3.000    .0470    -.0253
H( 9), H(12)    2.9466    2.8005    3.000    .0470    -.0432    *
H( 9), H(13)    2.3533    2.2812    3.000    .0470     .4682    *
H( 9), H(14)    4.1321    4.0452    3.000    .0470    -.0169
H( 9), H(15)    5.2339    5.0532    3.000    .0470    -.0046
H( 9), H(16)    4.1371    3.9969    3.000    .0470    -.0181
H( 9), H(17)    3.2778    3.2063    3.000    .0470    -.0495
H( 9), H(18)    5.1896    5.0240    3.000    .0470    -.0048
H( 9), H(19)    4.7181    4.5797    3.000    .0470    -.0083
H(10), H(11)    1.7865
H(10), H(12)    4.1369    4.0325    3.000    .0470    -.0172
H(10), H(13)    5.1867    5.0048    3.000    .0470    -.0049
H(10), H(14)    2.5948    2.4897    3.000    .0470     .1022    *
H(10), H(15)    2.3793    2.3004    3.000    .0470     .4168    *
H(10), H(16)    5.2860    5.1055    3.000    .0470    -.0043
H(10), H(17)    4.5225    4.4061    3.000    .0470    -.0104
H(10), H(18)    4.2988    4.1522    3.000    .0470    -.0146
H(10), H(19)    3.7292    3.6149    3.000    .0470    -.0306
H(11), H(12)    2.6843    2.7151    3.000    .0470    -.0260
H(11), H(13)    4.1370    4.0325    3.000    .0470    -.0172
H(11), H(14)    3.1140    2.9475    3.000    .0470    -.0544    *
H(11), H(15)    2.5974    2.4917    3.000    .0470     .1001    *
H(11), H(16)    4.2753    4.1808    3.000    .0470    -.0141
H(11), H(17)    4.0985    3.9821    3.000    .0470    -.0185
H(11), H(18)    3.7560    3.6351    3.000    .0470    -.0298
H(11), H(19)    2.5284    2.5318    3.000    .0470     .0646
H(12), H(13)    1.7866
H(12), H(14)    4.0986    3.9822    3.000    .0470    -.0185
H(12), H(15)    4.2752    4.1807    3.000    .0470    -.0141
H(12), H(16)    2.5975    2.4918    3.000    .0470     .1001    *
H(12), H(17)    3.1140    2.9475    3.000    .0470    -.0544    *
H(12), H(18)    3.7559    3.6350    3.000    .0470    -.0298
H(12), H(19)    2.5282    2.5317    3.000    .0470     .0647
H(13), H(14)    4.5227    4.4063    3.000    .0470    -.0104
```

FIG. 2.3.4. (*continued*)

```
H(13), H(15)   5.2860   5.1055   3.000   .0470   -.0043
H(13), H(16)   2.3792   2.3004   3.000   .0470    .4169    *
H(13), H(17)   2.5948   2.4897   3.000   .0470    .1022    *
H(13), H(18)   4.2988   4.1521   3.000   .0470   -.0146
H(13), H(19)   3.7291   3.6148   3.000   .0470   -.0306
H(14), H(15)   1.7893
H(14), H(16)   3.8165   3.6934   3.000   .0470   -.0275
H(14), H(17)   2.6454   2.6406   3.000   .0470   -.0003
H(14), H(18)   2.5238   2.4274   3.000   .0470    .1753    *
H(14), H(19)   3.1107   2.9440   3.000   .0470   -.0543    *
H(15), H(16)   4.2959   4.1508   3.000   .0470   -.0147
H(15), H(17)   3.8165   3.6934   3.000   .0470   -.0275
H(15), H(18)   2.4125   2.3280   3.000   .0470    .3509    *
H(15), H(19)   2.5002   2.4056   3.000   .0470    .2068    *
H(16), H(17)   1.7893
H(16), H(18)   2.4124   2.3280   3.000   .0470    .3510    *
H(16), H(19)   2.5002   2.4056   3.000   .0470    .2067    *
H(17), H(18)   2.5238   2.4274   3.000   .0470    .1752    *
H(17), H(19)   3.1107   2.9440   3.000   .0470   -.0543    *
H(18), H(19)   1.7861
```

BOND ANGLES, BENDING AND STRETCH-BEND ENERGIES   (36 ANGLES)

$$EB = 0.021914(KB)(DT)(DT)(1+SF*DT**4)$$
$$DT = THETA-TZERO$$
$$SF = .00700E-5$$

$$ESB(J) = 2.51124(KSB(J))(DT)(DR1+DR2)$$
$$DR(I) = R(I) - RO(I)$$

```
KSB(1) =  .120   X-F-Y   F = 1ST ROW ATOM
KSB(2) =  .250   X-S-Y   S = 2ND ROW ATOM
KSB(3) =  .090   X-F-H   (DR2 = 0)
KSB(4) = -.400   X-S-H   (DR2 = 0)
                 (X,Y = F OR S)
```

```
      A T O M S       THETA    TZERO     KB      EB     KSB    ESB
C( 2)- C( 1)- C( 3) 124.535  122.000                    .12    .0088
  IN-PLN  2- 1- 3   124.535  122.000   .550   .0775
  OUT-PL  1- 8- 1      .040            .050   .0000
C( 2)- C( 1)- H( 8) 118.443  120.000                    .09   -.0012
  IN-PLN  2- 1- 8   118.443  120.000   .360   .0191
  OUT-PL  1- 3- 1      .029            .050   .0000
C( 3)- C( 1)- H( 8) 117.022  118.200                    .09   -.0022
  IN-PLN  3- 1- 8   117.022  118.200   .360   .0110
  OUT-PL  1- 2- 1      .033            .050   .0000
C( 1)- C( 2)- C( 4) 124.535  122.000                    .12    .0087
  IN-PLN  1- 2- 4   124.535  122.000   .550   .0775
  OUT-PL  2- 9- 2      .043            .050   .0000
C( 1)- C( 2)- H( 9) 118.443  120.000                    .09   -.0012
  IN-PLN  1- 2- 9   118.443  120.000   .360   .0191
  OUT-PL  2- 4- 2      .032            .050   .0000
C( 4)- C( 2)- H( 9) 117.021  118.200                    .09   -.0021
  IN-PLN  4- 2- 9   117.021  118.200   .360   .0110
  OUT-PL  2- 1- 2      .036            .050   .0000
C( 1)- C( 3)- C( 5) 112.760  109.500   .450   .1048     .12    .0192
C( 1)- C( 3)- H(10) 108.010  109.410   .360   .0155     .09   -.0026
```

FIG. 2.3.4.   (*continued*)

```
C( 1)- C( 3)- H(11) 109.603 109.410   .360  .0003  .09  .0004
C( 5)- C( 3)- H(10) 109.020 109.410   .360  .0012  .09 -.0010
C( 5)- C( 3)- H(11) 110.820 109.410   .360  .0157  .09  .0037
H(10)- C( 3)- H(11) 106.385 109.400   .320  .0637
C( 2)- C( 4)- C( 6) 112.757 109.500   .450  .1046  .12  .0192
C( 2)- C( 4)- H(12) 109.602 109.410   .360  .0003  .09  .0004
C( 2)- C( 4)- H(13) 108.008 109.410   .360  .0155  .09 -.0026
C( 6)- C( 4)- H(12) 110.820 109.410   .360  .0157  .09  .0037
C( 6)- C( 4)- H(13) 109.017 109.410   .360  .0012  .09 -.0010
H(12)- C( 4)- H(13) 106.394 109.400   .320  .0634
C( 3)- C( 5)- C( 7) 114.063 109.500   .450  .2053  .12  .0330
C( 3)- C( 5)- H(14) 110.058 109.410   .360  .0033  .09  .0017
C( 3)- C( 5)- H(15) 108.021 109.410   .360  .0152  .09 -.0036
C( 7)- C( 5)- H(14) 110.341 109.410   .360  .0068  .09  .0026
C( 7)- C( 5)- H(15) 107.541 109.410   .360  .0276  .09 -.0053
H(14)- C( 5)- H(15) 106.479 109.400   .320  .0598
C( 4)- C( 6)- C( 7) 114.062 109.500   .450  .2053  .12  .0330
C( 4)- C( 6)- H(16) 108.020 109.410   .360  .0152  .09 -.0036
C( 4)- C( 6)- H(17) 110.059 109.410   .360  .0033  .09  .0017
C( 7)- C( 6)- H(16) 107.540 109.410   .360  .0276  .09 -.0053
C( 7)- C( 6)- H(17) 110.342 109.410   .360  .0068  .09  .0026
H(16)- C( 6)- H(17) 106.480 109.400   .320  .0598
C( 5)- C( 7)- C( 6) 114.964 109.500   .450  .2944  .12  .0412
C( 5)- C( 7)- H(18) 108.105 109.410   .360  .0134  .09 -.0037
C( 5)- C( 7)- H(19) 109.539 109.410   .360  .0001  .09  .0004
C( 6)- C( 7)- H(18) 108.106 109.410   .360  .0134  .09 -.0037
C( 6)- C( 7)- H(19) 109.539 109.410   .360  .0001  .09  .0004
H(18)- C( 7)- H(19) 106.210 109.400   .320  .0714
```

DIHEDRAL ANGLES, TORSIONAL ENERGY (ET)   (52 ANGLES)

$$ET = (V1/2)(1+COS(W))+(V2/2)(1-COS(2W))+(V3/2)(1+COS(3W))$$

SIGN OF ANGLE A-B-C-D : WHEN LOOKING THROUGH B TOWARD C,
IF D IS COUNTERCLOCKWISE FROM A, NEGATIVE.

```
      A T O M S        OMEGA    V1      V2      V3     ET
C( 1) C( 2) C( 4) C( 6)  60.989  -.440   .240    .060   -.143
C( 1) C( 2) C( 4) H(12) -62.966  0.000  0.000   -.240   -.001
C( 1) C( 2) C( 4) H(13)-178.485  0.000  0.000   -.240   -.000
C( 1) C( 3) C( 5) C( 7)  76.698   .170   .270    .093    .377
C( 1) C( 3) C( 5) H(14) -47.935  0.000  0.000    .500    .048
C( 1) C( 3) C( 5) H(15)-163.806  0.000  0.000    .500    .085
C( 2) C( 1) C( 3) C( 5) -61.004  -.440   .240    .060   -.143
C( 2) C( 1) C( 3) H(10) 178.463  0.000  0.000   -.240   -.000
C( 2) C( 1) C( 3) H(11)  62.953  0.000  0.000   -.240   -.001
C( 2) C( 4) C( 6) C( 7) -76.708   .170   .270    .093    .377
C( 2) C( 4) C( 6) H(16) 163.798  0.000  0.000    .500    .085
C( 2) C( 4) C( 6) H(17)  47.926  0.000  0.000    .500    .048
C( 3) C( 1) C( 2) C( 4)    .013  -.100 15.000   0.000   -.100
C( 3) C( 1) C( 2) H( 9) 179.881  0.000 15.000   0.000    .000
C( 3) C( 5) C( 7) C( 6) -66.613   .200   .270    .093    .370
C( 3) C( 5) C( 7) H(18) 172.533  0.000  0.000    .267    .010
C( 3) C( 5) C( 7) H(19)  57.213  0.000  0.000    .267    .001
C( 4) C( 2) C( 1) H( 8)-179.866  0.000 15.000   0.000    .000
C( 4) C( 6) C( 7) C( 5)  66.622   .200   .270    .093    .370
```

FIG. 2.3.4. (*continued*)

```
C( 4)  C( 6)  C( 7)  H(18)-172.524   0.000   0.000    .267    .010
C( 4)  C( 6)  C( 7)  H(19) -57.204   0.000   0.000    .267    .001
C( 5)  C( 3)  C( 1)  H( 8) 118.876   0.000   0.000    .010    .010
C( 5)  C( 7)  C( 6)  H(16)-173.613   0.000   0.000    .267    .007
C( 5)  C( 7)  C( 6)  H(17) -57.861   0.000   0.000    .267    .001
C( 6)  C( 4)  C( 2)  H( 9)-118.880   0.000   0.000    .010    .010
C( 6)  C( 7)  C( 5)  H(14)  57.869   0.000   0.000    .267    .001
C( 6)  C( 7)  C( 5)  H(15) 173.620   0.000   0.000    .267    .007
C( 7)  C( 5)  C( 3)  H(10)-163.350   0.000   0.000    .267    .048
C( 7)  C( 5)  C( 3)  H(11) -46.586   0.000   0.000    .267    .032
C( 7)  C( 6)  C( 4)  H(12)  46.573   0.000   0.000    .267    .032
C( 7)  C( 6)  C( 4)  H(13) 163.346   0.000   0.000    .267    .048
H( 8)  C( 1)  C( 2)  H( 9)   .002   0.000  15.000   0.000    .000
H( 8)  C( 1)  C( 3)  H(10)  -1.656   0.000   0.000    .520    .519
H( 8)  C( 1)  C( 3)  H(11)-117.166   0.000   0.000    .520    .517
H( 9)  C( 2)  C( 4)  H(12) 117.164   0.000   0.000    .520    .517
H( 9)  C( 2)  C( 4)  H(13)   1.646   0.000   0.000    .520    .519
H(10)  C( 3)  C( 5)  H(14)  72.017   0.000   0.000    .237    .023
H(10)  C( 3)  C( 5)  H(15) -43.853   0.000   0.000    .237    .040
H(11)  C( 3)  C( 5)  H(14)-171.219   0.000   0.000    .237    .012
H(11)  C( 3)  C( 5)  H(15)  72.911   0.000   0.000    .237    .026
H(12)  C( 4)  C( 6)  H(16) -72.921   0.000   0.000    .237    .026
H(12)  C( 4)  C( 6)  H(17) 171.207   0.000   0.000    .237    .012
H(13)  C( 4)  C( 6)  H(16)  43.852   0.000   0.000    .237    .040
H(13)  C( 4)  C( 6)  H(17) -72.020   0.000   0.000    .237    .023
H(14)  C( 5)  C( 7)  H(18) -62.985   0.000   0.000    .237    .001
H(14)  C( 5)  C( 7)  H(19)-178.305   0.000   0.000    .237    .000
H(15)  C( 5)  C( 7)  H(18)  52.765   0.000   0.000    .237    .008
H(15)  C( 5)  C( 7)  H(19) -62.555   0.000   0.000    .237    .001
H(16)  C( 6)  C( 7)  H(18) -52.759   0.000   0.000    .237    .008
H(16)  C( 6)  C( 7)  H(19)  62.561   0.000   0.000    .237    .001
H(17)  C( 6)  C( 7)  H(18)  62.993   0.000   0.000    .237    .001
H(17)  C( 6)  C( 7)  H(19) 178.313   0.000   0.000    .237    .000
```

DIPOLE INTERACTION ENERGY (DIELECTRIC CONSTANT = 1.50)

```
    DIPOLE(1)    MU(1)     DIPOLE(2)    MU(2)    R12(A)   E(KCAL*DC)
+ C( 3)- C( 1)   .300   + C( 4)- C( 2)   .300   2.19363    .1081
```

FINAL STERIC ENERGY IS   9.9330 KCAL.

```
      COMPRESSION        .2711
      BENDING           1.6461
      STRETCH-BEND       .1414
      VANDERWAALS
        1,4 ENERGY      5.2729
        OTHER          -1.3917
      TORSIONAL         3.8850
      DIPOLE             .1081
```

FIG. 2.3.4. (*continued*)

COORDINATES TRANSLATED TO NEW ORIGIN WHICH IS CENTER OF MASS
```
        C( 1)   -1.49481      .67216     -.18995    ( 2)
        C( 2)   -1.49665     -.66822     -.18982    ( 2)
        C( 3)    -.39917     1.52396      .39251    ( 1)
        C( 4)    -.40321    -1.52291      .39255    ( 1)
        C( 5)     .95695     1.29354     -.28751    ( 1)
        C( 6)     .95342    -1.29603     -.28763    ( 1)
        C( 7)    1.66035     -.00222      .14150    ( 1)
        H( 8)   -2.35022     1.19922     -.64800    ( 5)
        H( 9)   -2.35351    -1.19303     -.64774    ( 5)
        H(10)    -.69118     2.59428      .26845    ( 5)
        H(11)    -.32792     1.34260     1.49028    ( 5)
        H(12)    -.33138    -1.34169     1.49031    ( 5)
        H(13)    -.69804    -2.59244      .26840    ( 5)
        H(14)     .83855     1.32169    -1.39688    ( 5)
        H(15)    1.62773     2.14577     -.01862    ( 5)
        H(16)    1.62195    -2.15008     -.01892    ( 5)
        H(17)     .83482    -1.32369    -1.39698    ( 5)
        H(18)    2.68610     -.00359     -.30161    ( 5)
        H(19)    1.79934     -.00245     1.24885    ( 5)
```

MOMENT OF INERTIA WITH THE PRINCIPAL AXES (UNIT = 10**(-39) GM*CM**2)

        IX= 26.1463   IY= 26.5662   IZ= 46.7831

DIPOLE MOMENT =   .494 D

  COMPONENTS WITH PRINCIPAL AXES

        X=   .4363   Y=   .0006   Z=   .2322

HEAT OF FORMATION AND STRAIN ENERGY CALCULATIONS   (UNITS ARE KCAL.)

BOND ENTHALPY (BE) AND STRAINLESS BOND ENTHALPY (SBE) CONSTANTS AND SUMS

#	BOND OR STRUCTURE	--- NORMAL ---		--STRAINLESS--	
4	C-C SP3-SP3	-.004	-.02	.493	1.97
10	C-H ALIPHATIC	-3.205	-32.05	-3.125	-31.25
2	C-C SP2-SP3	.170	.34	-.105	-.21
1	C=C SP2-SP2	22.840	22.84	22.940	22.94
2	C-H OLEFINIC	-3.205	-6.41	-3.125	-6.25
			------		------
		BE =	-15.30	SBE =	-12.80

PARTITION FUNCTION CONTRIBUTION (PFC)
    CONFORMATIONAL POPULATION INCREMENT (POP)   0.00
    TORSIONAL CONTRIBUTION (TOR)   0.00
    TRANSLATION/ROTATION TERM (T/R)   2.40

                                          PFC =   2.40

FIG. 2.3.4. (*continued*)

```
STERIC ENERGY (E)                                                    9.93

SIGMA-STRETCHING (ECPI)                                              0.00

CORRECTED STERIC ENERGY (EC) = E-ECPI                               9.93

ENERGY FROM PLANAR SCF CALCULATION   (ESCF)                          0.00

HEAT OF FORMATION = EC + BE + PFC + ESCF                           -2.96

STRAINLESS HEAT OF FORMATION FOR SIGMA SYSTEM (HFS)
         HFS = SBE + T/R + ESCF - ECPI                            -10.40

INHERENT SIGMA STRAIN (SI) = E + BE - SBE                           7.43

SIGMA STRAIN ENERGY (S) = POP + TOR + SI                            7.43

(RESONANCE ENERGY IS NOT CALCULATED.)

------------------------------------------------------------------------

   END OF CYCLOHEPTENE


        TOTAL ELAPSED TIME IS      4.71 SEC.
```

FIG. 2.3.4.   (*continued*)

program

$$V = 71.94(K_s(r - r_0)^2(1 - 2.0(r - r_0)))\ ,$$

is given at the head of the table. The parameters K_s and r_0 are then given for each bond. The actual bond length in the initial geometry corresponds to r in the above expression. In this case there are no changes made to r_0 for electronegativity or anomeric effects, so these two columns are empty. In principle it would be possible to detect a very poor starting geometry from the presence of a high energy in this table, but the block diagonal Newton–Raphson method is remarkably insensitive to poor starting geometries (within reason), so that it is seldom necessary to check on the input structure. The program does, however, have an option to perform a single energy calculation without geometry optimization (requested by typing a "1" in column 72 of the first card). Note that the force field parameters for the two olefinic CH bonds (C_1–H_8 and C_2–H_9) are different from those for the aliphatic CHs.

The next table gives the van der Waals energies, the other quantity that depends only on the interatomic distances. As outlined in Section 2.1 only those interactions between atoms that are bonded neither to each other nor to a common atom are included. The expressions used for calculation of the van der

Waals energy are

$$V = K_v(2.9 \times 10^5 e^{(-12.5r/r_v)}) - 2.25(r_v/r)^6$$

for $r_v/r < 3.311$, and

$$V = 336.176(K_v(r_v/r)^2$$

for $r_v/r \geqslant 3.311$, where r is the interatomic distance, r_v is the sum of the van der Waals radii of the two atoms (from the force field), and K_v is a constant determining the steepness of the curve. The values stored for each atom in the force field are combined for each pair of atoms, so that

$$K_v = \sqrt{E_i \cdot E_j} \ ,$$

where E_i is the "hardness" parameter for atom i and E_j that for atom j. Thus for two atoms of the same type $K_v = E_i$; otherwise it is the geometric mean of the two hardness parameters. The use of a different expression at $r_v/r \geqslant 3.311$ (i.e. at very short interatomic distances) is necessary because the exponential minus 6 expression becomes binding at very small r, so that if this change were not made, atoms would otherwise fuse with each other. The "R#" value listed for interactions involving hydrogens is the interatomic distance obtained when the centers of the hydrogens are shifted 8.5% along the CH bonds to approximate the center of electron density, as described in Section 2.1. This is the value of r actually used in evaluating the van der Waals energy. Note that the 1,3 distances (e.g. C_1–C_4) are listed, but that no van der Waals energies are calculated for these atom pairs, and that a 1,4 interaction is marked by an asterisk at the end of the line. In this example there are 116 van der Waals energies to be considered.

The next table gives both the angle-bending energies (EB) and the energies from the stretch–bend cross terms (ESB). The bending energies are given by the expression

$$E_B = 0.021914(K_B(\theta - \theta_0)^2(1 + (7 \times 10^{-8}(\theta - \theta_0)^4))) \ ,$$

where K_B is a parameter determining the steepness of the curve, θ is the bond angle, and θ_0 is the equilibrium bond angle (TZERO in the table).

The stretch–bend energy is given by

$$E_{SB} = 2.51124(K_{SB}(\theta - \theta_0)((r_1 - r_0(1)) + (r_2 - r_0(2)))) \ ,$$

where K_{SB} is a constant that depends on the types of atom, as given in the output, r_1 and r_2 are the lengths of the two bonds defining the bond angle θ, and $r_0(1)$ and $r_0(2)$ are the equilibrium lengths for these two bonds. Note that this expression may give a negative contribution to the steric energy. The table of bond angles and bending energies also contains in-plane and out-of-plane entries for all angles involving the sp^2 carbons, C_1 and C_2. The out-of-plane angles are a measure of the nonplanarity of the sp^2 center, and are defined as the angles between the individual bonds to the sp^2 center and the projections of these bonds onto the plane of the three atoms bonded to the atom in question, as shown in the program manual.

The program then lists the dihedral angles and the torsional energies, which are given by

$$V_T = (V_1/2)(1 + \cos\omega) + (V_2/2)(1 - \cos 2\omega) + (V_3/3)(1 + \cos 3\omega) ,$$

where ω is the dihedral angle and V_1, V_2, and V_3 are the one-, two-, and threefold torsional barriers, respectively. The torsional energy may also be negative. Because each dihedral angle about a given type of bond is counted the V_1, V_2, and V_3 terms constitute a fraction of the total barrier. There are nine torsional interactions per sp^3–sp^3 bond, five per sp^3–sp^2 bond, and three for each sp^2–sp^2 linkage, giving a total of 52 for the CC bonds in cycloheptene.

The final table contains the dipole–dipole interaction term. This energy is given by

$$E_\mu = 14.39418(\mu_A \cdot \mu_B(\cos X - 3\cos a_A \cdot \cos a_B))r^3/1.5 ,$$

where μ_A and μ_B are the dipole moments of the two bonds, X is the angle between them, r is the distance between their midpoints, and a_A and a_B are the angles between the bonds and the line joining their midpoints. The constant 1.5 is a dielectric constant parameter appropriate to the gas phase. In this example nonzero dipole moments are assigned to only the two sp^3–sp^2 bonds (C_1–C_3 and C_2–C_4). All other CC bonds are either sp^2–sp^2 or sp^3–sp^3, and CH bonds are not included.

The program then prints the total steric energy, which is simply the sum of all the contributions given in the tables above, and the sums of the individual types of energy. As emphasized above, these energies do not necessarily have any physical significance. The dipole moment printed is the vector sum of the bond moments for the C_1–C_3 and C_2–C_4 bonds. Part 1 of the output then ends with a message giving the amount of computer time used to obtain the initial geometry and energy. This section of the output would normally be much shorter, but the full-print option was used here to demonstrate the individual features of the force field and the energy calculation.

Part 2 of the output summarizes the geometry optimization. The initial steric energy (identical to that given at the end of Part 1) is printed and the program then moves to the first optimization cycle. The message "(CH)−MOVEMENT = 1" indicates that in this optimization cycle the attached atoms are moved together with those to which they are bonded. The program then performs a total of 30 Newton–Raphson iterations, printing the steric energy every five iterations. Note that this energy decreases very rapidly from 40 to 10 kcal mol^{-1}, but much more slowly after that. In Cycle 2 of the optimization process the attached atoms are allowed to move independently ("(CH)$-$MOVEMENT $= 0$") in order to fully optimize the structure. At this stage the average movement of the atoms is very small (0.00005 Å), and the steric energy (printed as "TOTAL ENERGY") is constant. The geometry is therefore considered to be optimized. The program then prints the optimized Cartesian coordinates in the same format as was used in the input. This concludes the optimization section of the output, so the program moves on to Part 3, the analysis of the optimized structure.

Part 3 of the output begins with another box in which to draw the the molecule, and then continues with an analysis of the final structure exactly analogous to that given in Part 1 for the starting geometry. The optimized geometry is shown in Fig. 2.3.5. The program then prints a new set of coordinates that have been shifted so that the center of mass lies at the origin of the Cartesian coordinate system. The principal moments of inertia are then given and the dipole moment is printed with its x, y, and z components.

The program then moves on to the heat of formation and strain-energy calculations, which are performed exactly as outlined in Section 2.1. The bond increments for the structure are summed to give a bond enthalpy, which is used to convert the steric energy to a heat of formation, and a strainless bond enthalpy, which is used to calculate the heat of formation of a strain-free isomer. This table of bond increments presents a good opportunity to check

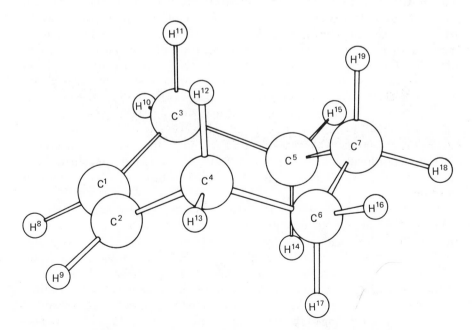

FIG. 2.3.5. MMP2 optimized geometry for cycloheptene.

that all is well with the calculation. If the numbers of the different types of bond do not correspond exactly to the molecule you intended to calculate, an error has been made, probably in the options used to define the coordinate calculations, but possibly by erroneous definition of an atom type in the coordinate table.

The program next calculates a partition function contribution to the heat of formation, which refers to 25°C. The translation and rotation term is the same for all molecules: the program adds a value of $2.4 \, kcal \, mol^{-1}$. The conformational population increment is intended to correct for the Boltzmann population of other energetically accessible conformations. Because the program knows nothing about the existence of such conformations, which would have to be calculated separately, this increment must be read in with the input if it is not zero. In the case of cycloheptene only one conformation appears to be important, so that POP is zero. The torsional contribution, TOR, is intended to account for the energy contribution due to the population of low-lying energy levels associated with an internal rotation. This correction need not be applied for methyl groups, because it is automatically taken into account in the group increments, but is important for n-alkanes and similar molecules. There are no internal rotations in cycloheptene, and so TOR is zero.

The steric energy and the other contributions to the heat of formation are now summed to give a calculated value of $-2.96 \, kcal \, mol^{-1}$, compared with the value of -2.25 ± 0.22 determined experimentally. The two terms "ECPI" and "ESCF" refer to the MO part of an MMP2 calculation, and therefore have values of zero in this example. The strainless heat of formation is calculated to be $-10.40 \, kcal \, mol^{-1}$, giving a strain energy of $7.43 \, kcal \, mol^{-1}$. The final two messages in the output give the computer time used and the time at completion of the job. Figure 2.3.6 shows the file written by this job to channel 7 (the punch file). It contains the revised input after completion of the coordinate calculations and also the optimized coordinates. Note that the options in cards 1 and 2 are not identical to those given in the input, and that the SCF information card is missing.

The next example, which uses the data written to the above file, illustrates the use of coordinate calculations to substitute groups in a molecule, in this case to give 5-^{t}butyl-cycloheptene with the alkyl substituent in the equatorial position.

To do this, we first use the revised atom connection and attached atom lists from Fig. 2.3.6 along with the optimized coordinates, which are also written to this file. We now wish to replace hydrogen 18 (see Fig. 2.3.5) by a ^{t}butyl group. The input for this calculation is shown in Fig. 2.3.7. The first three lines indicate that we intend to read in the coordinates for 19 atoms, with one atom connection list, 12 attached atom lists, and four coordinate calculation cards. The options for calculating the heat of formation (a "1" in column 65 of card 3) and for minimal printing of the results (a "4" in column 67 of card 1) have also been used. The atom connection list (card 4) and the 12 attached atom lists

```
CYCLOHEPTENE                                                  19    0    0
  1                       0   12    0
  1     2     4     6     7     5     3     1
  1     8     2     9     3    10     3    11    4    12    4    13    5    14    5    15
  6    16     6    17     7    18     7    19
   .67000    .00000    .00000    2        -.67000    .00000    .00000    2
  1.49000   1.10000    .60000    1       -1.49000   1.10000    .60000    1
  1.29000   2.45000   -.10000    1       -1.29000   2.45000   -.10000    1
  -.00000   3.10000    .00000    1        1.09164   -.89193   -.48651    5
 -1.09164   -.89193   -.48651    5        2.54934    .81199    .53038    5
  1.20639   1.19788   1.65829    5       -1.20639   1.19788   1.65829    5
 -2.54934    .81199    .53038    5        1.47309   2.31687  -1.17645    5
  2.01681   3.16747    .30863    5       -2.01681   3.16747    .30863    5
 -1.47309   2.31687  -1.17645    5        -.00000   3.86462   -.79079    5
  -.00000   3.59157    .98405    5
CYCLOHEPTENE                                                  19    0    0
   .67042   -.05553    .09876    2        -.66996   -.05639    .10131    2     2
  1.52380   1.09712    .55526    1       -1.52306   1.09527    .56078    1     4
  1.29293   2.37001   -.26998    1       -1.29664   2.36834   -.26542    1     6
  -.00173   3.11741    .08088    1        1.19625   -.95667   -.26289    5     8
 -1.19600   -.95821   -.25831    5        2.59375    .79243    .46243    5    10
  1.34436   1.28929   1.63875    5       -1.33992   1.28777   1.64359    5    12
 -2.59297    .78930    .47172    5        1.31910   2.12981  -1.35953    5    14
  2.14597   3.06578   -.07822    5       -2.14988   3.06306   -.07077    5    16
 -1.32628   2.12797  -1.35485    5        -.00332   4.08796   -.47280    5    18
   .00002   3.37784   1.16610    5        0.0       0.0       0.0            LAST
```

FIG. 2.3.6. The information written to the punch file by MMP2 for the cycloheptene calculation shown in Figs. 2.3.1 and 2.3.4.

```
!.........!.........!.........!.........!.........!.........!.........!.........!
5-T-BUTYL-CYCLOHEPTENE                                        19  4  0      1.0

  1                       0   12    0    4                    1
  1     2     4     6     7     5     3     1
  1     8     2     9     3    10     3    11    4    12    4    13    5    14    5    15
  6    16     6    17     7    18     7    19
   .67042   -.05553    .09876    2        -.66996   -.05639    .10131    2     2
  1.52380   1.09712    .55526    1       -1.52306   1.09527    .56078    1     4
  1.29293   2.37001   -.26998    1       -1.29664   2.36834   -.26542    1     6
  -.00173   3.11741    .08088    1        1.19625   -.95667   -.26289    5     8
 -1.19600   -.95821   -.25831    5        2.59375    .79243    .46243    5    10
  1.34436   1.28929   1.63875    5       -1.33992   1.28777   1.64359    5    12
 -2.59297    .78930    .47172    5        1.31910   2.12981  -1.35953    5    14
  2.14597   3.06578   -.07822    5       -2.14988   3.06306   -.07077    5    16
 -1.32628   2.12797  -1.35485    5        -.00332   4.08796   -.47280    5    18
   .00002   3.37784   1.16610    5        0.0       0.0       0.0            LAST
  7    18     7     6
  7    20     8     7
  7    21     8     7
  7    22     8     7
.........................................................................
```

FIG. 2.3.7. One possible MMP2 input for 5-ᵗbutyl-cycloheptene using coordinate calculation options.

(cards 5 and 6) are then given, followed by the optimized coordinates for cycloheptene. The four coordinate calculation cards (the last four) then instruct the program to add a ʹbutyl group in place of atom 18. The coordinate calculation option to add a methyl group is denoted by a "7". In the first stage of the substitution we replace hydrogen 18 by a methyl group using this option. This line therefore means that option **7**, substitution by a methyl group, will be performed on atom **18**, which is bound to atom **7**, which in turn is bound to atom **6**. The parameters used to define this option fully are given in the program manual. To proceed further we must know exactly what the program does when carrying out this substitution. Figure 2.3.8 shows the molecule as it is after the program has finished this first substitution. Note that the carbon atom of the methyl group has been given the next number after that of the last connected atom (in this case 8), and that the three hydrogens are added at the end of the atoms list (giving them numbers 20–22). What the program actually does is add one more atom connection list to make the methyl carbon itself a connected atom (hence the number 8). Note that the program produces a molecular definition corresponding to Fig. 2.3.8 before it moves on to the next coordinate calculation option (i.e., the next substitution is carried out on the renumbered molecule).

The last three lines of Fig. 2.3.7 give the coordinate calculation options for replacing the three methyl hydrogens by three methyl groups. Because the connected atoms that have already been defined are not renumbered during a substitution all three methyl groups can be added using option **7** to replace atoms **20**, **21**, and **22**, which are attached to atom **8**, which in turn is attached to atom **7**. An extract from the output for this calculation is shown in Fig. 2.3.9. Note that the three methyl carbons of the ʹbutyl group have been added as connected atoms 9, 10, and 11, and that there are now four atom connection lists. The final structure is shown in Fig. 2.3.10. The axial isomer could now be calculated by substituting for hydrogen 19 instead of hydrogen 18 in the first

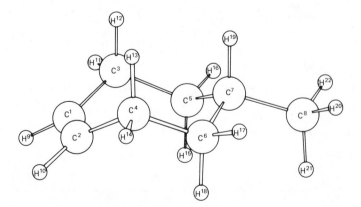

FIG. 2.3.8. Molecular structure and numbering after performance of the first coordinate calculation specified in Fig. 2.3.7.

ADDITIONS AND/OR REPLACEMENTS WILL BE MADE TO 4 OF THESE ATOMS.

CONFORMATIONAL ENERGY, PART 1: GEOMETRY AND STERIC ENERGY
                                OF INITIAL CONFORMATION.

CONNECTED ATOMS
    1- 2- 4- 6- 7- 5- 3- 1-

ATTACHED ATOMS
    1- 8, 2- 9, 3-10, 3-11, 4-12, 4-13, 5-14, 5-15,
    6-16, 6-17, 7-18, 7-19,

INITIAL ATOMIC COORDINATES

ATOM	X	Y	Z	TYPE
C(1)	.67042	-.05553	.09876	(2)
C(2)	-.66996	-.05639	.10131	(2)
C(3)	1.52380	1.09712	.55526	(1)
C(4)	-1.52306	1.09527	.56078	(1)
C(5)	1.29293	2.37001	-.26998	(1)
C(6)	-1.29664	2.36834	-.26542	(1)
C(7)	-.00173	3.11741	.08088	(1)
H(8)	1.19625	-.95667	-.26289	(5)
H(9)	-1.19600	-.95821	-.25831	(5)
H(10)	2.59375	.79243	.46243	(5)
H(11)	1.34436	1.28929	1.63875	(5)
H(12)	-1.33992	1.28777	1.64359	(5)
H(13)	-2.59297	.78930	.47172	(5)
H(14)	1.31910	2.12981	-1.35953	(5)
H(15)	2.14597	3.06578	-.07822	(5)
H(16)	-2.14988	3.06306	-.07077	(5)
H(17)	-1.32628	2.12797	-1.35485	(5)
H(18)	-.00332	4.08796	-.47280	(5)
H(19)	.00002	3.37784	1.16610	(5)

----
4 COORDINATE OPTIONS USED

  NEW HEADER CARD

  5-T-BUTYL-CYCLOHEPTENE   31   0   0   (PUNCHED)

      5   0   20   0   (PUNCHED)

  NEW CONNECTED ATOM LISTS
      1- 2- 4- 6- 7- 5- 3- 1-   (PUNCHED)
      7- 8-   (PUNCHED)
      8- 9-   (PUNCHED)
      8-10-   (PUNCHED)
      8-11-   (PUNCHED)

  NEW ATTACHED ATOM LISTS
      1-12, 2-13, 3-14, 3-15, 4-16, 4-17, 5-18, 5-19,   (PUNCHED)
      6-20, 6-21, 7-22, 9-23, 9-24, 9-25,10-26,10-27,   (PUNCHED)
     10-28,11-29,11-30,11-31,

FIG. 2.3.9. Extract from the MMP2 output generated using the input shown in Fig. 2.3.7. The omitted section of the output is indicated by "+ + + + + + +."

```
NEW ATOMIC COORDINATES
  ATOM        X             Y             Z        TYPE
      1       .67042      -.05553       .09876       2      (PUNCHED)
      2      -.66996      -.05639       .10131       2      (PUNCHED)
      3      1.52380      1.09712       .55526       1      (PUNCHED)
      4     -1.52306      1.09527       .56078       1      (PUNCHED)
      5      1.29293      2.37001      -.26998       1      (PUNCHED)
      6     -1.29664      2.36834      -.26542       1      (PUNCHED)
      7      -.00173      3.11741       .08088       1      (PUNCHED)
      8      -.00391      4.45071      -.67974       1      (PUNCHED)
      9      1.28054      5.22568      -.35444       1      (PUNCHED)
     10     -1.22406      5.28018      -.25614       1      (PUNCHED)
     11      -.07032      4.17935     -2.18910       1      (PUNCHED)
     12      1.19625      -.95667      -.26289       5      (PUNCHED)
     13     -1.19600      -.95821      -.25831       5      (PUNCHED)
     14      2.59375       .79243       .46243       5      (PUNCHED)
     15      1.34436      1.28929      1.63875       5      (PUNCHED)
     16     -1.33992      1.28777      1.64359       5      (PUNCHED)
     17     -2.59297       .78930       .47172       5      (PUNCHED)
     18      1.31910      2.12981     -1.35953       5      (PUNCHED)
     19      2.14597      3.06578      -.07822       5      (PUNCHED)
     20     -2.14988      3.06306      -.07077       5      (PUNCHED)
     21     -1.32628      2.12797     -1.35485       5      (PUNCHED)
     22       .00002      3.37784      1.16610       5      (PUNCHED)
     23      1.27897      6.18114      -.89951       5      (PUNCHED)
     24      2.15485      4.63118      -.65797       5      (PUNCHED)
     25      1.32807      5.42004       .72721       5      (PUNCHED)
     26     -1.22422      6.23519      -.80199       5      (PUNCHED)
     27     -1.17774      5.47539       .82541       5      (PUNCHED)
     28     -2.14443      4.72432      -.48837       5      (PUNCHED)
     29      -.07181      5.13563     -2.73273       5      (PUNCHED)
     30      -.99031      3.62312     -2.42199       5      (PUNCHED)
     31       .80357      3.58494     -2.49406       5      (PUNCHED)


INITIAL STERIC ENERGY IS     34.4270 KCAL.

     COMPRESSION        .9593
     BENDING           1.7001
     STRETCH-BEND       .1065
     VANDERWAALS
       1,4 ENERGY      9.7299
       OTHER          16.5197
     TORSIONAL         5.3034
     DIPOLE             .1081

DIPOLE MOMENT =        .494 D


   COMPONENTS WITH PRINCIPAL AXES

        X=    .0001    Y=   .4593    Z=   .1826


   INITIAL ENERGY REQUIRES    .23 SECONDS.

++++++++++++++++++++++++++++++++++++++++++++++++++++++++++++++++++++++++++++++++++++++
```

FIG. 2.3.9. *(continued)*

FINAL STERIC ENERGY IS   18.3022 KCAL.

```
        COMPRESSION      1.6397
        BENDING          3.2641
        STRETCH-BEND      .4049
        VANDERWAALS
          1,4 ENERGY     7.7814
           OTHER         -.8335
        TORSIONAL        5.9350
        DIPOLE            .1106
```

```
COORDINATES TRANSLATED TO NEW ORIGIN WHICH IS CENTER OF MASS
     C( 1)   -2.85275     .66809    -.27015   ( 2)
     C( 2)   -2.85218    -.67085    -.26797   ( 2)
     C( 3)   -1.80919    1.49677     .42633   ( 1)
     C( 4)   -1.80650   -1.49656     .42873   ( 1)
     C( 5)    -.39835    1.28021    -.14338   ( 1)
     C( 6)    -.39878   -1.27512    -.14672   ( 1)
     C( 7)     .33091     .00175     .33047   ( 1)
     C( 8)    1.86132    -.00043    -.01247   ( 1)
     C( 9)    2.56923    1.25471     .55728   ( 1)
     C(10)    2.57718   -1.21589     .62948   ( 1)
     C(11)    2.11899    -.04275   -1.53340   ( 1)
     H(12)   -3.65276    1.20145    -.81295   ( 5)
     H(13)   -3.65158   -1.20658    -.80934   ( 5)
     H(14)   -2.09686    2.56914     .30700   ( 5)
     H(15)   -1.83123    1.29125    1.52194   ( 5)
     H(16)   -1.82613   -1.29003    1.52416   ( 5)
     H(17)   -2.09117   -2.56981     .31027   ( 5)
     H(18)    -.43976    1.31856   -1.25645   ( 5)
     H(19)     .18397    2.17012     .18408   ( 5)
     H(20)     .18905   -2.16577     .16847   ( 5)
     H(21)    -.44892   -1.30506   -1.25973   ( 5)
     H(22)     .27756     .00258    1.44553   ( 5)
     H(23)    3.67682    1.18257     .46339   ( 5)
     H(24)    2.29655    2.18755     .01603   ( 5)
     H(25)    2.33909    1.39536    1.63813   ( 5)
     H(26)    3.68347   -1.14892     .51794   ( 5)
     H(27)    2.35990   -1.28629    1.71976   ( 5)
     H(28)    2.29937   -2.18319     .15610   ( 5)
     H(29)    3.21044    -.02984   -1.75587   ( 5)
     H(30)    1.70780    -.96473   -2.00134   ( 5)
     H(31)    1.67279     .83256   -2.05574   ( 5)
```

MOMENT OF INERTIA WITH THE PRINCIPAL AXES (UNIT = 10^{-39} GM*CM**2)

     IX= 43.8863   IY= 118.7579   IZ= 138.1383

DIPOLE MOMENT =   .501 D

  COMPONENTS WITH PRINCIPAL AXES

        X=   .4169   Y=   .0006   Z=   .2780

FIG. 2.3.9. (*continued*)

59

HEAT OF FORMATION AND STRAIN ENERGY CALCULATIONS   (UNITS ARE KCAL.)

BOND ENTHALPY (BE) AND STRAINLESS BOND ENTHALPY (SBE) CONSTANTS AND SUMS

#	BOND OR STRUCTURE	--- NORMAL ---		--STRAINLESS--	
8	C-C SP3-SP3	-.004	-.03	.493	3.94
18	C-H ALIPHATIC	-3.205	-57.69	-3.125	-56.25
2	C-C SP2-SP3	.170	.34	-.105	-.21
1	C=C SP2-SP2	22.840	22.84	22.940	22.94
2	C-H OLEFINIC	-3.205	-6.41	-3.125	-6.25
1	ISO (ALKANE)	.078	.08	-.073	-.07
1	NEO (ALKANE)	-.707	-.71	-1.120	-1.12
3	C(SP3)-METHYL	-1.510	-4.53	-1.575	-4.73
			BE = -46.11		SBE = -41.74

PARTITION FUNCTION CONTRIBUTION (PFC)
    CONFORMATIONAL POPULATION INCREMENT (POP)   0.00
    TORSIONAL CONTRIBUTION (TOR)   0.00
    TRANSLATION/ROTATION TERM (T/R)   2.40

                              PFC = 2.40

STERIC ENERGY (E)   18.30

SIGMA-STRETCHING (ECPI)   0.00

CORRECTED STERIC ENERGY (EC) = E-ECPI   18.30

ENERGY FROM PLANAR SCF CALCULATION   (ESCF)   0.00

HEAT OF FORMATION = EC + BE + PFC + ESCF   -25.41

STRAINLESS HEAT OF FORMATION FOR SIGMA SYSTEM (HFS)
        HFS = SBE + T/R + ESCF - ECPI   -39.34

INHERENT SIGMA STRAIN (SI) = E + BE - SBE   13.94

SIGMA STRAIN ENERGY (S) = POP + TOR + SI   13.94

(RESONANCE ENERGY IS NOT CALCULATED.)

    END OF 5-T-BUTYL-CYCLOHEPTENE

        TOTAL ELAPSED TIME IS   9.68 SEC.

        THIS JOB COMPLETED AT 12.00.00.

FIG. 2.3.9. *(continued)*

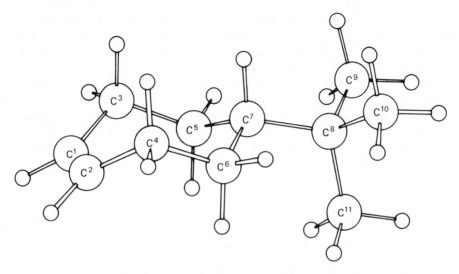

FIG. 2.3.10. The MMP2 optimized structure for 5-^{t}butyl-cycloheptene. The hydrogen atom numbering has been omitted for clarity.

coordinate calculation. The input would otherwise be identical to that shown in Fig. 2.3.7.

The third example illustrates the use of the MMP2 facility for conjugated systems. This can be illustrated by using 1,3-cycloheptadiene as an example. Figure 2.3.11 shows one possible input for this molecule. The same starting coordinates have been used as for cycloheptene, but the conjugated system has been redefined. The first change in the input compared with that in Fig. 2.3.1 is in the second card, which gives the options for the SCF calculation. The first 60 columns of this card define the atoms to be included in the SCF calculation (the π atoms). Each atom is assigned one column and it is necessary only to type a

```
!........!........!........!........!........!........!........!........!
CYCLOHEPTA-1,3-DIENE                                      7 0           1.0
TTTT                                              1
  1                                          7          1         0
  1    2    4    6    7   5   3   1
  0.70000   0.00000   0.00000   2      -0.70000   0.00000   0.00000   2
  1.45000   1.28000   0.20000   2      -1.45000   1.28000   0.20000   2
  1.26000   2.75000  -0.20000   1      -1.260000  2.75000  -0.20000   1
  0.00000   3.61000   0.00000   1
  4    1    2    3
  4    2    1    4
  4    3    1    5
  4    4    2    6
  2    5    3    7
  2    6    4    7
  2    7    5    6
```

FIG. 2.3.11. One possible MMP2 input for 1,3-cycloheptadiene.

"T" (for "true") in each column representing a π atom. In this case atoms 1–4 of the structure shown in Fig. 2.3.2 form the butadiene system, so that the four "T"s in columns 1–4 define these atoms as the π system. The "1" in column 62 indicates that the program should treat the conjugated system as nonplanar. Omission of this option would lead to an incorrect result, but not to a program error. The only other differences in the input compared with that for cycloheptene are that atoms 3 and 4 are now defined as type 2 (sp^2), and that the coordinate calculation cards for these two atoms have been changed correspondingly. The output for this calculation is shown in Fig. 2.3.12.

```
CYCLOHEPTA-1,3-DIENE


                                                        DATE   84/02/29.

        ADDITIONS AND/OR REPLACEMENTS WILL BE MADE TO   7 OF THESE ATOMS.

    CONFORMATIONAL ENERGY, PART 1:   GEOMETRY AND STERIC ENERGY
                                     OF INITIAL CONFORMATION.

    CONNECTED ATOMS
        1- 2- 4- 6- 7- 5- 3- 1-

    INITIAL ATOMIC COORDINATES
          ATOM        X           Y           Z        TYPE
          C( 1)      .70000     0.00000     0.00000    ( 2)    PI-ATOM
          C( 2)     -.70000     0.00000     0.00000    ( 2)    PI-ATOM
          C( 3)     1.45000     1.28000      .20000    ( 2)    PI-ATOM
          C( 4)    -1.45000     1.28000      .20000    ( 2)    PI-ATOM
          C( 5)     1.26000     2.75000     -.20000    ( 1)
          C( 6)    -1.26000     2.75000     -.20000    ( 1)
          C( 7)     0.00000     3.61000     0.00000    ( 1)


    -----------------------------
    7 COORDINATE OPTIONS USED
    -----------------------------


      NEW HEADER CARD

      CYCLOHEPTA-1,3-DIENE                            17    0   0   (PUNCHED)

         1                          0   10    0                     (PUNCHED)


      NEW CONNECTED ATOM LISTS
          1- 2- 4- 6- 7- 5- 3- 1-                                   (PUNCHED)


      NEW ATTACHED ATOM LISTS
          1- 8, 2- 9, 3-10, 4-11, 5-12, 5-13, 6-14, 6-15,   (PUNCHED)
          7-16, 7-17,
```

FIG. 2.3.12. The MMP2 output for 1,3-cycloheptadiene generated from the input shown in Fig. 2.3.11.

NEW ATOMIC COORDINATES

ATOM	X	Y	Z	TYPE	
1	.70000	.00000	.00000	2	(PUNCHED)
2	-.70000	.00000	.00000	2	(PUNCHED)
3	1.45000	1.28000	.20000	2	(PUNCHED)
4	-1.45000	1.28000	.20000	2	(PUNCHED)
5	1.26000	2.75000	-.20000	1	(PUNCHED)
6	-1.26000	2.75000	-.20000	1	(PUNCHED)
7	-.00000	3.61000	.00000	1	(PUNCHED)
8	1.19329	-.97140	-.15178	5	(PUNCHED)
9	-1.19329	-.97140	-.15178	5	(PUNCHED)
10	2.36404	1.09525	.78343	5	(PUNCHED)
11	-2.36404	1.09525	.78343	5	(PUNCHED)
12	1.45241	2.77477	-1.28276	5	(PUNCHED)
13	2.06022	3.29113	.32614	5	(PUNCHED)
14	-2.06022	3.29113	.32614	5	(PUNCHED)
15	-1.45241	2.77477	-1.28276	5	(PUNCHED)
16	-.00000	4.43235	-.73058	5	(PUNCHED)
17	-.00000	4.02559	1.01847	5	(PUNCHED)

SCF CALCULATION FOR NONPLANAR SYSTEM

```
    ITER  1   V(0) =  -1.718999   DIFF =  0.00000000
    ITER  2   V(0) =  -1.718999   DIFF =   .00000008
```

SCF CALCULATION FOR PLANAR SYSTEM

```
    ITER  1   V(0) =  -1.758423   DIFF =  0.00000000
    ITER  2   V(0) =  -1.758805   DIFF =   .00038200
    ITER  3   V(0) =  -1.758936   DIFF =   .00013117
    ITER  4   V(0) =  -1.758981   DIFF =   .00004475
    ITER  5   V(0) =  -1.758996   DIFF =   .00001521
    ITER  6   V(0) =  -1.759001   DIFF =   .00000516
```

SCF SEQUENCE REQUIRES   .06 SECONDS.

AROMATIC TORSIONAL CONSTANTS ARE FACTORED BY THE RATIO
   SUM(PIJ(0)BIJ(0)-PIJ(W)BIJ(W))/SUM(PIJ(0)BIJ(0)*(1-COS(W)*COS(W))
WHERE THE SUM IS OVER ALL DIHEDRAL ANGLES WITH CENTRAL PI-ATOMS I & J.
THE CALCULATED RATIO IS 1.0085/ 1.9408. THE PROGRAM USES .5196

INITIAL STERIC ENERGY IS   56.8731 KCAL.

```
    COMPRESSION     20.8118
    BENDING         14.4989
    STRETCH-BEND     1.4453
    VANDERWAALS
      1,4 ENERGY     5.0234
      OTHER         -1.0239
    TORSIONAL       16.0736
    DIPOLE            .0440
```

DIPOLE MOMENT =   .595 D

FIG. 2.3.12. *(continued)*

COMPONENTS WITH PRINCIPAL AXES

X=   .0000   Y=   .5745   Z=   .1563

NONPLANAR

F-MATRIX (PUNCHED)
  -.20555
  -.12972   -.20555
  -.16677   -.00000   -.20555
   .00000   -.16677   .03729   -.20555

MOLECULAR ORBITALS
   .60157   .60157   .37164   .37164
   .37164   -.37164   .60157   -.60157
  -.37164   -.37164   .60157   .60157
  -.60157   .60157   .37164   -.37164

EIGENVALUES
  -.43827   -.34586   -.06524   .02717

DENSITY MATRIX
  1.00000   .44753   .89427   -.00000
   .44753   1.00000   .00000   .89427
   .89427   .00000   1.00000   -.44753
  -.00000   .89427   -.44753   1.00000

ELECTRON DENSITIES
  1.0000   1.0000   1.0000   1.0000

PLANAR

F-MATRIX (PUNCHED)
  -.20555
  -.12269   -.20555
  -.18095   -.00000   -.20555
   .00000   -.18095   .03307   -.20555

H-CORE MATRIX
 -1.14272
  -.06759 -1.14272
  -.05818   0.00000 -1.03172
  0.00000   -.05818   0.00000 -1.03172

MOLECULAR ORBITALS
   .59095   .59095   .38831   .38831
   .38831   -.38831   .59095   -.59095
  -.38831   -.38831   .59095   .59095
  -.59095   .59095   .38831   -.38831

EINGENVALUES
  -.44740   -.35764   -.05346   .03630

DENSITY MATRIX
  1.00000   .39687   .91787   -.00000
   .39687   1.00000   .00000   .91787
   .91787   .00000   1.00000   -.39687
  -.00000   .91787   -.39687   1.00000

FIG. 2.3.12.   (*continued*)

ELECTRON DENSITIES
     1.0000   1.0000   1.0000   1.0000

BOND ORDERS (P) AND RESONANCE INTEGRALS (B) FOR PI-BONDS
          ---NONPLANAR GEOMETRY---   ----PLANAR GEOMETRY----
    BOND   P(W)   B(W)   P*B(W)   P(O)   B(O)   P*B(O)
    1- 2   .4475   .8954   .4007   .3969   .8954   .3554
    1- 3   .8943   .6247   .5587   .9179   .7707   .7074
    2- 4   .8943   .6247   .5587   .9179   .7707   .7074

BOND ENERGY FROM SCF CALCULATION
        SIGMA-BOND   -260.67
        PI-BOND   -74.34
        TOTAL ENERGY -335.01

INITIAL ENERGY REQUIRES   .24 SECONDS.

CYCLOHEPTA-1,3-DIENE

CONFORMATIONAL ENERGY, PART 2: ENERGY MINIMIZATION

INITIAL ENERGY CALCD FROM SYMMETRIZED COORDS:

TOTAL ENERGY IS   56.8731 KCAL.
    COMPRESS   20.8118   VANDERWAALS   TORSION   16.0736
    BENDING   14.4989   1,4   5.0234
    STR-BEND   1.4453   OTHER   -1.0239   DIPL/CHG   .0440

* * * * * * * * * * * * *   C Y C L E   1   * * * * * * * * * * * * *

                    (CH)-MOVEMENT = 1

        ITERATION   1   AVG. MOVEMENT = .06145 A
        ITERATION   2   AVG. MOVEMENT = .03973 A
        ITERATION   3   AVG. MOVEMENT = .02581 A
        ITERATION   4   AVG. MOVEMENT = .02279 A
        ITERATION   5   AVG. MOVEMENT = .01719 A

TOTAL ENERGY IS   9.8671 KCAL.
    COMPRESS   .2313   VANDERWAALS   TORSION   .1580
    BENDING   5.7277   1,4   4.6392
    STR-BEND   .1018   OTHER   -1.0304   DIPL/CHG   .0394

DELTA(T) =   .68 SEC.   ELAPSED TIME =   .92 SEC.

        ITERATION   6   AVG. MOVEMENT = .01198 A
        ITERATION   7   AVG. MOVEMENT = .00757 A
        ITERATION   8   AVG. MOVEMENT = .00502 A
        ITERATION   9   AVG. MOVEMENT = .00321 A
        ITERATION   10   AVG. MOVEMENT = .00210 A

FIG. 2.3.12. (*continued*)

```
TOTAL ENERGY IS      8.2363 KCAL.
   COMPRESS      .1111      VANDERWAALS                  TORSION        -.8403
   BENDING      5.1091       1,4           4.6862
   STR-BEND      .0932      OTHER          -.9617        DIPL/CHG        .0387

DELTA(T) =      .68 SEC.           ELAPSED TIME =    1.59 SEC.

        ITERATION  11      AVG. MOVEMENT =   .00135 A
        ITERATION  12      AVG. MOVEMENT =   .00088 A
        ITERATION  13      AVG. MOVEMENT =   .00057 A
        ITERATION  14      AVG. MOVEMENT =   .00038 A
        ITERATION  15      AVG. MOVEMENT =   .00028 A

TOTAL ENERGY IS      8.2129 KCAL.
   COMPRESS      .1111      VANDERWAALS                  TORSION        -.8055
   BENDING      5.0267       1,4           4.6980
   STR-BEND      .0936      OTHER          -.9495        DIPL/CHG        .0386

DELTA(T) =      .68 SEC.           ELAPSED TIME =    2.27 SEC.

        ITERATION  16      AVG. MOVEMENT =   .00021 A
        ITERATION  17      AVG. MOVEMENT =   .00018 A
        ITERATION  18      AVG. MOVEMENT =   .00015 A
        ITERATION  19      AVG. MOVEMENT =   .00014 A
        ITERATION  20      AVG. MOVEMENT =   .00013 A

TOTAL ENERGY IS      8.2123 KCAL.
   COMPRESS      .1111      VANDERWAALS                  TORSION        -.8020
   BENDING      5.0204       1,4           4.6990
   STR-BEND      .0936      OTHER          -.9484        DIPL/CHG        .0385

DELTA(T) =      .68 SEC.           ELAPSED TIME =    2.95 SEC.




* * * * * * * * * * * * *   C Y C L E   2  * * * * * * * * * * * * *



                    (CH)-MOVEMENT = 0

        ITERATION  21      AVG. MOVEMENT =   .00009 A

TOTAL ENERGY IS      8.2123 KCAL.
   COMPRESS      .1111      VANDERWAALS                  TORSION        -.8020
   BENDING      5.0207       1,4           4.6989
   STR-BEND      .0936      OTHER          -.9485        DIPL/CHG        .0385

DELTA(T) =      .15 SEC.           ELAPSED TIME =    3.10 SEC.

* * * * * ENERGY IS MINIMIZED WITHIN  .0014 KCAL * * * * *

      * * * * * ENERGY IS     8.2123 KCAL * * * * *
```

FIG. 2.3.12. (*continued*)

BECAUSE OF THE MOVEMENT DURING THE MINIMIZATION SEQUENCE, ANOTHER
VESCF-MINIMIZATION CYCLE MUST BE DONE TO INSURE COMPLETE CONVERGENCE.

SCF CALCULATION FOR NONPLANAR SYSTEM

```
ITER  1   V(0) =   -1.817087    DIFF =  0.00000000
ITER  2   V(0) =   -1.821467    DIFF =   .00437929
ITER  3   V(0) =   -1.822653    DIFF =   .00118615
ITER  4   V(0) =   -1.822965    DIFF =   .00031269
ITER  5   V(0) =   -1.823047    DIFF =   .00008141
ITER  6   V(0) =   -1.823068    DIFF =   .00002107
ITER  7   V(0) =   -1.823073    DIFF =   .00000544
```

SCF CALCULATION FOR PLANAR SYSTEM

```
ITER  1   V(0) =   -1.821194    DIFF =  0.00000000
ITER  2   V(0) =   -1.823137    DIFF =   .00194312
ITER  3   V(0) =   -1.823653    DIFF =   .00051515
ITER  4   V(0) =   -1.823787    DIFF =   .00013433
ITER  5   V(0) =   -1.823822    DIFF =   .00003475
ITER  6   V(0) =   -1.823831    DIFF =   .00000896
```

SCF SEQUENCE REQUIRES   .08 SECONDS.

AROMATIC TORSIONAL CONSTANTS ARE FACTORED BY THE RATIO
  SUM(PIJ(0)BIJ(0)-PIJ(W)BIJ(W))/SUM(PIJ(0)BIJ(0)*(1-COS(W)*COS(W))
WHERE THE SUM IS OVER ALL DIHEDRAL ANGLES WITH CENTRAL PI-ATOMS I & J.
THE CALCULATED RATIO IS   .0204/   .0611. THE PROGRAM USES   .3342

INITIAL ENERGY CALCD FROM SYMMETRIZED COORDS:

```
TOTAL ENERGY IS       8.5423 KCAL.
  COMPRESS     .5436    VANDERWAALS              TORSION      -.8779
  BENDING     5.0207    1,4           4.6989
  STR-BEND     .0670    OTHER         -.9485     DIPL/CHG      .0385
```

```
* * * * * * * * * * * *   C Y C L E   1   * * * * * * * * * * * * *
```

(CH)-MOVEMENT = 1

```
ITERATION    1    AVG. MOVEMENT =   .00376 A
ITERATION    2    AVG. MOVEMENT =   .00201 A
ITERATION    3    AVG. MOVEMENT =   .00097 A
ITERATION    4    AVG. MOVEMENT =   .00068 A
ITERATION    5    AVG. MOVEMENT =   .00032 A
```

```
TOTAL ENERGY IS       8.0597 KCAL.
  COMPRESS     .1136    VANDERWAALS              TORSION      -.8811
  BENDING     4.9488    1,4           4.6776
  STR-BEND     .0948    OTHER         -.9325     DIPL/CHG      .0385
```

FIG. 2.3.12.   (*continued*)

```
DELTA(T) =     .80 SEC.         ELAPSED TIME =    3.90 SEC.

       ITERATION   6    AVG. MOVEMENT =   .00019 A
       ITERATION   7    AVG. MOVEMENT =   .00011 A
       ITERATION   8    AVG. MOVEMENT =   .00007 A
       ITERATION   9    AVG. MOVEMENT =   .00005 A
       ITERATION  10    AVG. MOVEMENT =   .00003 A


TOTAL ENERGY IS      8.0594 KCAL.
   COMPRESS    .1134     VANDERWAALS                TORSION     -.8777
   BENDING    4.9432     1,4           4.6788
   STR-BEND    .0947     OTHER        -.9315        DIPL/CHG     .0385

DELTA(T) =     .67 SEC.         ELAPSED TIME =    4.57 SEC.




* * * * * * * * * * * * *   C Y C L E   2  * * * * * * * * * * * * *


                     (CH)-MOVEMENT = 0

       ITERATION  11    AVG. MOVEMENT =   .00002 A

TOTAL ENERGY IS      8.0594 KCAL.
   COMPRESS    .1134     VANDERWAALS                TORSION     -.8777
   BENDING    4.9432     1,4           4.6787
   STR-BEND    .0947     OTHER        -.9315        DIPL/CHG     .0385

DELTA(T) =     .15 SEC.         ELAPSED TIME =    4.72 SEC.


* * * * * ENERGY IS MINIMIZED WITHIN  .0014 KCAL * * * * *

      * * * * * ENERGY IS    8.0594 KCAL * * * * *



-------------------------------------------------------------------------
BECAUSE OF THE MOVEMENT DURING THE MINIMIZATION SEQUENCE, ANOTHER
VESCF-MINIMIZATION CYCLE MUST BE DONE TO INSURE COMPLETE CONVERGENCE.
-------------------------------------------------------------------------

SCF CALCULATION FOR NONPLANAR SYSTEM

       ITER  1   V(O) =  -1.827143    DIFF =  0.00000000
       ITER  2   V(O) =  -1.827203    DIFF =   .00006005
       ITER  3   V(O) =  -1.827218    DIFF =   .00001506
       ITER  4   V(O) =  -1.827222    DIFF =   .00000377

SCF CALCULATION FOR PLANAR SYSTEM

       ITER  1   V(O) =  -1.828624    DIFF =  0.00000000
       ITER  2   V(O) =  -1.828689    DIFF =   .00006549
       ITER  3   V(O) =  -1.828706    DIFF =   .00001637
       ITER  4   V(O) =  -1.828710    DIFF =   .00000408
```

FIG. 2.3.12. (*continued*)

```
-----------------------------------------------------------------
AROMATIC TORSIONAL CONSTANTS ARE FACTORED BY THE RATIO
 SUM(PIJ(O)BIJ(O)-PIJ(W)BIJ(W))/SUM(PIJ(O)BIJ(O)*(1-COS(W)*COS(W))
WHERE THE SUM IS OVER ALL DIHEDRAL ANGLES WITH CENTRAL PI-ATOMS I & J.
THE CALCULATED RATIO IS    .0386/    .1233.   THE PROGRAM USES    .3134
-----------------------------------------------------------------
```

INITIAL ENERGY CALCD FROM SYMMETRIZED COORDS:

```
TOTAL ENERGY IS        8.0270 KCAL.
   COMPRESS     .1135      VANDERWAALS              TORSION      -.9059
   BENDING     4.9432        1,4         4.6787
   STR-BEND     .0904      OTHER        -.9315      DIPL/CHG      .0385
```

* * * * * * * * * * * *   C Y C L E   1   * * * * * * * * * * * *

(CH)-MOVEMENT = 1

```
      ITERATION    1     AVG. MOVEMENT =   .00081 A
      ITERATION    2     AVG. MOVEMENT =   .00052 A
      ITERATION    3     AVG. MOVEMENT =   .00026 A
      ITERATION    4     AVG. MOVEMENT =   .00019 A
      ITERATION    5     AVG. MOVEMENT =   .00009 A
```

```
TOTAL ENERGY IS        8.0198 KCAL.
   COMPRESS     .1140      VANDERWAALS              TORSION      -.8974
   BENDING     4.9250        1,4         4.6722
   STR-BEND     .0948      OTHER        -.9274      DIPL/CHG      .0385
```

DELTA(T) =   .78 SEC.   ELAPSED TIME =   5.49 SEC.

```
      ITERATION    6     AVG. MOVEMENT =   .00005 A
      ITERATION    7     AVG. MOVEMENT =   .00003 A
      ITERATION    8     AVG. MOVEMENT =   .00002 A
      ITERATION    9     AVG. MOVEMENT =   .00002 A
      ITERATION   10     AVG. MOVEMENT =   .00001 A
```

```
TOTAL ENERGY IS        8.0197 KCAL.
   COMPRESS     .1140      VANDERWAALS              TORSION      -.8962
   BENDING     4.9232        1,4         4.6726
   STR-BEND     .0948      OTHER        -.9271      DIPL/CHG      .0385
```

DELTA(T) =   .68 SEC.   ELAPSED TIME =   6.17 SEC.

* * * * * * * * * * * *   C Y C L E   2   * * * * * * * * * * * *

(CH)-MOVEMENT = 0

      ITERATION   11   AVG. MOVEMENT =   .00001 A

```
TOTAL ENERGY IS        8.0197 KCAL.
   COMPRESS     .1140      VANDERWAALS              TORSION      -.8962
   BENDING     4.9231        1,4         4.6726
   STR-BEND     .0948      OTHER        -.9271      DIPL/CHG      .0385
```

FIG. 2.3.12.   (*continued*)

```
     DELTA(T) =      .15 SEC.          ELAPSED TIME =     6.32 SEC.


       * * * * * ENERGY IS MINIMIZED WITHIN  .0014 KCAL * * * * *

            * * * * * ENERGY IS    8.0197 KCAL * * * * *


     .71634     .30548     .39869     2       -.74551     .30761     .37127     2
    1.57809    1.26652     .01373     2      -1.59098    1.27490    -.03370     2
    1.27560    2.68594    -.37403     1      -1.27340    2.69652    -.40092     1
     -.00311    3.26533     .23367     1       1.17171    -.66579     .66056     5
   -1.21253    -.66471     .60749     5       2.64268     .98141    -.06779     5
   -2.65315     .99370    -.15223     5       1.26722    2.76777   -1.48614     5
    2.12845    3.30899    -.01164     5      -2.12987    3.32057    -.04897     5
   -1.24379    2.78825   -1.51192     5        .00295    4.36868     .05533     5
     -.01526    3.12264    1.34090     5


--------------------------------------------------------------------------



CYCLOHEPTA-1,3-DIENE

                                        C( 7) H(10)


       * M  M  P 2 * * * * * * * * * * * * 1 9 8 2 *         DATE   84/02/29.
       *                                             *
       *                                             *
       *                                             *
       *                                             *
       *                                             *
       *                                             *
       *                                             *
       *                                             *
       *                                             *
       *                                             *
     * * * * * * * * * * * * * * * * * * * * *

CONFORMATIONAL ENERGY, PART 3:   GEOMETRY AND STERIC ENERGY
                                 OF FINAL CONFORMATION.

CONNECTED ATOMS
     1- 2- 4- 6- 7- 5- 3- 1-

ATTACHED ATOMS
     1- 8, 2- 9, 3-10, 4-11, 5-12, 5-13, 6-14, 6-15,
     7-16, 7-17,

FINAL ATOMIC COORDINATES AND BONDED ATOM TABLE
     ATOM        X          Y          Z      TYPE    BOUND TO ATOMS    PI #
     C( 1)     .71634     .30548     .39869    ( 2)    2    3    8        1
     C( 2)    -.74551     .30761     .37127    ( 2)    1    4    9        2
     C( 3)    1.57809    1.26652     .01373    ( 2)    1    5   10        3
     C( 4)   -1.59098    1.27490    -.03370    ( 2)    2    6   11        4
     C( 5)    1.27560    2.68594    -.37403    ( 1)    3    7   12   13
     C( 6)   -1.27340    2.69652    -.40092    ( 1)    4    7   14   15
     C( 7)    -.00311    3.26533     .23367    ( 1)    5    6   16   17
```

FIG. 2.3.12.   (*continued*)

```
       H( 8)     1.17171    -.66579     .66056    ( 5)    1
       H( 9)    -1.21253    -.66471     .60749    ( 5)    2
       H(10)     2.64268     .98141    -.06779    ( 5)    3
       H(11)    -2.65315     .99370    -.15223    ( 5)    4
       H(12)     1.26722    2.76777   -1.48614    ( 5)    5
       H(13)     2.12845    3.30899    -.01164    ( 5)    5
       H(14)    -2.12987    3.32057    -.04897    ( 5)    6
       H(15)    -1.24379    2.78825   -1.51192    ( 5)    6
       H(16)      .00295    4.36868     .05533    ( 5)    7
       H(17)     -.01526    3.12264    1.34090    ( 5)    7
```

NONPLANAR

F-MATRIX (PUNCHED)
```
   -.20555
   -.09889   -.20555
   -.20999    .00000   -.20555
    .00000   -.21001    .02136   -.20555
```

MOLECULAR ORBITALS
```
    .56483    .56494    .42525    .42540
    .42540   -.42525    .56494   -.56483
   -.42540   -.42525    .56494    .56483
   -.56483    .56494    .42525   -.42540
```

EINGENVALUES
```
   -.46280   -.38514   -.02596    .05170
```

DENSITY MATRIX
```
   1.00000    .27639    .96104   -.00000
    .27639   1.00000   -.00000    .96104
    .96104   -.00000   1.00000   -.27639
   -.00000    .96104   -.27639   1.00000
```

ELECTRON DENSITIES
```
   1.0000    1.0000    1.0000    1.0000
```

PLANAR

F-MATRIX (PUNCHED)
```
   -.20555
   -.09880   -.20555
   -.21042    .00000   -.20555
   -.00000   -.21042    .02130   -.20555
```

H-CORE MATRIX
```
  -1.15061
   -.06140  -1.15065
   -.07426   0.00000  -1.03382
   0.00000   -.07425   0.00000  -1.03377
```

MOLECULAR ORBITALS
```
    .56474    .56470    .42557    .42553
   -.42553    .42557   -.56470    .56474
   -.42553   -.42557    .56470    .56474
   -.56474    .56470    .42557   -.42553
```

FIG. 2.3.12. (*continued*)

EINGENVALUES
  -.46317 -.38553 -.02557   .05207

DENSITY MATRIX
   1.00000   .27564   .96126   .00000
    .27564   1.00000   -.00000   .96126
    .96126   -.00000   1.00000   -.27564
    .00000   .96126   -.27564   1.00000

ELECTRON DENSITIES
   1.0000   1.0000   1.0000   1.0000

BOND ORDERS (P) AND RESONANCE INTEGRALS (B) FOR PI-BONDS

	---NONPLANAR GEOMETRY---			----PLANAR GEOMETRY----		
BOND	P(W)	B(W)	P*B(W)	P(O)	B(O)	P*B(O)
1- 2	.2764	.8131	.2247	.2756	.8134	.2242
1- 3	.9610	.9785	.9404	.9613	.9837	.9456
2- 4	.9610	.9788	.9406	.9613	.9837	.9456

BOND ENERGY FROM SCF CALCULATION
     SIGMA-BOND   -239.01
     PI-BOND   -118.10
     TOTAL ENERGY -357.11

AROMATIC TORSIONAL CONSTANTS ARE FACTORED BY   .3134

BOND LENGTHS AND STRETCHING ENERGY   (17 BONDS)

        ENERGY = 71.94(KS)(DR)(DR)(1+(CS)(DR))
                 DR = R-RO
                 CS = -2.000

					R(O)-CORR.	
BOND	LENGTH	R(O)	K(S)	ENERGY	E-NEG	ANOM
C(1)- C(2)	1.4621	1.4571	6.2714	.0111		
C(1)- C(3)	1.3470	1.3435	9.4208	.0084		
C(1)- H(8)	1.1042	1.1010	4.6000	.0034		
C(2)- C(4)	1.3470	1.3435	9.4208	.0085		
C(2)- H(9)	1.1042	1.1010	4.6000	.0034		
C(3)- C(5)	1.5022	1.4970	4.4000	.0085		
C(3)- H(10)	1.1051	1.1010	4.6000	.0056		
C(4)- C(6)	1.5022	1.4970	4.4000	.0086		
C(4)- H(11)	1.1051	1.1010	4.6000	.0056		
C(5)- C(7)	1.5297	1.5230	4.4000	.0142		
C(5)- H(12)	1.1151	1.1130	4.6000	.0015		
C(5)- H(13)	1.1166	1.1130	4.6000	.0043		
C(6)- C(7)	1.5297	1.5230	4.4000	.0139		
C(6)- H(14)	1.1166	1.1130	4.6000	.0043		
C(6)- H(15)	1.1152	1.1130	4.6000	.0016		
C(7)- H(16)	1.1177	1.1130	4.6000	.0072		
C(7)- H(17)	1.1165	1.1130	4.6000	.0039		

NON-BONDED DISTANCES, VAN DER WAALS ENERGY
     89 VDW INTERACTIONS (1,3 EXCLUDED)

FIG. 2.3.12. *(continued)*

$$\text{ENERGY} = KV*(2.90(10**5)\text{EXP}(-12.50/P) - 2.25(P**6))$$
$$RV = RVDW(I) + RVDW(K)$$
$$KV = \text{SQRT}(EPS(I)*EPS(K))$$
$$P = (RV/R) \text{ OR } (RV/R\#)$$
$$(\text{IF P.GT.3.311, ENERGY} = KV(336.176)(P**2))$$

IN THE VDW CALCULATIONS THE HYDROGEN ATOMS ARE RELOCATED
SO THAT THE ATTACHED HYDROGEN DISTANCE IS REDUCED BY .915

IN THE VDW CALCULATIONS THE DEUTERIUM ATOMS ARE RELOCATED
SO THAT THE ATTACHED DEUTERIUM DISTANCE IS REDUCED BY .915

ATOM PAIR	R	R#	RV	KV	ENERGY	(1,4)
C(1), C(4)	2.5398					
C(1), C(5)	2.5645					
C(1), C(6)	3.2118		3.840	.0440	.0784	*
C(1), C(7)	3.0505		3.840	.0440	.2274	*
C(1), H(9)	2.1692					
C(1), H(10)	2.0941					
C(1), H(11)	3.4829	3.3989	3.440	.0455	-.0529	*
C(1), H(12)	3.1494	3.0884	3.440	.0455	-.0190	*
C(1), H(13)	3.3442	3.2703	3.440	.0455	-.0476	*
C(1), H(14)	4.1704	4.0858	3.440	.0455	-.0317	
C(1), H(15)	3.6955	3.6436	3.440	.0455	-.0490	
C(1), H(16)	4.1396	4.0464	3.440	.0455	-.0332	
C(1), H(17)	3.0593	3.0427	3.440	.0455	-.0055	
C(2), C(3)	2.5390					
C(2), C(5)	3.2089		3.840	.0440	.0803	*
C(2), C(6)	2.5655					
C(2), C(7)	3.0526		3.840	.0440	.2248	*
C(2), H(8)	2.1695					
C(2), H(10)	3.4823	3.3984	3.440	.0455	-.0529	*
C(2), H(11)	2.0938					
C(2), H(12)	3.6815	3.6304	3.440	.0455	-.0495	
C(2), H(13)	4.1731	4.0881	3.440	.0455	-.0317	
C(2), H(14)	3.3423	3.2686	3.440	.0455	-.0474	*
C(2), H(15)	3.1541	3.0928	3.440	.0455	-.0202	*
C(2), H(16)	4.1415	4.0484	3.440	.0455	-.0331	
C(2), H(17)	3.0656	3.0486	3.440	.0455	-.0074	
C(3), C(4)	3.1694		3.880	.0440	.1360	*
C(3), C(6)	3.2168		3.840	.0440	.0751	*
C(3), C(7)	2.5581					
C(3), H(8)	2.0778					
C(3), H(9)	3.4453	3.3636	3.440	.0455	-.0522	*
C(3), H(11)	4.2433	4.1514	3.440	.0455	-.0294	
C(3), H(12)	2.1448					
C(3), H(13)	2.1155					
C(3), H(14)	4.2393	4.1505	3.440	.0455	-.0294	
C(3), H(15)	3.5505	3.5096	3.440	.0455	-.0526	
C(3), H(16)	3.4794	3.3965	3.440	.0455	-.0529	*
C(3), H(17)	2.7830	2.7470	3.440	.0455	.2150	*
C(4), C(5)	3.2131		3.840	.0440	.0775	*
C(4), C(7)	2.5602					
C(4), H(8)	3.4468	3.3652	3.440	.0455	-.0522	*
C(4), H(9)	2.0776					
C(4), H(10)	4.2440	4.1520	3.440	.0455	-.0294	
C(4), H(12)	3.5366	3.4965	3.440	.0455	-.0528	

FIG. 2.3.12. (continued)

73

C(4), H(13)	4.2394	4.1503	3.440	.0455	-.0295	
C(4), H(14)	2.1155					
C(4), H(15)	2.1438					
C(4), H(16)	3.4814	3.3985	3.440	.0455	-.0529	*
C(4), H(17)	2.7904	2.7541	3.440	.0455	.2057	*
C(5), C(6)	2.5492					
C(5), H(8)	3.5093	3.4253	3.340	.0460	-.0528	*
C(5), H(9)	4.2873	4.1951	3.340	.0460	-.0243	
C(5), H(10)	2.2064					
C(5), H(11)	4.2835	4.1918	3.340	.0460	-.0244	
C(5), H(14)	3.4793	3.3959	3.340	.0460	-.0534	*
C(5), H(15)	2.7663	2.7309	3.340	.0460	.1394	*
C(5), H(16)	2.1530					
C(5), H(17)	2.1904					
C(6), H(8)	4.2907	4.1985	3.340	.0460	-.0242	
C(6), H(9)	3.5098	3.4258	3.340	.0460	-.0528	*
C(6), H(10)	4.2882	4.1964	3.340	.0460	-.0243	
C(6), H(11)	2.2057					
C(6), H(12)	2.7636	2.7284	3.340	.0460	.1421	*
C(6), H(13)	3.4784	3.3950	3.340	.0460	-.0534	*
C(6), H(16)	2.1525					
C(6), H(17)	2.1905					
C(7), H(8)	4.1251	4.0331	3.340	.0460	-.0297	
C(7), H(9)	4.1289	4.0368	3.340	.0460	-.0295	
C(7), H(10)	3.5082	3.4238	3.340	.0460	-.0529	*
C(7), H(11)	3.5117	3.4272	3.340	.0460	-.0528	*
C(7), H(12)	2.1952					
C(7), H(13)	2.1461					
C(7), H(14)	2.1462					
C(7), H(15)	2.1941					
H(8), H(9)	2.3848	2.3064	3.000	.0470	.4018	*
H(8), H(10)	2.3254	2.2422	3.000	.0470	.5876	*
H(8), H(11)	4.2478	4.1026	3.000	.0470	-.0157	
H(8), H(12)	4.0505	3.9143	3.000	.0470	-.0203	
H(8), H(13)	4.1432	4.0069	3.000	.0470	-.0179	
H(8), H(14)	5.2245	5.0517	3.000	.0470	-.0046	
H(8), H(15)	4.7418	4.6045	3.000	.0470	-.0080	
H(8), H(16)	5.2037	5.0202	3.000	.0470	-.0048	
H(8), H(17)	4.0279	3.9383	3.000	.0470	-.0196	
H(9), H(10)	4.2460	4.1008	3.000	.0470	-.0157	
H(9), H(11)	2.3244	2.2413	3.000	.0470	.5906	*
H(9), H(12)	4.7238	4.5879	3.000	.0470	-.0082	
H(9), H(13)	5.2284	5.0550	3.000	.0470	-.0046	
H(9), H(14)	4.1418	4.0056	3.000	.0470	-.0179	
H(9), H(15)	4.0516	3.9155	3.000	.0470	-.0203	
H(9), H(16)	5.2074	5.0238	3.000	.0470	-.0048	
H(9), H(17)	4.0392	3.9487	3.000	.0470	-.0194	
H(10), H(11)	5.2965	5.1157	3.000	.0470	-.0043	
H(10), H(12)	2.6636	2.5497	3.000	.0470	.0511	*
H(10), H(13)	2.3844	2.3045	3.000	.0470	.4065	*
H(10), H(14)	5.3150	5.1344	3.000	.0470	-.0042	
H(10), H(15)	4.5227	4.4067	3.000	.0470	-.0104	
H(10), H(16)	4.2962	4.1481	3.000	.0470	-.0147	
H(10), H(17)	3.6924	3.5818	3.000	.0470	-.0320	
H(11), H(12)	4.5051	4.3903	3.000	.0470	-.0106	
H(11), H(13)	5.3145	5.1336	3.000	.0470	-.0042	
H(11), H(14)	2.3872	2.3069	3.000	.0470	.4005	*

FIG. 2.3.12. *(continued)*

H(11), H(15)	2.6562	2.5431	3.000	.0470	.0559	*
H(11), H(16)	4.2998	4.1516	3.000	.0470	-.0146	
H(11), H(17)	3.7041	3.5926	3.000	.0470	-.0316	
H(12), H(13)	1.7913					
H(12), H(14)	3.7298	3.6105	3.000	.0470	-.0308	
H(12), H(15)	2.5112	2.5145	3.000	.0470	.0791	
H(12), H(16)	2.5568	2.4559	3.000	.0470	.1388	*
H(12), H(17)	3.1246	2.9572	3.000	.0470	-.0546	*
H(13), H(14)	4.2585	4.1132	3.000	.0470	-.0154	
H(13), H(15)	3.7275	3.6086	3.000	.0470	-.0309	
H(13), H(16)	2.3760	2.2952	3.000	.0470	.4304	*
H(13), H(17)	2.5416	2.4418	3.000	.0470	.1562	*
H(14), H(15)	1.7913					
H(14), H(16)	2.3787	2.2975	3.000	.0470	.4242	*
H(14), H(17)	2.5382	2.4389	3.000	.0470	.1599	*
H(15), H(16)	2.5512	2.4509	3.000	.0470	.1449	*
H(15), H(17)	3.1241	2.9567	3.000	.0470	-.0546	*
H(16), H(17)	1.7904					

BOND ANGLES, BENDING AND STRETCH-BEND ENERGIES   (30 ANGLES)

$$EB = 0.021914(KB)(DT)(DT)(1+SF*DT**4)$$
$$DT = THETA-TZERO$$
$$SF = \quad .00700E-5$$

$$ESB(J) = 2.51124(KSB(J))(DT)(DR1+DR2)$$
$$DR(I) = R(I) - RO(I)$$
$$KSB(1) = \quad .120 \quad X-F-Y \quad F = 1ST \; ROW \; ATOM$$
$$KSB(2) = \quad .250 \quad X-S-Y \quad S = 2ND \; ROW \; ATOM$$
$$KSB(3) = \quad .090 \quad X-F-H \quad (DR2 = 0)$$
$$KSB(4) = -.400 \quad X-S-H \quad (DR2 = 0)$$
$$(X,Y = F \; OR \; S)$$

A T O M S		THETA	TZERO	KB	EB	KSB	ESB
C(2)- C(1)- C(3)		129.290	120.000			.12	.0238
IN-PLN	2- 1- 3	129.439	120.000	.430	.8401		
OUT-PL	1- 8- 1	2.556		.050	.0072		
C(2)- C(1)- H(8)		114.711	120.000			.09	-.0060
IN-PLN	2- 1- 8	114.847	120.000	.360	.2095		
OUT-PL	1- 3- 1	2.095		.050	.0048		
C(3)- C(1)- H(8)		115.564	120.000			.09	-.0035
IN-PLN	3- 1- 8	115.713	120.000	.360	.1450		
OUT-PL	1- 2- 1	1.930		.050	.0041		
C(1)- C(2)- C(4)		129.363	120.000			.12	.0241
IN-PLN	1- 2- 4	129.505	120.000	.430	.8519		
OUT-PL	2- 9- 2	2.493		.050	.0068		
C(1)- C(2)- H(9)		114.681	120.000			.09	-.0060
IN-PLN	1- 2- 9	114.811	120.000	.360	.2125		
OUT-PL	2- 4- 2	2.044		.050	.0046		
C(4)- C(2)- H(9)		115.542	120.000			.09	-.0036
IN-PLN	4- 2- 9	115.684	120.000	.360	.1470		
OUT-PL	2- 1- 2	1.883		.050	.0039		
C(1)- C(3)- C(5)		128.251	122.000			.12	.0164
IN-PLN	1- 3- 5	128.264	122.000	.550	.4730		
OUT-PL	3-10- 3	.784		.050	.0007		
C(1)- C(3)- H(10)		116.956	120.000			.09	-.0024
IN-PLN	1- 3-10	116.970	120.000	.360	.0724		
OUT-PL	3- 5- 3	.577		.050	.0004		

FIG. 2.3.12. (*continued*)

```
C( 5)- C( 3)- H(10)  114.753 118.200                      .09  -.0041
   IN-PLN  5- 3-10   114.766 118.200   .360   .0930
   OUT-PL  3- 1- 3     .643            .050   .0005
C( 2)- C( 4)- C( 6)  128.342 122.000                      .12   .0168
   IN-PLN  2- 4- 6   128.356 122.000   .550   .4870
   OUT-PL  4-11- 4     .799            .050   .0007
C( 2)- C( 4)- H(11)  116.922 120.000                      .09  -.0025
   IN-PLN  2- 4-11   116.937 120.000   .360   .0740
   OUT-PL  4- 6- 4     .588            .050   .0004
C( 6)- C( 4)- H(11)  114.693 118.200                      .09  -.0041
   IN-PLN  6- 4-11   114.706 118.200   .360   .0963
   OUT-PL  4- 2- 4     .655            .050   .0005
C( 3)- C( 5)- C( 7)  115.065 109.500   .450   .3054       .12   .0200
C( 3)- C( 5)- H(12)  109.164 109.410   .360   .0005       .09  -.0003
C( 3)- C( 5)- H(13)  106.837 109.410   .360   .0522       .09  -.0030
C( 7)- C( 5)- H(12)  111.229 109.410   .360   .0261       .09   .0028
C( 7)- C( 5)- H(13)  107.348 109.410   .360   .0335       .09  -.0031
H(12)- C( 5)- H(13)  106.766 109.400   .320   .0487
C( 4)- C( 6)- C( 7)  115.217 109.500   .450   .3223       .12   .0205
C( 4)- C( 6)- H(14)  106.839 109.410   .360   .0522       .09  -.0030
C( 4)- C( 6)- H(15)  109.086 109.410   .360   .0008       .09  -.0004
C( 7)- C( 6)- H(14)  107.361 109.410   .360   .0331       .09  -.0031
C( 7)- C( 6)- H(15)  111.141 109.410   .360   .0236       .09   .0026
H(14)- C( 6)- H(15)  106.762 109.400   .320   .0488
C( 5)- C( 7)- C( 6)  112.862 109.500   .450   .1115       .12   .0136
C( 5)- C( 7)- H(16)  107.816 109.410   .360   .0200       .09  -.0024
C( 5)- C( 7)- H(17)  110.773 109.410   .360   .0147       .09   .0021
C( 6)- C( 7)- H(16)  107.782 109.410   .360   .0209       .09  -.0025
C( 6)- C( 7)- H(17)  110.785 109.410   .360   .0149       .09   .0021
H(16)- C( 7)- H(17)  106.527 109.400   .320   .0579
```

DIHEDRAL ANGLES, TORSIONAL ENERGY (ET)   (42 ANGLES)

ET = (V1/2)(1+COS(W))+(V2/2)(1-COS(2W))+(V3/2)(1+COS(3W))

SIGN OF ANGLE A-B-C-D : WHEN LOOKING THROUGH B TOWARD C,
IF D IS COUNTERCLOCKWISE FROM A, NEGATIVE.

```
        A T O M S        OMEGA     V1     V2     V3      ET
C( 1) C( 2) C( 4) C( 6)   -9.041  -.270  4.445  0.000  -.159
C( 1) C( 2) C( 4) H(11)  168.463  0.000  4.445 -1.060   .084
C( 1) C( 3) C( 5) C( 7)   25.816  -.440   .240   .060  -.336
C( 1) C( 3) C( 5) H(12) -100.040  0.000  0.000  -.240  -.180
C( 1) C( 3) C( 5) H(13)  144.858  0.000  0.000  -.240  -.152
C( 2) C( 1) C( 3) C( 5)    9.428  -.270  4.445  0.000  -.149
C( 2) C( 1) C( 3) H(10) -168.123  0.000  4.445 -1.060   .089
C( 2) C( 4) C( 6) C( 7)  -24.974  -.440   .240   .060  -.339
C( 2) C( 4) C( 6) H(14) -144.126  0.000  0.000  -.240  -.156
C( 2) C( 4) C( 6) H(15)  100.814  0.000  0.000  -.240  -.184
C( 3) C( 1) C( 2) C( 4)    -.689  -.930  1.054  0.000  -.930
C( 3) C( 1) C( 2) H( 9)  171.505  0.000  1.054 -1.060  -.029
C( 3) C( 5) C( 7) C( 6)  -73.171   .170   .270   .093   .368
C( 3) C( 5) C( 7) H(16)  167.906  0.000  0.000   .500   .048
C( 3) C( 5) C( 7) H(17)   51.708  0.000  0.000   .500   .023
C( 4) C( 2) C( 1) H( 8) -172.691  0.000  1.054 -1.060  -.021
C( 4) C( 6) C( 7) C( 5)   72.540   .170   .270   .093   .366
```

FIG. 2.3.12. (*continued*)

```
C( 4) C( 6) C( 7) H(16)-168.517  0.000  0.000  .500  .044
C( 4) C( 6) C( 7) H(17) -52.333  0.000  0.000  .500  .020
C( 5) C( 3) C( 1) H( 8)-178.627  0.000  4.445  0.000  .003
C( 5) C( 7) C( 6) H(14)-168.598  0.000  0.000  .267  .023
C( 5) C( 7) C( 6) H(15) -52.184  0.000  0.000  .267  .011
C( 6) C( 4) C( 2) H( 9) 178.820  0.000  4.445  0.000  .002
C( 6) C( 7) C( 5) H(12)  51.612  0.000  0.000  .267  .013
C( 6) C( 7) C( 5) H(13) 168.071  0.000  0.000  .267  .025
C( 7) C( 5) C( 3) H(10)-156.589  0.000  0.000  .010  .003
C( 7) C( 6) C( 4) H(11) 157.476  0.000  0.000  .010  .003
H( 8) C( 1) C( 2) H( 9)   -.496  0.000  1.054  0.000  .000
H( 8) C( 1) C( 3) H(10)   3.822  0.000  4.445  0.000  .020
H( 9) C( 2) C( 4) H(11)  -3.676  0.000  4.445  0.000  .018
H(10) C( 3) C( 5) H(12)  77.556  0.000  0.000  .520  .102
H(10) C( 3) C( 5) H(13) -37.546  0.000  0.000  .520  .160
H(11) C( 4) C( 6) H(14)  38.324  0.000  0.000  .520  .150
H(11) C( 4) C( 6) H(15) -76.736  0.000  0.000  .520  .094
H(12) C( 5) C( 7) H(16) -67.311  0.000  0.000  .237  .009
H(12) C( 5) C( 7) H(17) 176.491  0.000  0.000  .237  .002
H(13) C( 5) C( 7) H(16)  49.148  0.000  0.000  .237  .019
H(13) C( 5) C( 7) H(17) -67.050  0.000  0.000  .237  .008
H(14) C( 6) C( 7) H(16) -49.655  0.000  0.000  .237  .017
H(14) C( 6) C( 7) H(17)  66.529  0.000  0.000  .237  .007
H(15) C( 6) C( 7) H(16)  66.759  0.000  0.000  .237  .007
H(15) C( 6) C( 7) H(17)-177.056  0.000  0.000  .237  .001
```

DIPOLE INTERACTION ENERGY (DIELECTRIC CONSTANT = 1.50)

```
   DIPOLE(1)    MU(1)      DIPOLE(2)    MU(2)    R12(A)  E(KCAL*DC)
 + C( 5)- C( 3)   .300   + C( 6)- C( 4)   .300   2.85929   .0385
```

FINAL STERIC ENERGY IS   8.0197 KCAL.

```
     COMPRESSION      .1140
     BENDING         4.9231
     STRETCH-BEND     .0948
     VANDERWAALS
       1,4 ENERGY    4.6726
       OTHER         -.9271
     TORSIONAL       -.8962
     DIPOLE           .0385
```

```
COORDINATES TRANSLATED TO NEW ORIGIN WHICH IS CENTER OF MASS
     C( 1)     .75147   -1.44815     .11368   ( 2)
     C( 2)    -.71047   -1.46897     .10665   ( 2)
     C( 3)    1.59094    -.41826    -.10775   ( 2)
     C( 4)   -1.57820    -.46181    -.11050   ( 2)
     C( 5)    1.25896    1.04120    -.23571   ( 1)
     C( 6)   -1.28997    1.00797    -.22616   ( 1)
     C( 7)    -.01883    1.48048     .48151   ( 1)
     H( 8)    1.22690   -2.44149     .19450   ( 5)
     H( 9)   -1.15759   -2.47635     .17440   ( 5)
     H(10)    2.65855    -.66437    -.25240   ( 5)
```

FIG. 2.3.12. (*continued*)

77

```
         H(11)    -2.63745    -.73793    -.26240   ( 5)
         H(12)     1.23086    1.31690   -1.31587   ( 5)
         H(13)     2.10739    1.60698    .21917    ( 5)
         H(14)    -2.15059    1.54414    .24151    ( 5)
         H(15)    -1.28023    1.29394   -1.30400   ( 5)
         H(16)     -.03373    2.59792    .49996    ( 5)
         H(17)     -.01035    1.14535    1.54644   ( 5)
```

MOMENT OF INERTIA WITH THE PRINCIPAL AXES (UNIT = 10**(-39) GM*CM**2)

    IX= 24.5807   IY= 25.3176   IZ= 46.2217

DIPOLE MOMENT =   .587 D

  COMPONENTS WITH PRINCIPAL AXES

       X=   .0087   Y=   .5850   Z=   .0487

HEAT OF FORMATION AND STRAIN ENERGY CALCULATIONS   (UNITS ARE KCAL.)

BOND ENTHALPY (BE) AND STRAINLESS BOND ENTHALPY (SBE) CONSTANTS AND SUMS

#	BOND OR STRUCTURE	--- NORMAL ---		--STRAINLESS--	
2	C-C SP3-SP3	-.004	-.01	.493	.99
6	C-H ALIPHATIC	-3.205	-19.23	-3.125	-18.75
2	C-C SP2-SP3	.170	.34	-.105	-.21
4	C-H OLEFINIC	-3.205	-12.82	-3.125	-12.50
3	C-C DELOCALIZED	152.750	458.25	152.850	458.55
2	SEC-DELOCALIZED	-28.540	-57.08	-29.167	-58.33
			BE = 369.45		SBE = 369.74

PARTITION FUNCTION CONTRIBUTION (PFC)
    CONFORMATIONAL POPULATION INCREMENT (POP)   0.00
    TORSIONAL CONTRIBUTION (TOR)   0.00
    TRANSLATION/ROTATION TERM (T/R)   2.40
                                          PFC = 2.40

STERIC ENERGY (E)   8.02

SIGMA-STRETCHING (ECPI)   .03

CORRECTED STERIC ENERGY (EC) = E-ECPI   7.99

ENERGY FROM PLANAR SCF CALCULATION (ESCF)   -357.11

HEAT OF FORMATION = EC + BE + PFC + ESCF   22.73

STRAINLESS HEAT OF FORMATION FOR SIGMA SYSTEM (HFS)
      HFS = SBE + T/R + ESCF - ECPI   15.00

FIG. 2.3.12. (*continued*)

INHERENT SIGMA STRAIN (SI) = E + BE - SBE   7.73

SIGMA STRAIN ENERGY (S) = POP + TOR + SI   7.73

(RESONANCE ENERGY IS NOT CALCULATED.)

   END OF CYCLOHEPTA-1,3-DIENE

     TOTAL ELAPSED TIME IS   6.75 SEC.

     THIS JOB COMPLETED AT 12.00.00.

FIG. 2.3.12. (*continued*)

The beginning of the output is very similar to that for cycloheptene, except that the π atoms are marked as such in the table of initial coordinates. After working out the complete table of atomic coordinates, including those for the hydrogen atoms the program moves on to a MO calculation for the π system. This is divided into two parts because the conjugated system is nonplanar. The program must perform an SCF calculation on the corresponding planar system to determine the rotation barriers, which are dependent on the planar bond orders. The SCF calculation is an iterative procedure in which the energy [$V(0)$] is minimized and the difference between consecutive density matrices (DIFF) should approach zero. The values of $V(0)$ and DIFF are printed for each cycle. The program assigns bond-stretching and torsional parameters to the conjugated system on the basis of the calculated π bond orders, and then calculates an initial energy using this force field, exactly as for cycloheptene. A summary of the results of the initial energy calculation is printed, followed by the results of the SCF calculation. The *F*-matrix, which can be used as input for another calculation on the same system to speed up convergence, is similar to a Hückel matrix, and gives the overlap integrals for the different bonds. The molecular orbitals are given as coefficients of the *p*-orbitals assigned to each carbon in horizontal rows. Thus the first molecular orbital has coefficients of 0.60157 on C_1 and C_2 and 0.37164 on C_3 and C_4. Because all the coefficients are positive, this orbital is bonding overall. The remaining three orbitals complete the well-known pattern for the π orbitals of 1,3-butadiene. Note that only one orbital is used per atom (i.e., this is a π-only calculation), and that only the four π atoms are included. The eigenvalues represent the energies of the molecular orbitals, the lower two of which are occupied in this case. The density matrix is formed from the orbital coefficients and their occupancies, as outlined in Section 4.4, and the electron densities printed are the π orbital charges for atoms 1–4. These tables are repeated for the planar calculation, in which the interatomic distances between bonded atoms are retained, but the entire system is assumed to be planar. In this case the results are fairly similar to those of the nonplanar calculation. The program then gives the π bond orders and overlap integrals for the planar and nonplanar systems. These are the values used to determine the potentials for the force field. In this example the formal double bonds (C_1–C_3 and C_2–C_4) have a nonplanar π bond order of 0.8943 and the single C_1–C_2 bond an order of 0.4475.

The program next uses the force field determined from this SCF calculation (with the normal force field used for the other parts of the molecule) for a geometry optimization as before. At the end of this optimization, however, the geometry of the optimized system has changed significantly from that on which the SCF calculation was performed, so that the force field must be redetermined after another SCF calculation on the new geometry. The criterion actually used to decide whether a new SCF is necessary is the number of Newton–Raphson iterations in the optimization cycle. If less than 10 iterations were necessary, the optimization is considered to be converged. The new force field is then used to optimize the geometry once more. In this example the SCF–optimization cycle was repeated three times before a minimum was found. The program terminates automatically after three SCF-optimization cycles, even if the geometry has not converged. In this case the message

* * * * TOTAL CONVERGENCE NOT ACCOMPLISHED IN THREE
VESCF−MINIMIZATION CYCLES
* * * * PROGRAM QUITS

is printed. The results of such a calculation are not reliable because the optimization is not complete. The information written to the punch file should then be used to restart the job. In this case, however, everything is in order, and the optimized Cartesian coordinates are printed at the end of Part 2.

Part 3 of the output then summarizes the SCF and force field calculations for the final geometry, as before. Note that the final bond orders have changed considerably from those found for the initial geometry (C_1–C_3 and C_2–C_4, 0.9610; C_1–C_2, 0.2764). The table of bond lengths and stretching energies indicates that the force field used has preferred bond lengths of 1.4571 Å for the C_1–C_2 bond and 1.3435 Å for the two double bonds, the constants for the stretching potentials being 6.2714 and 9.4208, respectively. The torsional constants V_1 and V_2 are set accordingly for the conjugated system. The remainder of the output is analogous to that for cycloheptene, except that the calculated energy from the SCF calculation (ESCF) must now be included in the expression for the heat of formation. The calculated structure is shown in Fig. 2.3.13. The calculated heat of formation (22.73 kcal mol^{-1}) compares with an experimental value of 22.5 ± 0.2.

The fourth example, norcaradiene, demonstrates the use of a different type of carbon atom for cyclopropane rings. The input is shown in Fig. 2.3.14. There are two important features to note: first that the program must be told that the molecule contains three- or four-membered rings (a "1" in column 25 of the

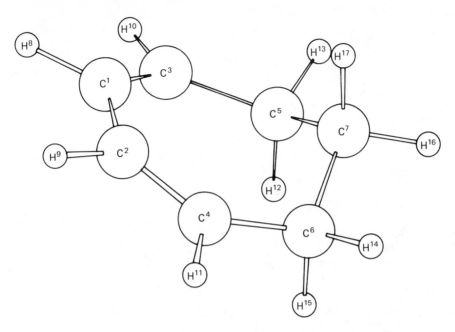

FIG. 2.3.13. The MMP2 optimized structure for 1,3-cycloheptadiene.

third card); and second that the atom type for a cyclopropane ring carbon is 22, rather than the 1 for a normal sp^3 carbon. Note also the extra atom connection list for the bridging $(C_5–C_6)$ bond. Although this atom connection list contains only two atoms, they cannot be defined as attached atoms because they also appear in the first atom connection list. The output for this job is exactly analogous to that for cycloheptadiene, so only the last section of Part 3 is shown in Fig. 2.3.15. Note in this output that the steric energy is only 7.63 kcal mol^{-1},

```
!........!..........!..........!..........!..........!..........!..........!..........!..........!
NORCARADIENE                                                        7 2              1.0
TTTT                                                             1
    2                       1              7                     1              0
    1    2    4    6    5    3    1
    5    7    6
 0.74000   0.07000   0.03000    2      -0.72000   0.07000   0.02000    2
 1.43000   1.22000  -0.05000    2      -1.42000   1.22000  -0.06000    2
 0.76000   2.55000  -0.13300   22      -0.75000   2.55000  -0.13000   22
 0.00000   3.00000   1.09000   22
    4    1    2    3
    4    2    1    4
    4    3    1    5
    4    4    2    6
    1    5    3    7
    1    6    4    7
    2    7    5    6
.............................................................................
```

FIG. 2.3.14. One possible MMP2 input for norcaradiene.

```
FINAL STERIC ENERGY IS      7.6301 KCAL.

        COMPRESSION        .2434
        BENDING           2.7153
        STRETCH-BEND      -.1589
        VANDERWAALS
          1,4 ENERGY      2.9419
          OTHER           -.6735
        TORSIONAL         2.5370
        DIPOLE             .0249

-------------------------------------------------------------------

COORDINATES TRANSLATED TO NEW ORIGIN WHICH IS CENTER OF MASS
        C( 1)   -1.47068     .73006     .12891    ( 2)
        C( 2)   -1.47067    -.73007     .12890    ( 2)
        C( 3)    -.34517    1.42555    -.12263    ( 2)
        C( 4)    -.34516   -1.42556    -.12262    ( 2)
        C( 5)     .95970     .75708    -.40603    (22)
        C( 6)     .95972    -.75710    -.40599    (22)
        C( 7)    1.59331     .00003     .73665    (22)
        H( 8)   -2.40956    1.26702     .34554    ( 5)
        H( 9)   -2.40956   -1.26703     .34550    ( 5)
        H(10)    -.35851    2.52841    -.09902    ( 5)
        H(11)    -.35850   -2.52842    -.09901    ( 5)
        H(12)    1.59367    1.28108   -1.11883    ( 5)
        H(13)    1.59371   -1.28112   -1.11875    ( 5)
        H(14)    2.67921     .00004     .79368    ( 5)
        H(15)    1.08580     .00004    1.69853    ( 5)

MOMENT OF INERTIA WITH THE PRINCIPAL AXES (UNIT = 10**(-39) GM*CM**2)

        IX=  18.6474    IY=  24.9718    IZ=  37.7897

DIPOLE MOMENT =        .268 D


COMPONENTS WITH PRINCIPAL AXES

        X=    .2622    Y=    .0000    Z=    .0569

-------------------------------------------------------------------

HEAT OF FORMATION AND STRAIN ENERGY CALCULATIONS   (UNITS ARE KCAL.)

BOND ENTHALPY (BE) AND STRAINLESS BOND ENTHALPY (SBE) CONSTANTS AND SUMS

    #   BOND OR STRUCTURE      --- NORMAL ---      --STRAINLESS--
    4   C-H OLEFINIC           -3.205   -12.82     -3.125   -12.50
    4   H-C CYCLOPROPANE       -3.205   -12.82     -3.125   -12.50
    3   C-C CYCLOPROPANE       -7.429   -22.29       .493     1.48
    2   C-C SP2-CYCLOPROPANE    3.800     7.60      3.800     7.60
    1   CYCLOPROPANE RINGS     47.625    47.63      0.000     0.00
    3   C-C DELOCALIZED       152.750   458.25    152.850   458.55
    2   SEC-DELOCALIZED       -28.540   -57.08    -29.167   -58.33
                             ---------------     ---------------
                              BE =  408.47        SBE =  384.30
```

FIG. 2.3.15. The last section of the output for norcaradiene generated from the input shown in Fig. 2.3.14.

```
PARTITION FUNCTION CONTRIBUTION (PFC)
     CONFORMATIONAL POPULATION INCREMENT (POP)   0.00
     TORSIONAL CONTRIBUTION (TOR)                0.00
     TRANSLATION/ROTATION TERM (T/R)             2.40
                                              ------------
                                       FFC =   2.40
```

STERIC ENERGY (E)   7.63

SIGMA-STRETCHING (ECPI)   .02

CORRECTED STERIC ENERGY (EC) = E-ECPI   7.61

ENERGY FROM PLANAR SCF CALCULATION   (ESCF)   -357.47

HEAT OF FORMATION = EC + BE + PFC + ESCF   61.01

STRAINLESS HEAT OF FORMATION FOR SIGMA SYSTEM (HFS)
        HFS = SBE + T/R + ESCF - ECPI   29.21

INHERENT SIGMA STRAIN (SI) = E + BE - SBE   31.80

SIGMA STRAIN ENERGY (S) = POP + TOR + SI   31.80

(RESONANCE ENERGY IS NOT CALCULATED.)

    END OF NORCARADIENE

    TOTAL ELAPSED TIME IS   3.94 SEC.

    THIS JOB COMPLETED AT   12.00.00.

FIG. 2.3.15.   (*continued*)

but the strain energy is 31.8 kcal mol^{-1}. This is because the cyclopropane carbons have a preferred bond angle of 60° in the force field, and therefore do not give rise to much angle strain. The strain energy is introduced *via* the bond increments and group increment for the cyclopropane ring. This once again demonstrates the model nature of the calculations. Norcaradiene is also a case in which considerable conjugation may exist between the diene system and the three-membered ring, so that this calculation represents a "classical" model in which such conjugation is considered only in the force field for the vinyl cyclopropane units.

The MO calculations used in MMP2 do not include interactions between atoms that are not bonded to each other. This is normally not critical, as 1,3 or 1,4 interactions are accounted for in the parametrization of the force field, and therefore need not be considered. In some cases, however, overlap between nonbonded atoms must be considered. Cyclohepta-1,3,5-triene is an example. The structure and energy of the molecule are influenced by 1,6 overlap between the termini of the hexatriene system:

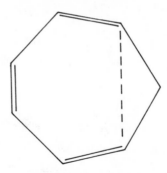

This interaction is not included in the force field parametrization, and would not normally be included in an MMP2 calculation. It can be added to the calculation by the use of a suitable option, as shown in the input given in Fig. 2.3.16(a). The "1" in column 77 of the second (SCF) card indicates that the next card to be read contains lists of π atoms that are to be allowed to overlap with each other, in this case atoms 5 and 6. The input then continues normally with the atom connection list. A further feature of this input is that the option to increase the number of SCF cycles has been used ("16" in columns 71 and 72 of the SCF card). Normally the program carries out only eight SCF cycles and prints a message if the SCF has failed to converge. This is rarely a significant problem because it usually occurs only in the first SCF–optimization cycle, so that the later cycles give reliable results. In this example the optimization does not terminate successfully in the first job, and so the calculation must be restarted using the information written to the punch file. The input for the continuation job is shown in Fig. 2.3.16(b). The relevant options and the SCF card have simply been added to the correct sections of the punch file. The F-matrices for the SCF calculation provide a good starting point for the calculation, and therefore speed up convergence.

The calculation optimizes successfully at the second attempt. The π bond order between C_5 and C_6 is calculated to be 0.1835, compared with 0.9488 for the terminal double bonds, 0.9190 for the central C_1–C_2 bond, and 0.2988 for the two single bonds of the conjugated system. The calculated heat of formation is 45.51 kcal mol^{-1}, compared with an experimental value of 43.7 ± 0.3 and a calculated value of 48.10 kcal mol^{-1} if the 1,6 interaction is not considered. The structures obtained for 1,3,5-cycloheptatriene and norcaradiene, its isomer, are shown in Fig. 2.3.17.

The final input example, shown in Fig. 2.3.18, illustrates the use of a nonstandard force field. The van der Waals parameters for a carbenium ion center (atom type 30) are included in the MMP2 program, but the stretching, bending, and torsional constants are not. The program lists the missing constants in any calculation, so that the simplest way to determine which constants are needed is to try to run the job with no extra parameters. In the

```
!........!..........!..........!..........!..........!..........!..........!..........!
CYCLOHEPTA-1,3,5-TRIENE                                            7 0              1.0
TTTTTT                                                           1          16    1
     5    6
     1                                    7                          1          0
     1    2    4    6    7    5    3    1
   0.77500   0.41500   0.63000    2      -0.67500   0.41500   0.63000    2
   1.54000   1.33000   0.00000    2      -1.54000   1.33000   0.00000    2
   1.26000   2.63000  -0.32000    2      -1.26000   2.63000  -0.32000    2
   0.00000   3.33000   0.00000    1
     4    1    2    3
     4    2    1    4
     4    3    1    5
     4    4    2    6
     4    5    3    7
     4    6    4    7
     2    7    5    6
```

(a)

```
!........!..........!..........!..........!..........!..........!..........!..........!
CYCLOHEPTA-1,3,5-TRIENE                                           15    0    0
TTTTTT                                                           1 2       16    1
     5    6
     1                      0    8    0                              1
     1    2    4    6    7    5    3    1
     1    8    2    9    3   10    4   11    5   12    6   13    7   14    7   15
    .65735    .46964    .75148    2      -.69552    .48496    .79146    2    2
   1.51337   1.33319   -.05359    2     -1.57775   1.36838    .03766    2    4
   1.22220   2.59504   -.42846    2     -1.28059   2.62355   -.35440    2    6
   -.00979   3.35009   -.02066    1      1.16784   -.33162   1.31344    5    8
  -1.18999   -.30485   1.38315    5      2.50115    .93034   -.33656    5   10
  -2.58931    .98818   -.18653    5      1.99162   3.16254   -.98162    5   12
  -2.06818   3.20876   -.86163    5      -.01458   4.32734   -.55875    5   14
    .02509   3.58114   1.06916    5      0.0       0.0       0.0      LAST
2 = NONPLANAR F-MATRIX          CYCLOHEPTA-1,3,5-TRIENE
 -.2055494 -.2029944 -.0917638  .0000000  .0000000  .0197804 -.2055494 -.0000000
 -.0914645  .0198589 -.0000000 -.2055494  .0088486 -.2081840 -.0000000 -.2055494
  .0000000 -.2082683 -.2055494 -.0381685 -.2055494
3 = PLANAR F-MATRIX             CYCLOHEPTA-1,3,5-TRIENE
 -.2055494 -.1993522 -.1078410 -.0000000 -.0000000  .0226945 -.2055494  .0000000
 -.1076589  .0227755  .0000000 -.2055494  .0077880 -.2063390 -.0000000 -.2055494
  .0000000 -.2064296 -.2055494 -.0331347 -.2055494
```

(b)

FIG. 2.3.16. (a) One possible MMP2 input for 1,3,5-cycloheptatriene including 1,6 overlap. (b) The input for the continuation of the optimization, as prepared from the punch file.

case of the ^{t}butyl cation, the C^+–$C(sp^3)$ stretching constant must be read in with the C–C–C^+ in- and out-of-plane bending constants, the C^+–C–H bending constant, and the C–C^+–C–H torsional constants V_1, V_2, and V_3. If a heat of formation calculation is required the bond increment for the C^+–$C(sp^3)$ bonds must also be read in. The "1" in column 75 of the first card indicates that additional constants for the force field are to be read in. The input then proceeds normally, with two atom connection lists, atomic coordinates for four

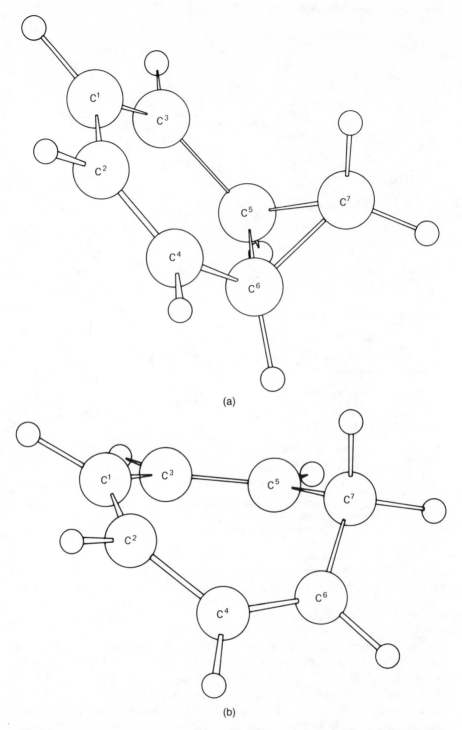

(a)

(b)

FIG. 2.3.17. MMP2 structures for (a) norcaradiene and (b) 1,3,5-cycloheptatriene.

```
!........!.........!.........!.........!.........!.........!.........!.........!
T-BUTYL CATION                                                    4 0        1 1.0

   2                                        3                      0          0
   1     2     3
   2     4
   1.45000   0.00000   0.00000   1        0.00000   0.00000   0.00000   30
  -0.80000   1.20000   0.00000   1       -0.80000  -1.20000   0.00000   1
   3     1     2
   3     3     2
   3     4     2
   1     1         3              1          0
   5     1    30     10.0        -3.0      -1.000
   1   304.4         1.497
   1    30     1     0.55        120.
   0    30     1     0.05          0.0
   5     1    30     0.36        109.5
   1    30 81.00     67.25          1      C(+)-C(SP3)
..........................................................................
```

FIG. 2.3.18. MMP2 input for the ^{t}butyl cation using a trial force field.

atoms (the carbons), and coordinate calculation cards to add three hydrogens to each of the three sp^3 carbons. The next card of the input gives the number of constants of each type to be read in; in this case 1 torsional parameter (column 5), 1 stretching parameter (column 10), 0 van der Waals parameters (column 15), 3 bending parameters (column 20), and 1 heat of formation increment card (column 40). This last number automatically requests that the heat of formation be calculated, even if the relevant option was not used in card 3. The force field parameters can be guessed for a first calculation, and later refined to fit experimental or *ab initio* data. In this case the force field has been adapted from that for olefinic carbons with a few changes. The torsional potential function is defined (line 12) for the H–C(sp^3)–C$^+$–C(sp^3) (5–1–30–1) dihedral angles *via* the V_1, V_2, and V_3 parameters. In this example V_1 has been set to zero, V_3 (−1.0) has been set equal to the corresponding parameter for the H–C(sp^3)–C(sp^2)–C(sp^3) torsional function, and a value of −3.0 has been estimated for V_2 to reflect the hyperconjugation between the carbenium-ion center and the CH bonds. A negative value for V_2 gives minima at $\omega = 90°$ and $\omega = 270°$. This parameter is an arbitrary guess at the correct potential function for C$^+$–C(sp^3) bonds.

Line 13 of the input gives the K_s and r_0 parameters for the C$^+$–C(sp^3) bond-stretch term. These parameters are taken directly from the alkene force field. The next three cards (14–16) define the three angle-bending functions needed. The first is the C(sp^3)–C$^+$–C(sp^3) in-plane bending function, as indicated by the three atom types (1–30–1). The second card, which begins at zero, gives the out-of-plane bending function for the carbenium center. These parameters are once again taken directly from the alkene force field. The last angle bending parameters are for the 5–1–30 bond angles [H–C(sp^3)–C$^+$], and are identical to those for H–C(sp^3)–C(sp^2) angles. The stretch-bend parameters are already defined in the program. The last card of the input would not normally be included in such a trial calculation, since it gives the bond increment for the C^+–$C(sp^3)$ bonds. This increment is needed to calculate the heat of formation, and has been included only to illustrate the adjustment of the calculated steric energy to give the correct heat of formation. This is possible only when the results of the calculation are already known. The two increments, 81.00 and 67.25 kcal mol^{-1}, are the normal and strain-free values, respectively, and have been chosen to give the correct heat of formation and zero strain energy. The "1" in column 35 indicates that a title for the bond increment will be read in from column 41 onwards.

An extract from the results is shown in Fig. 2.3.19, and the structure in Fig. 2.3.20. The planarity of the carbenium-ion center, the conformation of the hydrogen atoms, and the lengths of the CC bonds could now be adjusted by varying the force field parameters. Note that the negative torsional barriers result in a negative steric energy. This is then corrected to give the heat of formation *via* the C^+–$C(sp^3)$ bond increment, which would normally be determined after the force field had been optimized. It is easy to obtain the correct heat of formation for one molecule, as in this example. If a force field were to be determined for carbenium-ions, all the available data would be used with, perhaps, geometries from MO calculations to form a parametrization set.

```
T-BUTYL CATION


                                                                        DATE  84/02/29.


     ADDITIONS AND/OR REPLACEMENTS WILL BE MADE TO   3 OF THESE ATOMS.
     SOME CONSTANTS WILL BE READ IN.

CONFORMATIONAL ENERGY, PART 1:   GEOMETRY AND STERIC ENERGY
                                 OF INITIAL CONFORMATION.

CONNECTED ATOMS
     1- 2- 3-
     2- 4-

INITIAL ATOMIC COORDINATES
     ATOM          X              Y              Z          TYPE
     C( 1)       1.45000       0.00000        0.00000       ( 1)
    C+( 2)       0.00000       0.00000        0.00000       (30)
     C( 3)       -.80000       1.20000        0.00000       ( 1)
     C( 4)       -.80000      -1.20000        0.00000       ( 1)


-------------------------------
3 COORDINATE OPTIONS USED
-------------------------------


NEW HEADER CARD

T-BUTYL CATION                                        13    0   1   (PUNCHED)

     2                        0     9     0                       (PUNCHED)
```

FIG. 2.3.19. An extract from the MMP2 output for the ^{t}butyl cation obtained using the input shown in Fig. 2.3.18.

```
NEW CONNECTED ATOM LISTS
   1- 2- 3-                                                    (PUNCHED)
   2- 4-                                                       (PUNCHED)

NEW ATTACHED ATOM LISTS
   1- 5, 1- 6, 1- 7, 3- 8, 3- 9, 3-10, 4-11, 4-12,            (PUNCHED)
   4-13,

NEW ATOMIC COORDINATES
   ATOM        X            Y           Z       TYPE
     1      1.45000     -.00000     0.00000      1            (PUNCHED)
     2      -.00000     0.00000     0.00000     30            (PUNCHED)
     3      -.80000     1.20000     0.00000      1            (PUNCHED)
     4      -.80000    -1.20000     0.00000      1            (PUNCHED)
     5      1.81556     1.03748     0.00000      5            (PUNCHED)
     6      1.81713     -.51871     -.89786      5            (PUNCHED)
     7      1.81713     -.51871     .89786       5            (PUNCHED)
     8     -1.86601     .92867      0.00000      5            (PUNCHED)
     9      -.57206     1.79320     -.89786      5            (PUNCHED)
    10      -.57206     1.79320     .89786       5            (PUNCHED)
    11      -.13954    -2.07965     0.00000      5            (PUNCHED)
    12     -1.43524    -1.21774     -.89786      5            (PUNCHED)
    13     -1.43524    -1.21774     .89786       5            (PUNCHED)


     THE FOLLOWING  1 TORSIONAL PARAMETERS ARE READ IN
             (* FOR 4-MEMBERED RING)
          ATOM TYPE NOS.    V1       V2       V3
          5   1  30   1    0.000   -3.000   -1.000


     THE FOLLOWING  1 STRETCHING PARAMETERS ARE READ IN
          BOND TYPE     K(S)      L(0)
           1 - 30       4.40     1.4970


     THE FOLLOWING  3 BENDING PARAMETERS ARE READ IN

                    (* FOR 4-MEMBERED RING)
                    (+ FOR 3-MEMBERED RING)

             ATOM TYPES      K(B)    THETA(0)    ED. TYPE
              1 30  1        .550    120.000
                30  1        .050
              5  1 30        .360    109.500


     INITIAL STERIC ENERGY IS    -30.3997 KCAL.

             COMPRESSION      2.4567
             BENDING           .9993
             STRETCH-BEND      .0201
             VANDERWAALS
              1,4 ENERGY      2.2285
              OTHER           -.1130
             TORSIONAL      -35.9915

     INITIAL ENERGY REQUIRES    .11 SECONDS.
```

FIG. 2.3.19. (*continued*)

```
++++++++++++++++++++++++++++++++++++++++++++++++++++++++++++++++++++++++++++++++

     FINAL STERIC ENERGY IS    -40.4976 KCAL.

         COMPRESSION        .3090
         BENDING           1.5660
         STRETCH-BEND      -.0261
         VANDERWAALS
           1,4 ENERGY      1.1405
           OTHER            .6769
         TORSIONAL       -44.1639

     ------------------------------------------------------------------

     COORDINATES TRANSLATED TO NEW ORIGIN WHICH IS CENTER OF MASS
        C( 1)     1.41176    -.45535     .03946    ( 1)
        C+( 2)     .00009    -.00004    -.22017    (30)
        C( 3)     -.31156    1.45029     .03944    ( 1)
        C( 4)    -1.10029    -.99490     .03946    ( 1)
        H( 5)     1.97641     .27655     .65910    ( 5)
        H( 6)     1.94502    -.63104    -.92153    ( 5)
        H( 7)     1.44294   -1.37424     .66647    ( 5)
        H( 8)    -1.22729    1.57326     .65974    ( 5)
        H( 9)     -.42664    1.99980    -.92156    ( 5)
        H(10)      .46899    1.93692     .66585    ( 5)
        H(11)     -.74863   -1.85022     .65850    ( 5)
        H(12)    -1.51947   -1.36842    -.92154    ( 5)
        H(13)    -1.91133    -.56256     .66697    ( 5)

     MOMENT OF INERTIA WITH THE PRINCIPAL AXES (UNIT = 10**(-39) GM*CM**2)

           IX=  10.5957    IY=  10.5963    IZ=  19.2452


     ------------------------------------------------------------------


     HEAT OF FORMATION AND STRAIN ENERGY CALCULATIONS  (UNITS ARE KCAL.)

     THE FOLLOWING  1 STRUCTURAL AND BOND ENTHALPY PARAMETERS ARE READ IN.
        1   30    81.000    67.250  TITLE : C(+)-C(SP3)

     BOND ENTHALPY (BE) AND STRAINLESS BOND ENTHALPY (SBE) CONSTANTS AND SUMS

         #   BOND OR STRUCTURE        --- NORMAL ---     --STRAINLESS--
         9   C-H ALIPHATIC            -3.205  -28.85     -3.125  -28.13
         3   C(+)-C(SP3)              81.000  243.00     67.250  201.75
                                     ----------------   ----------------
                                     BE =  214.16       SBE =  173.63

     PARTITION FUNCTION CONTRIBUTION (PFC)
         CONFORMATIONAL POPULATION INCREMENT (POP)   0.00
         TORSIONAL CONTRIBUTION (TOR)                0.00
         TRANSLATION/ROTATION TERM (T/R)             2.40
                                                ------------
                                                PFC =  2.40
```

FIG. 2.3.19.   (*continued*)

```
STERIC ENERGY (E)                                              -40.50

SIGMA-STRETCHING (ECPI)                                          0.00

CORRECTED STERIC ENERGY (EC) = E-ECPI                          -40.50

ENERGY FROM PLANAR SCF CALCULATION  (ESCF)                      0.00

HEAT OF FORMATION = EC + BE + PFC + ESCF                       176.06

STRAINLESS HEAT OF FORMATION FOR SIGMA SYSTEM (HFS)
        HFS = SBE + T/R + ESCF - ECPI                          176.03

INHERENT SIGMA STRAIN (SI) = E + BE - SBE                        .03

SIGMA STRAIN ENERGY (S) = POP + TOR + SI                         .03


-----------------------------------------------------------------

    END OF T-BUTYL CATION


        TOTAL ELAPSED TIME IS     3.26 SEC.

        THIS JOB COMPLETED AT  12.00.00.
```

FIG. 2.3.19. (*continued*)

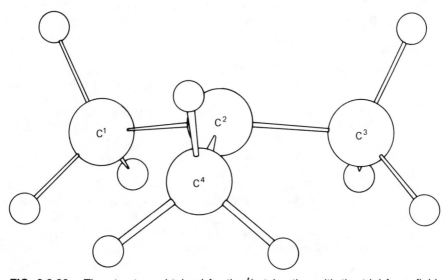

FIG. 2.3.20. The structure obtained for the ^{t}butyl cation with the trial force field.

The MMP2 program contains other options, such as one for symmetry or the dihedral driver, which will not be illustrated here. They are described in the program manuals and in reference 1. Calculations with MM2 or MMP2 may give erroneous results if the input is incorrect, but one can always detect this situation by checking that the numbers of each type of atom and of the bond group increments are correct. Location of transition states for conformational interconversions is not always straightforward with the dihedral driver. Two or more dihedral angles may have to be driven simultaneously if the correct reaction path is to be found. Because of the size of the molecules involved the use of a plotting program to produce a ball and stick picture, as in this chapter, is advantageous.

REFERENCES

1. U. Burkert and N. L. Allinger, "Molecular Mechanics," ACS Monograph 177, American Chemical Society, Washington, D.C., 1982.

2. K. Mislow, D. A. Dougherty, and H. B. Schlegel, *Tetrahedron*, 1978, **34**, 1441.

3. S. W. Benson, "Thermochemical Kinetics," 2nd ed., Wiley, New York, 1976; J. D. Cox and G. W. Pilcher, "Thermochemistry of Organic and Organometallic Compounds," Academic, New York, 1970; T. Clark and M. A. McKervey in "Comprehensive Organic Chemistry," vol. 1, p. 37, Pergamon, Oxford, 1979.

4. N. L. Allinger and J. T. Sprague, *J. Am. Chem. Soc.*, 1973, **95**, 3893; J. Kao and N. L. Allinger, *ibid.*, 1977, **99**, 975.

5. J. E. Williams, P. Stang, and P. v. R. Schleyer, *Ann. Rev. Phys. Chem.* , 1968, **19**, 591.

6. N. L. Allinger, *Adv. Phys. Org. Chem.*, 1976, **13**, 1.

7. D. H. Wertz and N. L. Allinger, *Tetrahedron*, 1974, **30**, 1579.

8. E. M. Engler, J. D. Andose, and P. v. R. Schleyer, *J. Am. Chem. Soc.*, 1973, **95**, 8005.

9. E. Osawa, J. B. Collins, and P. v. R. Schleyer, *Tetrahedron*, 1977, **33** , 2667.

10. T. Clark, T. M. Knox, M. A. McKervey, H. Mackle, and J. J. Rooney, *J. Am. Chem. Soc.*, 1979, **101**, 2404.

11. E. Osawa, *QCPE Bulletin*, 1983, **3**, (4), 87.

12. S. Profeta, Jr. and M. Rahman, unpublished work cited in ref. 1; H.-D. Beckhaus, unpublished.

13. H.-D. Beckhaus, EUCHEM conference on Organic Free Radicals, Schloss Elmau, W. Germany, 1983.

14. W. F. Maier and P. v. R. Schleyer, *J. Am. Chem. Soc.*, 1981, **103**, 1891.

15. See, for instance, H. A. Scheraga, *Adv. Phys. Org. Chem.*, 1968, **6**, 103; *Chem. Rev.*, 1971, **71**, 195; *Pure Appl. Chem.*, 1973, **36**, 1.

16. C. Altona and D. H. Faber, *Top. Curr. Chem.*, 1974, **45**, 1.

17. M. J. Browman, L. M. Carruthers, K. L. Kashuba, F. A. Momany, M. S. Pottle, S. P. Posen, S. M. Rumsey, and H. A. Scheraga, UNICEPP, QCPE program No. 361, 1978.

18. O. Ermer, "Aspekte von Kraftfeldrechnungen," Wolfgang Bauer Verlag, Munich, 1981.

19. E. Osawa and H. Musso, *Angew. Chem.*, 1983, **95**, 1.

CHAPTER 3

MOLECULAR ORBITAL THEORY

3.1. INTRODUCTION

The molecular orbital (MO) programs to be discussed in this section are all based on the LCAO–SCF type of procedure. This chapter gives a qualitative introduction to these and other terms commonly used in quantum chemistry. The object of all MO programs is to build a set of molecular orbitals to be occupied by the electrons assigned to the molecule. In principle this can be achieved by combining any number of different types of electron probability functions, or even by writing one extremely complex function to describe the electron density in each molecular orbital. A far more convenient procedure is to build up the molecular orbitals from sets of orbitals centered on the constituent atoms. The MO calculation then simply involves finding the combinations of these atomic orbitals that have the proper symmetries and that give the lowest (most negative) electronic energy. This is the *linear combination of atomic orbitals* (LCAO) formalism.

The simplest example illustrating this procedure is the hydrogen molecule. The atomic orbitals in this case can be limited to two $1s$ orbitals, one for each hydrogen atom. Two identical atoms must belong to the point group $D_{\infty h}$, for which all symmetry species must be either symmetric or antisymmetric with respect to a mirror plane midway between the two atoms and perpendicular to the connecting line between them (see Fig. 3.1.1). Therefore only two possible types of atomic-orbital combination fulfill these symmetry conditions, Ψ_1 and Ψ_2 in Fig. 3.1.1. In Ψ_1 the two hydrogen $1s$ orbitals are added to each other to give a symmetric, or bonding, combination. In Ψ_2 one orbital has changed phase (i.e., it is subtracted from the other) to give an antisymmetric, or antibonding, combination. The LCAO

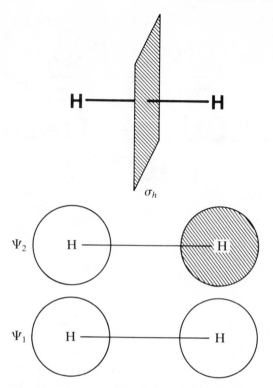

FIG. 3.1.1. The σ_h mirror plane and the molecular orbitals of H_2.

equations for these two molecular orbitals are

$$\Psi_1 = c\psi_1 + c\psi_2$$

and

$$\Psi_2 = c'\psi_1 - c'\psi_2 ,$$

where ψ_1 and ψ_2 are the atomic orbitals of H_1 and H_2 and c and c' are the *coefficients* of these atomic orbitals in the molecular orbitals Ψ_1 and Ψ_2. The coefficients are equal for ψ_1 and ψ_2 in this case because of the symmetry constraints. This is not the case for unsymmetrical molecules. The magnitude of the coefficients is determined by the *normalization* condition. This arises from the fact that the total probability of finding an electron in the molecular orbital must be 1, just as for each constituent atomic orbital. This probability is equal to the sum of the squares of the coefficients. Therefore both c in Ψ_1 and c' in Ψ_2 must be $1\sqrt{2}$, or 0.7071. In hydrogen, we can now fill Ψ_1 with two electrons to give a hydrogen–hydrogen bond, and leave Ψ_2 unoccupied. Such unoccupied orbitals are also known as *virtual orbitals*. Note that the composition of a virtual orbital does not affect the energy of the molecule, since it contains no electrons.

The effect of electronegativity can be illustrated by a similar example, lithium hydride. In this case the lithium atomic orbitals can be represented by 1s and 2s orbitals, although in most cases a set of 2p orbitals would be considered necessary for lithium. The lithium 1s orbital is a *core orbital*; that is, it belongs to a shell that is fully occupied in the isolated atom, and therefore does not take part in bonding. For this reason core orbitals and their electrons are often omitted from semiempirical calculations. This *frozen core approximation* (or *valence electron approximation*) will be discussed in connection with MINDO/3 and MNDO. Figure 3.1.2. shows the molecular orbitals obtained for lithium hydride when the lithium 1s orbital is ignored. Note that the coefficient of the hydrogen 1s orbital in Ψ_1 is now much larger than that of the lithium 2s. This occurs because the energy of an electron in a given molecular orbital is determined primarily by two factors: the bonding interactions between atomic orbitals (the *overlap*) and their original energies in the isolated atoms. In this case the 1s orbital in H. (ionization potential 13.4 eV) is far lower in energy than the 2s in Li. (ionization potential 5.4 eV). The lowest-energy bonding combination, Ψ_1, is a compromise between the need to maximize the bonding between the two atoms (which occurs when the two coefficients are equal) and the tendency for electrons to occupy the more stable of the two atomic orbitals, the hydrogen 1s. Therefore the greater the difference in energy between the interacting atomic orbitals is, the more unsymmetrical will be the bonding molecular orbital. The extreme case where one coefficient sinks to zero gives a purely ionic bond. Figure 3.1.2. shows typical coefficients for LiH. These coefficients can be squared to give an estimate of the charge on each atom from this molecular orbital. Because the orbital is doubly occupied (analogously to H_2) the total number of electrons that can be assigned to hydrogen by this procedure is ($c_h^2 \times 2 = 1.445$), but the neutral hydrogen atom has one electron, so that the calculated charge on hydrogen is −0.445. This simple procedure is called a *population analysis* and is carried out automatically by most MO programs. Note that the virtual orbital (Ψ_2 in Fig. 3.1.2.) mirrors the behavior of the occupied orbital Ψ_1.

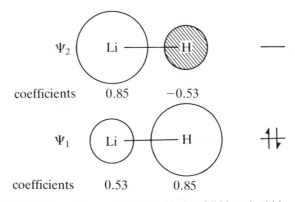

FIG. 3.1.2.   The molecular orbitals of lithium hydride.

One important point to note for the interpretation of molecular orbitals is that if p atomic orbitals are used to construct bonding molecular orbitals as in Fig. 3.1.3., the coefficients are of the same sign for a π orbital (Fig. 3.1.3(b)), but of opposite sign for a σ orbital (Fig. 3.1.3(a)).

The above discussion ignores interactions between electrons in different molecular orbitals. Many simple MO treatments, such as Hückel and extended Hückel theories, are based on this *one-electron* treatment. "One electron" in this case means that the electron is considered not to interact with the others in the molecule. One-electron theories give, for instance, the same orbital energies for benzene and its radical cation and anion. This is not realistic, because any one electron in benzene$^{-\cdot}$ is repelled by two more electrons than in benzene$^{+\cdot}$. Nevertheless, one-electron theories are used extensively, especially extended Hückel theory (EHT) for transition-metal complexes. The methods discussed in this book, however, take electron–electron repulsion into account by considering the interaction between an electron in a given orbital and the mean field of the other electrons in the molecule. This approach is known as the *self-consistent field* (SCF) method, and involves an iterative process in which the orbitals are improved from cycle to cycle until the electronic energy reaches a constant minimum value and the orbitals no longer change. This situation is described as "self consistent." At the SCF level the electron–electron repulsion is actually overestimated. The theory does not allow the electrons to avoid each other, but assumes that their instantaneous positions are independent of one

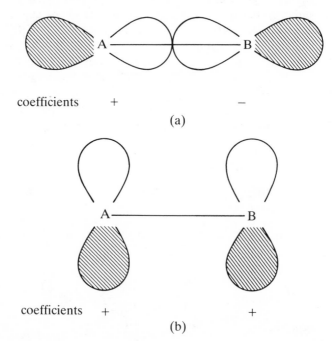

FIG. 3.1.3. Bonding combinations of (a) σ and (b) π molecular orbitals.

another. However, the error is reasonably consistent, so that its effects can be made to cancel by the use of proper comparisons. The SCF method is also known as *Hartree–Fock* or *single-determinant* theory.

Self-consistent field calculations can be used successfully as long as the number and type of electron pairs (bonds, lone pairs) does not change from one side of the equation to the other. This is because the amount by which the electron–electron repulsion is overestimated is approximately constant for each type of electron pair. This consideration leads to the use of *isodesmic equations* for evaluation of results of SCF calculations. An isodesmic equation is one in which there are equal numbers of each type of bond on each side. For instance, the equation

$$CH_3F + CH_3{}^+ \rightarrow CH_2F^+ + CH_4$$

can be used to evaluate the stabilization energy of the fluoromethyl cation. Each side of the equation has six CH bonds and one CF bond. Strictly speaking, the use of isodesmic equations should not be necessary for parametrized methods such as MINDO/3 or MNDO, as the overestimation of the electron–electron repulsion should be compensated by the parametrization, which is based on experimental results.

In real systems the movements of the electrons are not independent of each other, as assumed in SCF methods, but are correlated to a certain extent so as to minimize repulsions as much as possible. This *electron correlation* means, in effect, that if electron A is at one end of the molecule, electron B prefers to be at the other end. The methods used to calculate the *correlation energy* will be discussed in Chapter 5 on *ab initio* calculations. An important point to note is that MINDO/3 and MNDO are parametrized at the SCF level to fit experimental results, which already reflect correlation. Therefore any calculations using these methods and considering correlation do not give realistic heats of formation, because the correlation energy has been included twice, once *via* the parametrization and once explicitly.

One of the major problems in using the SCF–MO method is how to treat both open- and closed-shell molecules consistently (i.e., to calculate molecules with and without unpaired electrons at the same level of approximation). Closed-shell systems are almost always calculated using *restricted Hartree–Fock* (RHF) theory, also known as spin restricted Hartree–Fock theory. One set of molecular orbitals is calculated, but each orbital may only be doubly occupied or empty. The problem of electron spin therefore does not play a role because all spins are paired.

More possibilities exist for open-shell systems. The simplest model conceptually is that used in the original versions of MINDO/3 and MNDO, the *half-electron* (HE) method. A half-electron calculation for an open-shell system is simply an RHF calculation in which the singly occupied orbitals contain not one electron, that would have a spin, but rather two half electrons of opposite spin. A small correction (the half-electron correction) is then applied to compensate for the pairing energy of the two half electrons. The HE method has the advantage that it is on the same energy scale as RHF calculations for closed-shell molecules; this means, for instance, that the relative energies of singlet and triplet states can be compared directly if the singlet is calculated with the RHF and the triplet with the HE method. The disadvantages of the HE approach are that no information about spin densities is available and that the method is not well suited for optimizations using analytically calculated forces; thus HE optimizations may be very slow. Koopman's theorem (see below) ionization potentials are also not reliable in HE calculations. Because the energies of singly and doubly occupied orbitals are artificially separated, HE calculations do not reproduce Jahn–Teller distortions in open-shell species.

An alternative method for calculating open-shell systems is the *unrestricted Hartree–Fock* (UHF) formalism, which is most often used for open-shell calculations with the GAUSSIAN series of *ab initio* programs and by the newer versions of MINDO/3 and MNDO treated here (although the MOPAC program can also perform HE calculations). Unrestricted Hartree–Fock calculations determine two sets of molecular orbitals, one for each type of spin. These two sets of orbitals, named alpha and beta, are similar, but not identical. A doublet (radical), for instance, has one more alpha than beta electron, and a triplet has two more (there are always at least as many alpha electrons as beta). This is shown in Fig. 3.1.4. The UHF procedure is more flexible than the RHF because the paired alpha and beta orbitals, which correspond to doubly occupied molecular orbitals in the RHF formalism, need not be identical. This is both a strength and a weakness of the method. On the one hand, it allows for spin polarization, the process by which an unpaired electron perturbs formally paired spins, and therefore it gives realistic estimates of spin densities. On the other hand, this extra flexibility gives a more negative electronic energy than would be obtained with spin-restricted theory. It is therefore not possible to compare the energies of open- and closed-shell systems directly. A further problem that may arise with UHF calculations is that of *spin contamination*. The UHF

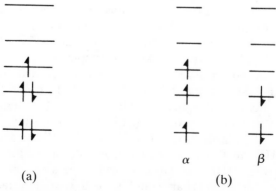

FIG. 3.1.4.   (a) Restricted and (b) unrestricted Hartree–Fock orbitals for a radical.

wave function is not limited to one pure electronic state (for instance a doublet radical), but can also mix in states of higher spin (in this case a quadruplet or even higher). In extreme cases the admixture of higher-spin states may distort the results, leading to too negative an energy and unrealistic spin densities. The degree of spin contamination can be seen from the $\langle S^2 \rangle$ parameter printed after the SCF calculation by most UHF programs. The expectation value for $\langle S^2 \rangle$ is $S(S + 1)$, where S is the total spin (i.e. $\langle S^2 \rangle = 0$ for a singlet, 0.75 for a doublet, and 2.0 for a triplet). In practice the value of $\langle S^2 \rangle$ will always be larger than the expectation value, but values up to about 10% larger usually are acceptable.

Spin contamination is not a problem for HE calculations or for the *open-shell restricted Hartree–Fock* (ROHF) procedure. In these the doubly occupied orbitals are restricted so as to be identical for alpha and beta spins, thus no admixture of higher-spin states is possible. Open-shell restricted Hartree–Fock calculations are possible with *ab initio* programs such as GAUSSIAN82, but not with the usual semiempirical programs. A more detailed treatment of open-shell systems, with examples, will be given for each method.

3.2. HOW THE PROGRAMS WORK

The basic steps involved in performing a typical *ab initio* or semiempirical calculation are very similar and can be treated together. The first task is to determine the type of calculation required. The programs are usually controlled by specific *keywords*, which request given types of calculation. In nonstandard calculations a more detailed specification of the steps to be taken by the program must be supplied. If the keywords are used, the program converts them to internal parameters, which then control the execution.

The next step is to read in a title for the job (used only for information, and printed out exactly as it is read in), the molecular charge, and the required multiplicity (singlet, triplet, etc.). The molecular geometry is then read in, usually in the form of a *Z-matrix* of atomic numbers, bond lengths, bond angles, and dihedral angles. Note that the *Z*-matrix is only a geometrical means of defining the positions of the atoms. It is not meant to tell the program where to put bonds or to represent a given electronic state. The program will (almost) always give the electronic configuration that is most stable for the geometrical arrangement of atoms defined by the *Z*-matrix. The information from the *Z*-matrix is used to calculate the Cartesian (x,y,z) coordinates of the atoms and, in conjunction with the charge, atomic numbers, and multiplicity, to work out the total number of electrons and the orbital occupancies. At this stage the nuclear repulsion energy may be calculated. Because this number depends only on the atomic numbers and the molecular geometry, it should be the same in different calculations on the same structure. This constancy can be used to check that, for instance, *ab initio* calculations with different basis sets have been performed on exactly the same geometry. The atomic orbitals are then assigned to each nucleus. Semiempirical programs use a set of predetermined parameters to define the forms and energies of the atomic orbitals. *Ab initio* programs may use an internally stored standard set of coefficients and exponents that define the orbitals (the *basis set*), or these may be read in with the input for a nonstandard basis. The *p*-orbitals are always oriented in the *x*, *y*, and *z* directions, although any set of mutually perpendicular axes would give identical results. *Ab initio* programs next calculate the various one- and two-electron integrals (often hundreds of thousands of them) required later in the calculation. The integrals are identified by indices that are uniquely assigned to them and with which they are written to magnetic disk files for later use.

Both semiempirical and *ab initio* programs must then produce an *initial guess*, a trial set of molecular orbitals used as a starting point for the SCF calculations. There are several possibilities for the initial guess. Semiempirical programs often simply divide the electrons evenly among the atomic orbitals and allow the SCF procedure to find more realistic molecular orbitals. The usual form of initial guess for *ab initio* programs is that obtained from an extended Hückel calculation on the molecule in question (i.e., the initial guess overlay is actually a simple extended Hückel program). The set of orbitals thus obtained may then be "projected" to give a good approximation of the orbitals expected for the basis set to be used in the *ab initio* calculation. At this stage the electronic configuration required may also be chosen, although the configuration given by the initial guess is usually the ground state.

The program uses the initial guess as the starting point for an iterative SCF calculation, the part of the program that actually delivers the results required. The solution to the SCF equations is improved cycle by cycle until the electronic energy is at a minimum and the density matrix does not change. At this stage the calculation is said to be converged, or to have reached self consistency, and the program proceeds to the next step. In some cases the SCF does not converge, but either oscillates between two possible solutions or diverges rapidly. The techniques used to overcome this problem will be discussed for the individual methods.

The next stage of the calculation depends on the type of job to be performed. For a single point, the program may either move directly to the *population analysis*, which calculates the atomic charges, overlaps, dipole moment, and so forth, or perform some sort of post-SCF correlation energy calculation (by a perturbational, configuration interaction, or density-functional method).

For a geometry optimization the atomic forces are then determined analytically and used to estimate the minimum-energy geometry for the molecular species being calculated. The above process is repeated for each new geometry until the atomic forces are close to zero and the total energy

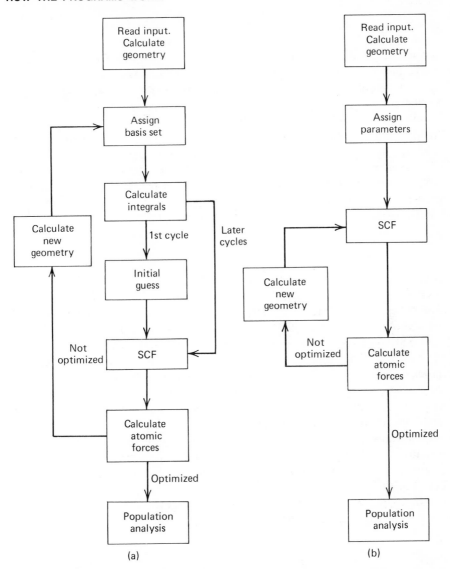

FIG. 3.2.1. Typical flow charts for (a) an *ab initio* optimization, and (b) a corresponding semi-empirical calculation.

does not change significantly from cycle to cycle. At this stage the optimization is complete and the program moves on to a population analysis of the optimized species. The procedures used to convert the atomic forces into changes in the geometry vary from program to program, and there may even be several options in one program, but the principles outlined above apply to them all. Figure 3.2.1 shows flow charts for some typical semiempirical and *ab initio* jobs.

3.3. GEOMETRIES: THE Z-MATRIX

The minimum possible input for a MO program consists of the molecular charge and multiplicity and some definition of the molecular structure (the types of atom used and their geometrical positions). The structure is conveniently defined by means of a Z-matrix, which is a method of defining a molecule atom by atom in terms of bond lengths, bond angles, and dihedral angles. All the MO programs treated here use Z-matrix input, although there are small differences in input between the different programs. It is important to note, as mentioned earlier, that the Z-matrix in no way defines the bonds to be formed in the calculation. It is simply a geometrical device used to define the positions and types of the atoms. The following examples illustrate the important principles.

Methylene

To facilitate interpretation of the results of the calculations, which are given in terms of p_x-, p_y-, and p_z-orbitals, it is desirable to orient the molecule so that it lies in one Cartesian plane and a second Cartesian plane coincides with the second molecular mirror plane, as shown in Fig. 3.3.1(a). GAUSSIAN82 chooses such an orientation automatically, but other programs do not. The required orientation is most easily achieved by use of a *dummy atom*. This is simply a point in space that is treated as an atom for the purposes of the geometry definition. At some stage in the calculation after the determination of the Cartesian coordinates the dummy atoms will be removed so that only the real molecule remains. Because the program automatically places the first two atoms along one of the Cartesian axes (usually z) and the first three in a Cartesian plane (usually xz), the desired orientation can be achieved by using a dummy as the first atom, as shown in Fig. 3.3.1(b).

The Z-matrix for the structure shown in Fig. 3.3.1(b) can now be built up, one line per atom, using the numbering shown.

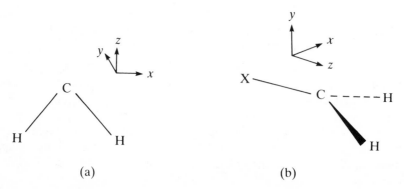

(a)   (b)

FIG. 3.3.1. (a) An ideal orientation for the methylene fragment. (b) The use of a dummy atom (X) to define the correct orientation.

First Atom. The first atom is always placed at the origin of the coordinate system, and therefore only its atomic number code need be given. A dummy atom is given the symbol **XX** or the number 99 in MOPAC, and the symbol X in GAUSSIAN82. Older programs from the GAUSSIAN series use either zero (GAUSSIAN70) or −1 (GAUSSIAN76) to indicate a dummy atom.

Second Atom. The second atom, the carbon in Fig. 3.3.1(b), must now be defined. Because the second atom is always placed on a predetermined axis, only the distance from atom 1 need be defined. Figure 3.3.2 shows the methylene Z-matrix for GAUSSIAN82. The second line indicates that a carbon atom **C** is bound to atom **1** at a distance of **1.0 Å** . The defined structure at this stage is also shown in Fig. 3.3.2.

Third Atom. The third atom, H_3 in Fig. 3.3.2, is defined using the distance from the carbon atom and the dummy–carbon–hydrogen angle. The program automatically places this atom in a predetermined Cartesian plane, so that no further definition is required. The third line in Fig. 3.3.2 indicates that a hydrogen atom, **H**, is bound to the carbon atom, number **2**, at a distance of **1.09 Å** and that it makes an angle with the dummy atom, number **1**, of **122.0°**. The defined structure at this stage is also shown.

Fourth Atom. The fourth atom (H_4) is defined exactly as the third, except that an extra parameter is needed to specify its position uniquely. The distance from C_2 and the dummy–carbon–hydrogen angle together define a circle on which atom 4 must lie:

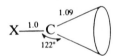

The exact position is defined using a dihedral angle to H_3. Imagine a Newman projection along the C_2-dummy bond. The dihedral angle is simply

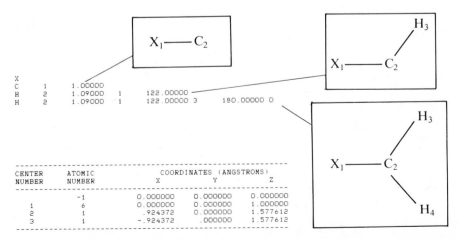

FIG. 3.3.2. A Z-matrix definition for the methylene structure shown in Fig. 3.3.1(b).

the angle between the two relevant bonds in this projection. Thus the fourth line in Fig. 3.3.2 defines a hydrogen atom, **H**, bound to the carbon, number **2**, at a distance of **1.09 Å**, making an angle with the dummy atom, number **1**, of **122.0°**, and a dihedral angle with hydrogen atom number **3** of **180.0°**. In general the dihedral angle is obtained by imagining a Newman projection along the bond between the atom to which the current atom is bonded and that with reference to which the angle is defined (the first two numbered atoms in the current card). The direction used for the dihedral angle is not important, as long as it is consistent throughout the Z-matrix.

The molecule is now completely defined. The Z-matrix and Cartesian coordinates obtained with GAUSSIAN82 are shown in Fig. 3.3.2. The carbene is symmetrically oriented in the coordinate system, and is thus ideally suited for interpretation of the results. The examples given in this chapter are all given in GAUSSIAN82 format, but the corresponding MOPAC Z-matrices are given in Appendix A for those who intend to use the semiempirical program. Note that in older MINDO/3 and MNDO programs the third atom *must* be bound to the second, not the first, atom. If the Z-matrix is defined differently, the programs assume that atom 3 is bound to atom 2 and thus change the structure. This restriction does not apply to MOPAC or to the GAUSSIAN programs.

Ethylene

The Z-matrix for D_{2h} ethylene is shown in Fig. 3.3.3. The first four atoms are defined exactly analogously to those in the previous example except that

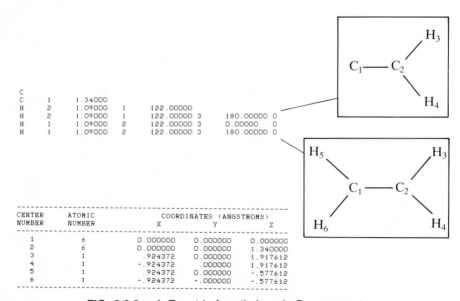

C							
C	1	1.34000					
H	2	1.09000	1	122.00000			
H	2	1.09000	1	122.00000	3	180.00000	0
H	1	1.09000	2	122.00000	3	0.00000	0
H	1	1.09000	2	122.00000	3	180.00000	0

CENTER NUMBER	ATOMIC NUMBER	COORDINATES (ANGSTROMS)		
		X	Y	Z
1	6	0.000000	0.000000	0.000000
2	6	0.000000	0.000000	1.340000
3	1	.924372	0.000000	1.917612
4	1	-.924372	.000000	1.917612
5	1	.924372	0.000000	-.577612
6	1	-.924372	-.000000	-.577612

FIG. 3.3.3. A Z-matrix for ethylene in D_{2h} symmetry.

the dummy atom has been replaced by carbon 1. The fifth atom, a hydrogen, **H**, is bound to carbon **1** at a distance of **1.09** Å and makes an angle with carbon **2** of **122.0°**. The dihedral angle with hydrogen **3** is obtained from a Newman projection along the C_1-C_2 bond, and is **0.0°**.

The sixth atom is defined identically, except that the dihedral angle to hydrogen 3 is now 180.0°. There are many possible ways of defining the last two dihedral angles (for instance 0° to atoms 3 and 4, respectively), but using one atom (in this case H_3) as a reference for the others is often the least confusing method. Note that atom 6 could, in principle at least, be defined as having a bond angle of $-122.0°$ and a dihedral angle of 0.0° to H_3. This definition would not be accepted by GAUSSIAN82, for which all bond angles must be between 0° and 180°, and would be very confusing when used with semiempirical programs. Beginners are often tempted to use negative bond angles (or even bond lengths) to define molecules. This leads at best to confusion, and more often to an error when the job is run. Bond angles and bond lengths should always be positive; the desired structure can always be obtained *via* the dihedral angles, which may be negative.

Staggered Ethane

Staggered (D_{3d}) ethane is an example of a molecule with threefold symmetry. Its Z-matrix is shown in Fig. 3.3.4. The definition of the first three atoms is exactly as in the previous examples, except that the appropriate bond lengths and bond angle for ethane have been used. The fourth atom, a hydrogen, **H**, is bound to carbon **2** at a distance of **1.09** Å, and makes an angle with carbon **1** of **110.0°**. The dihedral angle, defined with respect to hydrogen 3, is **120.0°**, as seen from the Newman projection along the C_2-C_1 bond. H_5 is defined analogously, except that the dihedral angle (taken from the same Newman projection) is 240°. In this case a dihedral angle of $-120°$ leads to exactly the same result. In general, dihedral angles of $-n°$ or $360 - n°$ are equivalent and either may be used without affecting the geometry. Once again we emphasize that this is the case only for dihedral angles, not for bond angles. The molecule at this stage of the Z-matrix is shown in Fig. 3.3.4.

The sixth atom, a hydrogen, **H**, bonded to carbon **1** at a distance of **1.09** Å, makes an angle with carbon **2** of **110.0°**. The dihedral angle is now defined with respect to a Newman projection along the C_1-C_2 bond, that is, in the opposite direction to that used for atoms 3–5, because atoms 1 and 2 appear in reverse order in the definition of H_6 as compared with that used for H_4 and H_5. H_6 is defined as making a dihedral angle with hydrogen **3** of **60.0°**. H_7 and H_8 are defined similarly, but with dihedral angles of 180° and 300°, respectively, as shown in Fig. 3.3.4.

Propyne

Propyne illustrates a common pitfall in defining molecules with linear portions: The fourth atom cannot be defined if the first three are linear (any

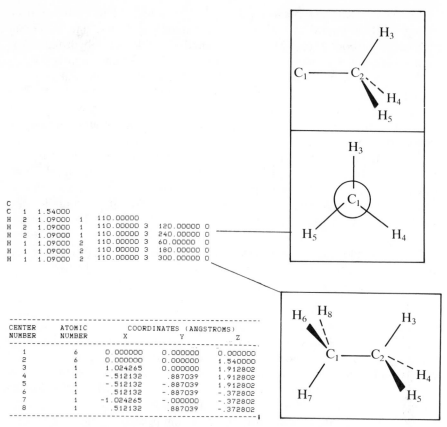

```
C
C   1   1.54000
H   2   1.09000   1   110.00000
H   2   1.09000   1   110.00000   3   120.00000   0
H   2   1.09000   1   110.00000   3   240.00000   0
H   1   1.09000   2   110.00000   3    60.00000   0
H   1   1.09000   2   110.00000   3   180.00000   0
H   1   1.09000   2   110.00000   3   300.00000   0
```

CENTER	ATOMIC	COORDINATES (ANGSTROMS)		
NUMBER	NUMBER	X	Y	Z
1	6	0.000000	0.000000	0.000000
2	6	0.000000	0.000000	1.540000
3	1	1.024265	0.000000	1.912802
4	1	-.512132	.887039	1.912802
5	1	-.512132	-.887039	1.912802
6	1	.512132	-.887039	-.372802
7	1	-1.024265	-.000000	-.372802
8	1	.512132	.887039	-.372802

FIG. 3.3.4. A Z-matrix for ethane in D_{3d} symmetry.

dihedral angle is in this case meaningless). The same problem may some-times occur during an optimization if a bond angle approaches 180°. Figure 3.3.5 shows one possible Z-matrix for propyne in which the problem has been solved by the use of dummy atoms and by defining the methyl group first.

The definition of the first five atoms in Fig. 3.3.5 is exactly the same as that for the first five atoms of ethane in the previous example. By starting at this end of the molecule, the problems of linear structures can be avoided; the dihedral angles can be defined with respect to hydrogen 3. The next problem arises with the definition of the C_2–C_1–C_7 angle. Strictly speaking, an atom that is defined with a 180° bond angle can only occupy one position, and therefore does not need a dihedral angle. The programs expect a dihedral angle, however, and so if a 180° bond angle were used an arbitrary dihedral angle would have to be defined. Figure 3.3.5 shows an alternative solution that will also work with the GAUSSIAN80 and -82 programs, which do not accept 180° bond angles for geometry optimizations. The

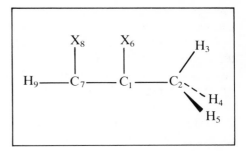

```
C
C   1  1.46000
H   2  1.09000  1  110.00000
H   2  1.09000  1  110.00000  3  120.00000  0
H   2  1.09000  1  110.00000  3  240.00000  0
X   1  1.00000  2   90.00000  3    0.00000  0
C   1  1.20000  6   90.00000  2  180.00000  0
X   7  1.00000  1   90.00000  6    0.00000  0
H   7  1.08000  8   90.00000  1  180.00000  0
```

CENTER	ATOMIC	COORDINATES (ANGSTROMS)		
NUMBER	NUMBER	X	Y	Z
1	6	0.000000	0.000000	0.000000
2	6	0.000000	0.000000	1.460000
3	1	1.024265	0.000000	1.832802
4	1	-.512132	.887039	1.832802
5	1	-.512132	-.887039	1.832802
	-1	1.000000	0.000000	.000000
6	6	.000000	.000000	-1.200000
	-1	1.000000	.000000	-1.200000
7	1	.000000	.000000	-2.280000

FIG. 3.3.5. A Z-matrix for propyne in C_{3v} symmetry.

C_2–C_1–C_7 angle is defined using dummy number 6 at right angles to the C_1–C_2 bond. This type of definition always works for both semiempirical and GAUSSIAN programs, and will therefore be used throughout. Atom 6, a dummy, X, is bound to carbon 1 at a distance of 1.0 Å and makes an angle with carbon 2 of 90.0°. The dihedral angle, defined with respect to hydrogen 3 is 0° (i.e., H_3, C_2, C_1 and dummy 6 are all in one plane). The seventh atom, a carbon, is bound to carbon 1 at a distance of 1.2 Å, makes an angle with dummy 6 of 90.0°, and forms a dihedral angle with carbon 2 of 180.0° (Newman projection along the dummy 6–C_1 vector). The definitions of X_8 and H_9 are exactly analogous to those of atoms 6 and 7.

Methanol

Methanol in the staggered conformation shown in Fig. 3.3.6 illustrates the use of a dummy atom to define a methylene group within C_s symmetry. This is a structure that arises particularly often in organic calculations. It is often incorrectly assumed that the methyl groups in low-symmetry molecules have local threefold symmetry, probably because the protons are usually chemically equivalent. This is of course not the case, and the symmetry of the methyl groups must correspond to the molecular point group.

The definition of the first four atoms is straightforward, as shown in Fig. 3.3.6. To define the two symmetrical protons of the methyl group, it is advantageous to define a dummy atom that lies on the bisector of the

H–C–H angle. This atom (the fifth) is defined as a dummy, **X**, bonded to carbon **1** at an arbitrary distance of **1.0 Å**. The dummy makes an angle with oxygen **2** of **125.0°** and a dihedral angle with H_3 of **0.0°** (Newman projection along the C_1–O_2 bond). The angle of **125°** is approximately correct, the corresponding angle in a tetrahedral molecule being 125.2645°. The structure at this stage of the definition is shown in Fig. 3.3.6. The two hydrogen atoms can now be defined relative to the dummy. To do this we can use the fact that the plane of the CH_2 group is perpendicular to the dummy–C_1–O_2 plane, as shown in Fig. 3.3.6. The dihedral angles of the two hydrogens to the oxygen atom (using the Newman projection along the C_1–dummy "bond") are therefore 90° and −90° (or 270°). The hydrogens, **H**, are thus defined as being bound to carbon **1** at a distance of **1.09 Å**. They make an angle with the dummy **5** equal to half the H–C–H angle, or **55.0°** (this is because the dummy lies on the H–C–H bisector). The dihedral angles to oxygen **2** are **90.0°** for H_6 and −90° for H_7. (If you are unable to visualize these dihedral angles molecular models can be helpful.) This method of defining methylene groups not only retains C_s symmetry, but also has advantages during geometry optimization (see below). What we have done, in effect, is to define an O–C–dummy angle and two equivalent half H–C–H angles rather than two equivalent O–C–H angles and two dihedral angles, one of which is the reverse of the other. We shall use this type of definition whenever methylene groups have a plane of symmetry. Modern programs can optimize dihedral angles so that a plane of symmetry is retained, but the above method often leads to faster optimization.

A few more complex examples will now be given to illustrate strategies

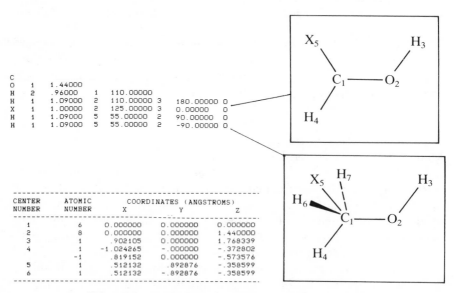

C							
O	1	1.44000					
H	2	.96000	1	110.00000			
H	1	1.09000	2	110.00000	3	180.00000	0
X	1	1.00000	2	125.00000	3	0.00000	0
H	1	1.09000	5	55.00000	2	90.00000	0
H	1	1.09000	5	55.00000	2	−90.00000	0

CENTER NUMBER	ATOMIC NUMBER	COORDINATES (ANGSTROMS)		
		X	Y	Z
1	6	0.000000	0.000000	0.000000
2	8	0.000000	0.000000	1.440000
3	1	.902105	0.000000	1.768339
4	1	−1.024265	−.000000	−.372802
	−1	.819152	0.000000	−.573576
5	1	.512132	.892876	−.358599
6	1	.512132	−.892876	−.358599

FIG. 3.3.6. A Z-matrix for methanol in C_s symmetry.

for writing Z-matrices. The definitions will not be discussed in detail, but the important features of each Z-matrix will be pointed out. A useful exercise for the reader would be to work through the Z-matrices to determine the structures.

Cyclopropane, D_{3h}

The most convenient way to define cyclopropane within D_{3h} symmetry is shown in Fig. 3.3.7. Two dummies (atoms 1 and 2) have been used to define the main (C_3) axis, and the positions of the carbon atoms (3–5) have been defined relative to the center of the ring (dummy 2). This is a case in which not the conventional chemical bonds but rather distances to a defined point (in this case the center of symmetry), are used to define the geometry. This is a fundamental difference from the input for molecular mechanics calculations, in which the bonds themselves must be defined.

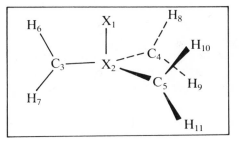

```
X
X   1  1.00000
C   2   .90000   1  90.00000
C   2   .90000   1  90.00000   3  120.00000  0
C   2   .90000   1  90.00000   3  240.00000  0
H   3  1.08000   2 122.00000   1    0.00000  0
H   3  1.08000   2 122.00000   1  180.00000  0
H   4  1.08000   2 122.00000   1    0.00000  0
H   4  1.08000   2 122.00000   1  180.00000  0
H   5  1.08000   2 122.00000   1    0.00000  0
H   5  1.08000   2 122.00000   1  180.00000  0
```

CENTER NUMBER	ATOMIC NUMBER	COORDINATES (ANGSTROMS)		
		X	Y	Z
	-1	0.000000	0.000000	0.000000
	-1	0.000000	0.000000	1.000000
1	6	.900000	0.000000	1.000000
2	6	-.450000	.779423	1.000000
3	6	-.450000	-.779423	1.000000
4	1	1.472313	0.000000	.084108
5	1	1.472313	.000000	1.915892
6	1	-.736156	1.275060	.084108
7	1	-.736156	1.275060	1.915892
8	1	-.736156	-1.275060	.084108
9	1	-.736156	-1.275060	1.915892

FIG. 3.3.7. A Z-matrix for cyclopropane in D_{3h} symmetry.

Oxirane, C_{2v}

Figure 3.3.8 shows one possible Z-matrix for oxirane within C_{2v} symmetry. There are several points to note about the geometry definition. Once again the main symmetry axis (the C_2) is defined using a dummy atom (number 2), but this time in conjunction with oxygen 1. Two half bond lengths (from dummy 2 to carbons 3 and 4) are used to define the CC bond. The methylene groups, that have C_s local symmetry, are defined using dummies

```
O
X   1   1.10000
C   2    .74000   1  90.00000
C   2    .74000   1  90.00000   3  180.00000  0
X   3   1.00000   2  130.00000  1  180.00000  0
X   4   1.00000   2  130.00000  1  180.00000  0
H   3   1.08000   5  55.00000   2  90.00000   0
H   3   1.08000   5  55.00000   2  -90.00000  0
H   4   1.08000   6  55.00000   2  90.00000   0
H   4   1.08000   6  55.00000   2  -90.00000  0
```

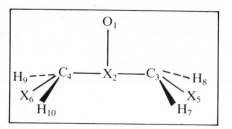

```
----------------------------------------------------------------------
CENTER        ATOMIC              COORDINATES (ANGSTROMS)
NUMBER        NUMBER          X            Y            Z
----------------------------------------------------------------------
   1             8        0.000000     0.000000     0.000000
                -1        0.000000     0.000000     1.100000
   2             6         .740000     0.000000     1.100000
   3             6        -.740000      .000000     1.100000
                -1        1.382788      .000000     1.866044
                -1       -1.382788      .000000     1.866044
   4             1        1.138183      .884684     1.574536
   5             1        1.138183     -.884684     1.574536
   6             1       -1.138183     -.884684     1.574536
   7             1       -1.138183      .884684     1.574536
----------------------------------------------------------------------
```

FIG. 3.3.8. A Z-matrix for oxirane in C_{2v} symmetry.

on the H–C–H bisectors, as shown above for methanol. This type of Z-matrix is often used for substituted three-membered rings or for complexes between Lewis acids and ethylene.

Twist Cyclopentane, C_2

At first sight the definition shown in Fig. 3.3.9 for twist cyclopentane may appear unnecessarily complicated. The Z-matrix is, however, written with the geometry optimization process in mind and incorporates several features that help to give a fast, problem-free optimization. The definition of the first three atoms is conventional, but the next two carbon atoms (5 and 6) are defined as being bound to dummy 1 rather than to carbon 2. This type of definition leads to faster optimization in the GAUSSIAN programs because the BERNY optimization procedure is based on estimated force constants, which can be considerably inaccurate for cyclic structures. Distances and angles measured relative to dummy atoms give fewer problems and faster optimization. The next two carbon atoms (9 and 10) are similarly defined with respect to dummy 8. Dummy 8 however, is defined with the help of the extra dummy 7 in order to avoid the 180° angle between C_2, X_1, and X_8. Note that molecules with C_n point groups can be defined using dihedral angles that are equal and have the same sign (e.g., for C_9 and C_{10}). Dihedral angles for atoms related to each other by a mirror plane have opposite signs.

Chair Cyclohexane, D_{3d}

As in the previous examples, the chair cyclohexane structure is best defined using the two dummy atoms (1 and 2 in Fig. 3.3.10) to first define the principal axis. The sets of carbon atoms, 3–5 and 6–8, are then defined in

```
X
C   1   .91000
H   2  1.08000  1  125.00000
H   2  1.08000  1  125.00000  3  180.00000  0
C   1  1.25000  2   90.00000  3   90.00000  0
C   1  1.25000  2   90.00000  4   90.00000  0
X   1  1.00000  2   90.00000  3    0.00000  0
X   1  1.41000  7   90.00000  2  180.00000  0
C   8   .77000  1   90.00000  6   29.60000  0
C   8   .77000  1   90.00000  5   29.60000  0
H   5  1.08000  2  110.00000  6  120.00000  0
H   6  1.08000  2  110.00000  5  120.00000  0
H   5  1.08000  2  110.00000  6 -120.0000  0
H   6  1.08000  2  110.00000  5 -120.0000  0
H   9  1.08000  5  110.00000 10  120.00000  0
H  10  1.08000  6  110.00000  9  120.00000  0
H   9  1.08000  5  110.00000 10 -120.0000  0
H  10  1.08000  6  110.00000  9 -120.0000  0
```

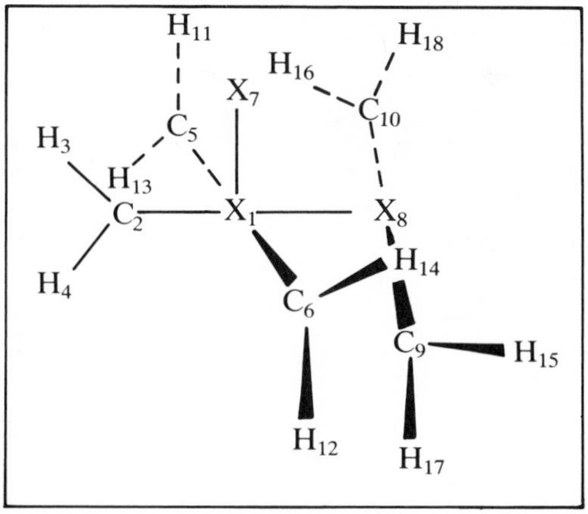

CENTER NUMBER	ATOMIC NUMBER	COORDINATES (ANGSTROMS)		
		X	Y	Z
	-1	0.000000	0.000000	0.000000
1	6	0.000000	0.000000	.910000
2	1	.884684	0.000000	1.529463
3	1	-.884684	.000000	1.529463
4	6	.000000	-1.250000	.000000
5	6	.000000	1.250000	.000000
	-1	1.000000	0.000000	.000000
	-1	.000000	.000000	-1.410000
6	6	.380335	.669511	-1.410000
7	6	-.380335	-.669511	-1.410000
8	1	.878901	-1.847282	.192836
9	1	-.878901	1.847282	.192836
10	1	-.878901	-1.847282	.192836
11	1	.878901	1.847282	.192836
12	1	.034767	1.565330	-.915540
13	1	-.034767	-1.565330	-.915540
14	1	1.434821	.769988	-1.620630
15	1	-1.434821	-.769988	-1.620630

FIG. 3.3.9. A Z-matrix for twist cyclopentane in C_2 symmetry.

two planes, and staggered by 60° relative to each other. The axial hydrogens, 9–14, are also defined relative to dummies 1 and 2, since this automatically gives the correct symmetry. The bond angles relative to the dummy–carbon vector can be estimated from models or from a "standard geometry" with all carbon atoms tetrahedral, in which case the axial hydrogens are all perpendicular to the C_3–C_4–C_5 and C_6–C_7–C_8 planes. The six equatorial hydrogens (15–20) are defined similarly, but with 0° dihedral angles relative to the two dummies and with the correct bond angles.

3.4. GEOMETRY OPTIMIZATION

The modern MO programs considered here are capable of automatically optimizing the molecular geometry (the bond lengths, angles, and dihedral angles in the Z-matrix) within the specified symmetry constraints. Faulty

```
X
X  1   .51000
C  2  1.45000   1   90.00000
C  2  1.45000   1   90.00000   3   120.00000   0
C  2  1.45000   1   90.00000   3   240.00000   0
C  1  1.45000   2   90.00000   3   180.00000   0
C  1  1.45000   2   90.00000   3    60.00000   0
C  1  1.45000   2   90.00000   3   -60.00000   0
H  3  1.08000   2   95.00000   1   180.00000   0
H  4  1.08000   2   95.00000   1   180.00000   0
H  5  1.08000   2   95.00000   1   180.00000   0
H  6  1.08000   1   95.00000   2   180.00000   0
H  7  1.08000   1   95.00000   2   180.00000   0
H  8  1.08000   1   95.00000   2   180.00000   0
H  3  1.08000   2  155.00000   1     0.00000   0
H  4  1.08000   2  155.00000   1     0.00000   0
H  5  1.08000   2  155.00000   1     0.00000   0
H  6  1.08000   1  155.00000   2     0.00000   0
H  7  1.08000   1  155.00000   2     0.00000   0
H  8  1.08000   1  155.00000   2     0.00000   0
```

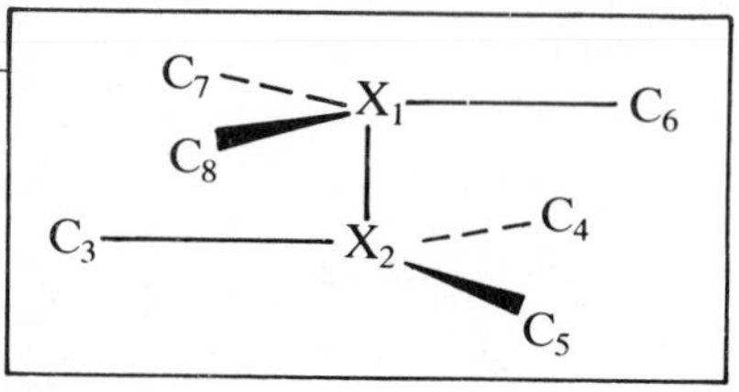

CENTER NUMBER	ATOMIC NUMBER	COORDINATES (ANGSTROMS)		
		X	Y	Z
	-1	0.000000	0.000000	0.000000
	-1	0.000000	0.000000	.510000
1	6	1.450000	0.000000	.510000
2	6	-.725000	1.255737	.510000
3	6	-.725000	-1.255737	.510000
4	6	-1.450000	-.000000	.000000
5	6	.725000	-1.255737	.000000
6	6	.725000	1.255737	.000000
7	1	1.544128	.000000	1.585890
8	1	-.772064	1.337254	1.585890
9	1	-.772064	-1.337254	1.585890
10	1	-1.544128	-.000000	-1.075890
11	1	.772064	-1.337254	-1.075890
12	1	.772064	1.337254	-1.075890
13	1	2.428812	0.000000	.053572
14	1	-1.214406	2.103413	.053572
15	1	-1.214406	-2.103413	.053572
16	1	-2.428812	-.000000	.456428
17	1	1.214406	-2.103413	.456428
18	1	1.214406	2.103413	.456428

FIG. 3.3.10. A Z-matrix for chair cyclohexane in D_{3d} symmetry.

specification of which parameters to optimize and which to set equal to each other is one of the most common sources of error for inexperienced users. The GAUSSIAN80 and -82 programs indicate whether the geometry has been fully or partly optimized within the symmetry of the Z-matrix, but the semiempirical programs do not, so that a careful check that the geometry has been properly optimized is necessary.

For complete optimization of a molecule consisting of N atoms without symmetry, a total of $3N - 6$ ($3N - 5$ for a diatomic molecule) independent parameters must be considered. This number represents the $3N$ degrees of freedom of the molecule minus the three translations and the three rotations (two for diatomics). The number of independent parameters may be reduced substantially by the use of symmetry, but should never be more than $3N - 6$. At first sight it seems unnecessary to warn against optimizing too many parameters, but this can easily happen for a large molecule with several dummy atoms. Usually this results only in wasted computer time, but occasionally it can lead to an error because the program attempts to optimize a parameter that has a gradient of exactly zero. A few examples based on the Z-matrices used earlier will help to illustrate the use of symmetry in geometry optimizations and outline some common errors. In the following Z-matrices the parameters to be optimized together will be indicated by an asterisk (*) bracketing them in the Z-matrix. The methods used to define equivalent parameters in the programs themselves vary, and will be described in Chapters 4 and 5.

Methylene C_{2v}

A full optimization of CH_2 involves $3N - 6 = 3$ independent parameters. Using the Z-matrix from Fig. 3.3.2, these would be the two bond lengths to nondummy atoms and *one* of the dummy–C–H angles. Optimizing the dummy-to-carbon distance would simply give a translation, and optimizing the second dummy–C–H angle, a rotation, as shown in Fig. 3.4.1. To retain C_{2v} symmetry, however, the two CH distances can be held equal. To keep the molecule properly oriented, the two dummy–C–H angles should also be kept equal, but optimized. The resulting Z-matrix, in which only two parameters are optimized, gives correctly oriented C_{2v} optimized methylene:

```
X
C   1   1.00
H   2   1.09 ⎫*    1   122.0 ⎫*
H   2   1.09 ⎭     1   122.0 ⎭     3   180 .
```

Ethylene and Ethane

Because the Z-matrix for ethylene was written without dummies (Fig. 3.3.3) all $3N - 6$ parameters may be optimized. Confining the molecule to D_{2h} symmetry, however, means that no dihedral angles need be optimized,

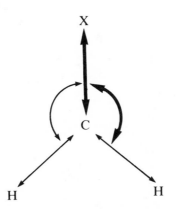

FIG. 3.4.1. The optimization of methylene. Optimization of the two parameters indicated by heavy arrows would lead to a translation and a rotation.

because they define the plane of the molecule. Similarly, all C–C–H angles and all CH bond lengths are equal, so that only 3 of the original 12 parameters remain independent. The Z-matrix from Fig. 3.3.3 can be optimized as follows:

```
C
C   1   1.34  *
H   2   1.09 ⎫*     1   122.0 ⎫*
H   2   1.09 |      1   122.0 |
H   1   1.09 |      2   122.0      3   180.0
H   1   1.09 ⎭      2   122.0 ⎭    3     0.0
                                   3   180.0
```

Exactly the same arguments apply to ethane (D_{3d}), so that only three parameters from the Z-matrix given in Fig. 3.3.4 need be optimized (in this case the dihedral angles are defined by the C_3 principal axis):

```
C
C   1   1.54  *
H   2   1.09 ⎫*     1   110.0 ⎫*
H   2   1.09 |      1   110.0 |
H   2   1.09 |      1   110.0      3   120.0
H   1   1.09 |      2   110.0      3   240.0
H   1   1.09 |      2   110.0      3    60.0
H   1   1.09 ⎭      2   110.0 ⎭    3   180.0
                                   3   300.0
```

Propyne

The two dummy atoms in the propyne Z-matrix (Fig. 3.3.5) are simply used to help define 180° angles, so none of the parameters associated with them should be optimized. The remaining 15 parameters for the "real" atoms could be optimized if symmetry were not used. However, because the two 180° angles (and hence the 90° X_6–C_1–C_7 and X_8–C_7–H_9 angles) and the methyl-group dihedral angles are defined by the C_{3v} symmetry, these need not be optimized. The C–C–H bond angles and the CH bond lengths of the methyl group are all equivalent, so only five independent parameters remain:

```
C
C   1   1.46  *
H   2   1.08 ⎫*     1   110.0 ⎫*
H   2   1.08 |      1   110.0 |
H   2   1.08 ⎭      1   110.0 ⎭    3   120.0
C   1   1.00        2    90.0      3   240.0
C   1   1.20  *     6    90.0      3     0.0
X   7   1.00        1    90.0      2   180.0
H   7   1.08  *     8    90.0      6     0.0
                                   1   180.0
```

Methanol

One of the most common mistakes in geometry optimization is to assume too much symmetry in the methyl group of a C_s molecule such as methanol (i.e., to make all three CH bond lengths and all three O–C–H bond angles equal). The methanol Z-matrix shown in Fig. 3.3.6 serves to illustrate the optimization of a C_s methylene group using a dummy on the H–C–H bisector. Figure 3.4.2 shows that optimizing the O_2–C_1–X_5 angle, the X_5–C_1–H angles symmetrically, and the C_1H_6 and C_1H_7 distances symmetrically is equivalent to optimizing the O–C–H angles and CH bond lengths symmetrically and the H–O–C–H_6 and H–O–C–H_7 dihedral angles antisym-

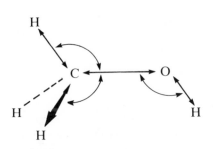

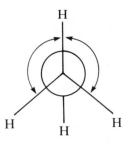

These bond lengths and
angles are equal for
the two equivalent
hydrogens.

These two dihedral
angles are optimized
antisymmetrically

(a)

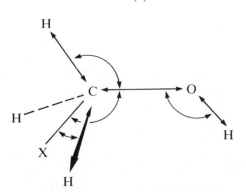

Two equivalent X–C–H angles and CH
bond lengths are optimized symmetrically.

(b)

FIG. 3.4.2. Two equivalent modes of optimization of C_s methanol: (a) using antisymmetric dihedral angles, (b) using a dummy atom on the H–C–H bisector.

metrically. Note that the number of parameters to be optimized in each case is three if the program has an option to optimize dihedral angles antisymmetrically; otherwise the use of the dummy simplifies the optimization. C_s methanol, as defined in Fig. 3.3.6, can therefore be optimized *via* eight independent parameters:

```
C
O   1   1.44  *
H   2   0.96  *      1   110.0  *
H   1   1.09  *      2   110.0  *    3   180.0
X   1   1.00         2   125.0  *    3     0.0
H   1   1.09 }*      5    55.0 }*    2    90.0
H   1   1.09         5    55.0       2   -90.0
```

Note that the 90° and −90° dihedral angles for H_6 and H_7 should not be optimized. They define the $H_6C_1H_7$ plane perpendicular to the $H_3O_2C_1H_4$ plane, as shown in Fig. 3.3.6.

Cyclopropane

D_{3h} cyclopropane is a good example of the use of symmetry in a molecule. All CC and CH bond lengths and all C–C–H bond angles in cyclopropane are equal, so that only three independent parameters need be optimized:

```
X
X   1   1.00
C   2   0.90 }*      1    90.0
C   2   0.90         1    90.0       3   120.0
C   2   0.90         1    90.0       3   240.0
H   3   1.08 }*      2   122.0 }*    1     0.0
H   3   1.08         2   122.0       1   180.0
H   4   1.08         2   122.0       1     0.0
H   4   1.08         2   122.0       1   180.0
H   5   1.08         2   122.0       1     0.0
H   5   1.08         2   122.0       1   180.0
```

In this case neither the CC bond lengths nor the C–C–H angles but rather distances and angles defined relative to the center of symmetry, are optimized directly. The 120° and 240° dihedral angles are defined by the C_3 axis and the 0° and 180° dihedrals for the hydrogens are defined by the σ_v planes. No dihedral angles need be optimized.

Oxirane

The optimization of oxirane in C_{2v} symmetry is relatively straightforward, except that it is easy to optimize too many parameters. The O–X–C angles should not be optimized, because this, in conjunction with optimization of

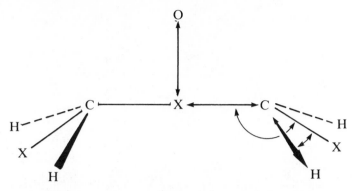

FIG. 3.4.3. Optimization of oxirane. Optimization of the O–X–C angles would be redundant. All XC and CH distances and X–C–X angles are equivalent.

the O–X and X–C, distances would amount to allowing the dummy to wander along the principal axis, as shown in Fig. 3.4.3. Oxirane in C_{2v} symmetry can be optimized using five independent parameters:

```
O
X  1  1.10 *
C  2  0.74 }*    1   90.0
C  2  0.74 ]     1   90.0    3   180.0
X  3  1.00       2  130.0 }* 1   180.0
X  4  1.00       2  130.0 ]  1   180.0
H  3  1.08 }*    5   55.0 }* 2    90.0
H  3  1.08 |     5   55.0 |  2   -90.0
H  4  1.08 |     6   55.0 |  2    90.0
H  4  1.08 ]     6   55.0 ]  2   -90.0
```

Twist Cyclopentane, C_2

The C_2 symmetry of twist cyclopentane presents an interesting optimization problem. Because the dihedral angles are related by a C_2 axis, they are all symmetric. The optimization is best performed by fixing the $H_3C_2H_4$ plane and optimizing the twist angles of the $C_2C_5C_6$ plane and the C_9C_{10} bond (see Fig. 3.4.4). As in the oxirane example above, the 90° angles defining C_5, C_6, C_9, and C_{10} should not be optimized. This leaves 20 independent parameters

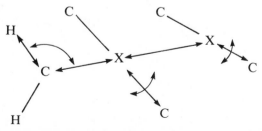

FIG. 3.4.4. Optimization of twist cyclopentane in C_2 symmetry. The remaining eight hydrogens have been omitted for clarity.

when using the Z-matrix given in Fig. 3.3.9:

```
X
C    1   0.91  *
H    2   1.08 }*   1   125.0 }*
H    2   1.08 ]    1   125.0 ]    3    180.0
C    1   1.25 }*   2    90.0     3     90.0 }*
C    1   1.25 ]    2    90.0     4     90.0 ]
X    1   1.00      2    90.0     3      0.0
X    1   1.41  *   7    90.0     2    180.0
C    8   0.77 }*   1    90.0     6     29.6 }*
C    8   0.77 ]    1    90.0     5     29.6 ]
H    5   1.08 }*   2   110.0 }*  6    120.0 }*
H    6   1.08 ]    2   110.0 ]   5    120.0 ]
H    5   1.08 }*   2   110.0 }*  6   -120.0 }*
H    6   1.08 ]    2   110.0 ]   5   -120.0 ]
H    9   1.08 }*   5   110.0 }*  10   120.0 }*
H   10   1.08 ]    6   110.0 ]   9    120.0 ]
H    9   1.08 }*   5   110.0 }*  10  -120.0 }*
H   10   1.08 ]    6   110.0 ]   9   -120.0 ]
```

Cyclohexane, D_{3d}

Cyclohexane, which can be fully optimized within the point group D_{3d} using only six independent parameters, represents another example of the use of symmetry to reduce the optimization task to manageable proportions. The Z-matrix given in Fig. 3.3.10 can be optimized as follows:

```
X
X    1   0.51  *
C    2   1.45 }*   1    90.0
C    2   1.45 |    1    90.0     3    120.0
C    2   1.45 |    1    90.0     3    240.0
C    1   1.45 |    2    90.0     3    180.0
C    1   1.45 |    2    90.0     3     60.0
C    1   1.45 ]    2    90.0     3    -60.0
H    3   1.08 }*   2    95.0 }*  1    180.0
H    4   1.08 |    2    95.0 |   1    180.0
H    5   1.08 |    2    95.0 |   1    180.0
H    6   1.08 |    1    95.0 |   2    180.0
H    7   1.08 |    1    95.0 |   2    180.0
H    8   1.08 ]    1    95.0 ]   2    180.0
H    3   1.08  *   2   155.0 }*  1      0.0
H    4   1.08 |    2   155.0 |   1      0.0
H    5   1.08 |    2   155.0 |   1      0.0
H    6   1.08 |    1   155.0 |   2      0.0
H    7   1.08 |    1   155.0 |   2      0.0
H    8   1.08 ]    1   155.0 ]   2      0.0
```

As with cyclopropane, the dihedral angles are defined by the principal axis and the σ_d planes.

3.5. POTENTIAL SURFACES

The nature of the potential surface is of prime importance in chemical calculations. The potential surface for a molecule with N atoms is $3N - 6$ dimensional. Such a surface is conceptually unmanageable, and potential-energy diagrams are often reduced to graphs of one or two dimensions, such as the simple reaction profile shown in Fig. 3.5.1(a) or the potential surface shown in Fig. 3.5.1(b). The nature of a given structure can, however, be defined by calculating the normal vibrations. This process is computationally expensive, but is becoming a prerequisite in high-quality work. To calculate the normal vibrations it is first necessary to calculate the second derivatives of the molecular energy with respect to the full set of geometrical parameters (either Cartesian coordinates or bond lengths, angles, etc.). The second derivatives (force constants) form a matrix that can then be diagonalized to give the vibrations. Newer programs calculate the second derivatives analytically, but earlier versions performed a series of geometry changes on the given structure, evaluated the first derivatives of the energy (the atomic forces) at each geometry, and then calculated the second derivatives by finite difference. The latter procedure resulted in very long calculations; so diagonalization of the force constant matrix is only now becoming a standard procedure.

Of the different types of definable points on the potential-energy surface only two interest us here: The *minimum*, for which all normal vibrations (eigenvalues of the force-constant matrix) are positive; and the *saddle point* or *transition state*, which has one negative eigenvalue. These two are *stationary points*; that is, the atomic forces (first derivatives of the energy) are all zero both at a minimum and at a transition state. At a minimum, any alteration of the geometry increases the energy, whereas at a saddle point, displacement along one normal vibration decreases the energy (and hence the vibration has a negative frequency), but all other displacements result in an energy increase. Structures with more than one negative eigenvalue are neither minima nor transition states. For instance, in Fig. 3.5.1(b) the points marked A are minima; the point B, which lies at the highest point in a "valley," is a transition state; and the structure C, which represents a "hilltop," has at least two negative eigenvalues.

Geometry optimization programs are usually designed to find minima, although some programs can also locate transition states. Optimization within a given symmetry may, however, arrive at a stationary point that is not a minimum, or even a transition state. For the methylene imine shown in Fig. 3.5.1(b), for instance, points A, B, and C can each be obtained by optimization within the appropriate symmetry. Furthermore, optimization starting from a symmetrical geometry may give a false minimum with the starting symmetry *even if no symmetry constraints are used in the optimiza-*

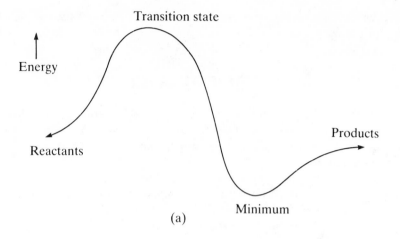

(a)

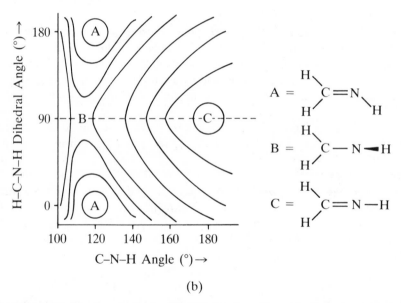

(b)

FIG. 3.5.1. (a) Simple reaction profile. (b) Contour diagram of a potential surface.

tion. This situation arises because the forces at a symmetrical transition state or hump are zero (distortions in opposite, but symmetrically equivalent, directions give the same energy change) and so the program sees no reason to try an unsymmetrical structure. In practice small numerical errors often eliminate the exact symmetry and give the unsymmetrical structure, but a good rule is never to use a starting geometry with more symmetry than the optimization itself uses. This problem disappears with optimization methods, such as BERNY, which check the nature of the stationary point, but most current methods do not check.

Geometry optimizations by MO programs also cannot cross energy barriers to find a deeper minimum, as shown in the section on molecular mechanics. The minimum found in a given optimization is that obtained by decreasing the energy from the starting point. There is therefore no guarantee that the structure obtained is the most stable, only that it is a stationary point. One of the problems often encountered in chemical calculations is that the chemist's imagination is not adequate to consider all the structural possibilities open to the molecule, and thus the global (most stable) minimum remains unnoticed. There is, unfortunately, no way to locate the global minimum automatically.

As with molecular mechanics, MO calculations refer to a hypothetical motionless state, without even zero-point energy. Thus vibrational corrections to the calculated energies may be necessary in very accurate work.

3.6. QUALITATIVE MOLECULAR ORBITAL THEORY

Let us assume that the program has given the right answer, has predicted a given type of behavior, or has reproduced the experimental trends. How can we now extract the reasons for this behavior from the output and communicate them to other chemists? The disadvantage of many modern methods is that they are too much like experiments. If the reasons for a particular type of behavior in an experimental system are complicated any calculation that reproduces this behavior is likely to be difficult to interpret. It is therefore necessary to translate the calculational results into a language that can be understood by other chemists. Traditionally this language has been valence bond theory and resonance structures. There is, however, little point in performing a MO calculation to determine that resonance structure is most important in the valence-bond description of the molecule.

The results of MO calculations are therefore usually discussed in terms of qualitative MO theories, such as the perturbational molecular orbital (PMO) treatment. Such qualitative theories provide easy-to-understand pictures of orbital interactions, but their value should not be overestimated. Their great strength is in interpreting the results of quantitative calculations, rather than in making predictions. To make real chemical predictions one must consider the relative importance of many possible interactions, so that in many cases only a quantitative theory can hope to be successful. Nevertheless, PMO theory is becoming the language in which we discuss chemistry, and into which we translate our results.

Group Orbitals

Molecular orbitals can be built up *via* a process equivalent to the LCAO approach used by the programs themselves. However, this is unnecessarily complicated, because the orbitals of a given group (methylene, methyl, amino, etc.) are largely transferable from molecule to molecule. It is therefore useful to build up the molecular orbitals of larger molecules from these *group orbitals*, rather than starting from atomic orbitals. The group orbitals are essentially the same as those found in the isolated fragments (carbenes, radicals, etc.), and so can be determined for these small species. Methylene (CH_2) can be used to illustrate the principles behind such determinations.

The "basis set" of atomic orbitals used to construct the group orbitals for CH_2 is shown in Fig. 3.6.1. Each hydrogen has a $1s$ orbital, but since the two hydrogens are equivalent, their orbitals have been combined to make the symmetric (ψ_1) and antisymmetric (ψ_2) combinations. It is important that all orbitals or combinations thereof, in this "basis set" have the proper symmetry for the species to be constructed. The orbitals of sets of equivalent atoms must therefore always be combined. The carbon atom has a $2s$- (ψ_3) and three $2p$- (ψ_4–ψ_6) orbitals. The carbon $1s$-orbital is part of a filled shell and has been omitted for simplicity. In the first step of the orbital combination process the hydrogen orbitals ψ_1 and ψ_2 are allowed to interact with the carbon orbitals ψ_3–ψ_6 to form six molecular orbitals, as shown in Fig. 3.6.2. Each hydrogen combination is allowed to combine with *one* carbon orbital of the same symmetry type. Thus ψ_1, which is symmetric with respect to both mirror planes, can combine with either the carbon s-orbital ψ_3 or the p_x-orbital ψ_4. In Fig. 3.6.2 the s-orbital, the one closest in energy to ψ_1, has been used, but the result would be the same if ψ_4 were used because the resulting orbitals mix at the next stage. Orbitals ψ_1 and ψ_3 are combined to form a bonding combination, labeled σ_{CH} in Fig. 3.6.2 to denote its CH sigma-bonding character, and an antibonding combination (σ^*_{CH} in Fig. 3.6.2). The antibonding combination is raised in energy above the average

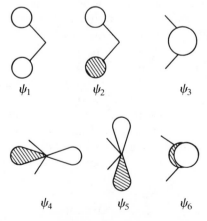

FIG. 3.6.1. The symmetry-adapted atomic orbitals used to construct the molecular orbitals of methylene.

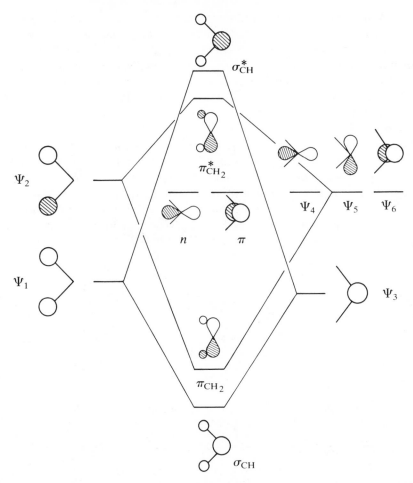

FIG. 3.6.2. Orbitals produced from a first-order interaction of the atomic orbitals shown in Fig. 3.6.1.

level of ψ_1 and ψ_3 by an amount slightly larger than than that by which the bonding combination lies below this level. In this case the energy splitting between σ_{CH} and σ^*_{CH} is very large because the overlap between the adjacent s-orbitals is large.

Orbital ψ_2, which is antisymmetric with respect to the mirror plane perpendicular to the molecular plane, can be combined only with the p_y-orbital, which also has this symmetry property. The two resulting orbitals, a bonding and an antibonding combination, are denoted as π_{CH_2} and $\pi^*_{CH_2}$ in Fig. 3.6.2. The π notation indicates the antisymmetric nature of the orbitals. The overlap between the p_y-orbital, which is not directed toward the hydrogens, and ψ_2 is smaller than that between ψ_1 and ψ_3, so that the energy difference between the π_{CH_2} and $\pi^*_{CH_2}$ orbitals is smaller than that between σ_{CH} and σ^*_{CH}. The remaining carbon orbitals, ψ_4 and ψ_6, are used unchanged in CH_2, at least in this first-order approximation, to give the two nonbonding orbitals denoted as n (in plane) and π in Fig. 3.6.2. The orbitals shown in Fig. 3.6.2 are not, however, the the final set. Three of them, σ_{CH}, n, and σ^*_{CH}, are symmetrical with respect to both mirror planes; that is, they have the same symmetry and can interact with each other. The π orbital is the only one of its type and the π_{CH_2} and $\pi^*_{CH_2}$ orbitals are derived from ψ_2 and ψ_5, so only the magnitudes of the coefficients can change with varying electronegativities, as outlined for LiH in section 3.1.

To decide how the σ_{CH}, n-, and σ^*_{CH} orbitals mix we must know how many electrons to put into the molecule. The carbon 1s-orbital, which is occupied by two electrons in the atom, was omitted from the orbitals in Fig. 3.6.1, so that the carbon has only four electrons. Each of the hydrogens has one electron, giving a total of six for CH_2. The first four electrons go into the σ_{CH} and π_{CH_2} orbitals, as shown in Fig. 3.6.3, but the next two could go

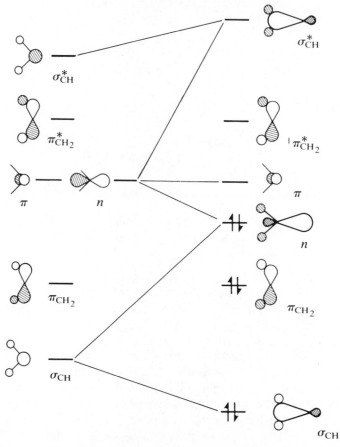

FIG. 3.6.3. Second-order interaction of the molecular orbitals shown in Fig. 3.6.2 to give the methylene molecular orbitals.

into either the n- or the π orbitals in a closed-shell configuration. The π orbital cannot be stabilized by mixing with other orbitals because it is the only one of its symmetry type. The n-orbital, however, can mix with the σ_{CH} and the σ^*_{CH} to make it more stable than the π, and therefore will be doubly occupied in the lowest closed-shell singlet state. Because the σ_{CH} and the n-orbitals are occupied and the σ^*_{CH} is unoccupied, the former two can be stabilized at the expense of the latter. This process is shown in Fig. 3.6.3, in which the first order orbitals from Fig. 3.6.2 are mixed to give a more realistic set of CH_2 molecular orbitals. The σ_{CH} orbital is made more strongly bonding by mixing in some n to direct the carbon contribution toward the hydrogens. The n-orbital is stabilized by mixing in some s-character from the σ_{CH} and σ^*_{CH}. This results in an energy lowering, because s-orbitals are more stable than p. A further consequence of this mixing is that the σ^*_{CH} orbital becomes more strongly antibonding and rises in energy. This does not affect the energy of the molecule, however, because the orbital is unoccupied. Later sections will deal with the changes that occur when the methylene orbitals are occupied differently, but the orbitals shown in Fig. 3.6.3 serve both as the molecular orbitals of the singlet carbene itself and as the group orbitals for CH_2 in larger molecules.

The group orbitals of other common organic fragments are shown in Fig. 3.6.4. They can all be constructed exactly as shown above for methylene, and can be used for any heavy atom A because the s-and-p "basis set" used above for carbon is appropriate for any main-group element.

FIG. 3.6.4. (a–e) The group orbitals of some common organic fragments. "A" represents any heavy atom.

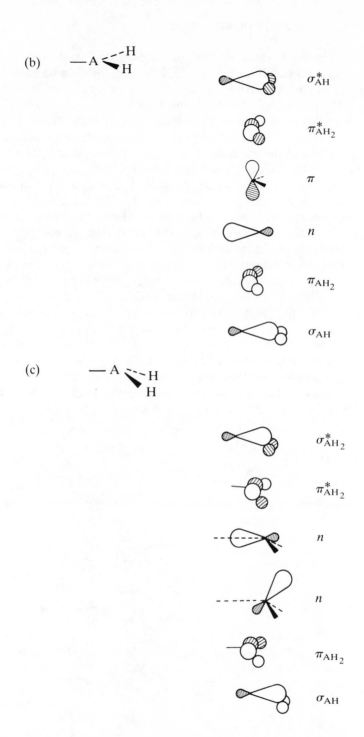

FIG. 3.6.4. (*continued*)

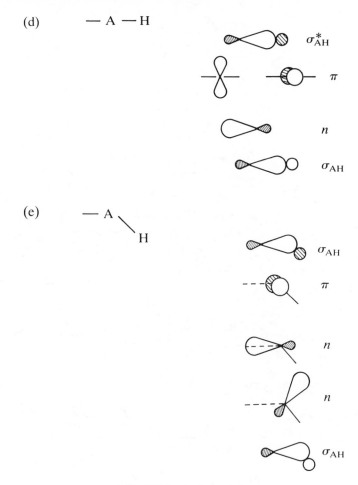

FIG. 3.6.4. (*continued*)

Molecular Orbitals from Group Orbitals

The methylene group orbitals can now be used to construct molecular orbitals for more complex molecules. The simplest example is the combination of two methylene units to form ethylene, as shown in Fig. 3.6.5. Ethylene is a particularly simple example because of its symmetry. Each orbital of one methylene group can interact with its counterpart in the other to give bonding and antibonding combinations. Thus the σ_{CH} and the π_{CH_2} orbitals form two new sets of ethylene orbitals, also designated σ_{CH} and π_{CH_2} in Fig. 3.6.5. The energy-splitting between these orbitals is not large, because their main lobes are not directed toward each other and overlap is therefore small. The n-orbitals, on the other hand, point directly at each other, so the overlap, and hence the energy difference between the bonding

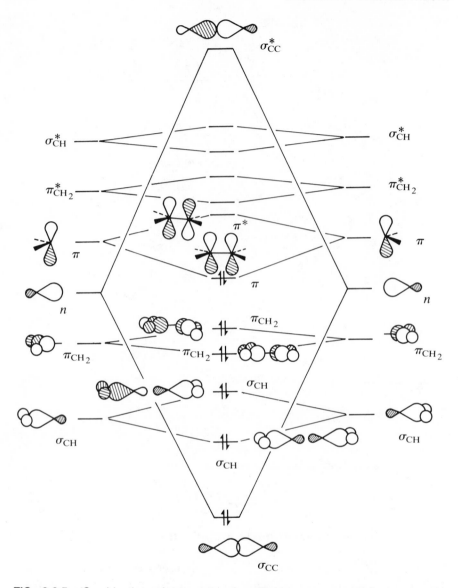

FIG. 3.6.5. Combination of two sets of methylene group orbitals to give the molecular orbitals of ethylene.

and antibonding combinations, is large. The symmetrical combination gives the new carbon–carbon sigma-bonding orbital denoted as σ_{CC} in Fig. 3.6.5, and therefore sinks low in the MO manifold for ethylene. The σ^*_{CC} orbital is strongly destabilized, as shown in Fig. 3.6.5. Similarly, the π orbitals of the two methylene units form the π and π^* molecular orbitals of ethylene. In this case the energy-splitting between the bonding and antibonding combinations is not as large as for the σ orbitals, because π type overlap is not as effective as σ overlap. The remaining antibonding orbitals (the $\pi^*_{CH_2}$ and σ^*_{CH}) form symmetric and antisymmetric combinations in the same manner as their bonding counterparts do.

It now remains to add the electrons to the ethylene molecular orbitals to complete the picture. Note that in building up the molecular orbitals of a molecule from group orbitals the electrons are not considered until after construction of the molecular orbitals. The methylene group orbitals are being used exactly as the basis set in an LCAO calculation. The molecular orbitals obtained from this basis set for ethylene are then occupied according to the Aufbau principle to give the ground-state molecule. Doubly occupying the lowest six valence molecular orbitals in Fig. 3.6.5 (2×4 electrons for the two carbons and one each for the four hydrogens) leads to the well-known picture for ethylene in which the π orbital is the *highest occupied molecular orbital* (HOMO) and the $\pi*$ the *lowest unoccupied molecular orbital* (LUMO). The orbitals of ethylene could now be used to to consider, for instance, the effect of substituting a methyl group for a hydrogen to make propene.

A more complex, but very well-known, example is cyclopropane, which can be constructed from three methylene groups, as first suggested by Walsh. Note the fundamental difference between MO and valence-bond theories in this respect. Molecular orbital theory uses the same methylene group orbitals, and hence the same atomic basis set, for construction of ethylene, cyclopropane, and even cyclohexane. Valence-bond theory considers the sp^2 hybridized carbon atoms in ethylene to be different to the sp^3 carbons of alkanes. It is desirable, but not always possible, to avoid terms like "hybrid" or "rehybridization" in discussing molecular orbitals. Once again, however, these concepts often help chemists who are more familiar with valence-bond terminology to understand the processes involved.

A partial orbital picture for cyclopropane is shown in Fig. 3.6.6. The methylene σ^*_{CH} and $\pi^*_{CH_2}$ orbitals have been omitted because they form only unoccupied molecular orbitals in cyclopropane. A cyclic, symmetrical combination of any three equivalent orbitals gives two degenerate and one unique molecular orbital in a fashion analogous to formation of the π orbitals of the cyclopropenium cation. The cyclopropane molecular orbitals consist of combinations of the methylene group orbitals in this way. Thus the σ_{CH} methylene orbital forms one all-bonding cyclopropane molecular orbital (Ψ_2) and one degenerate set (Ψ_3 and Ψ_4). Similarly, the π_{CH_2} methylene orbital forms Ψ_5 and the degenerate set Ψ_6 and Ψ_7. In both cases the splitting between energy levels is relatively small, because there is no direct sigma overlap. As seen previously for ethylene, combinations of the methylene n-orbital are strongly split because of the direct sigma overlap. The most stable combination, the all-symmetrical Ψ_1, sinks deep in the MO manifold, while the two antibonding orbitals, Ψ_{10} and Ψ_{11}, are considerably raised in energy above the level of the methylene n orbital. The final

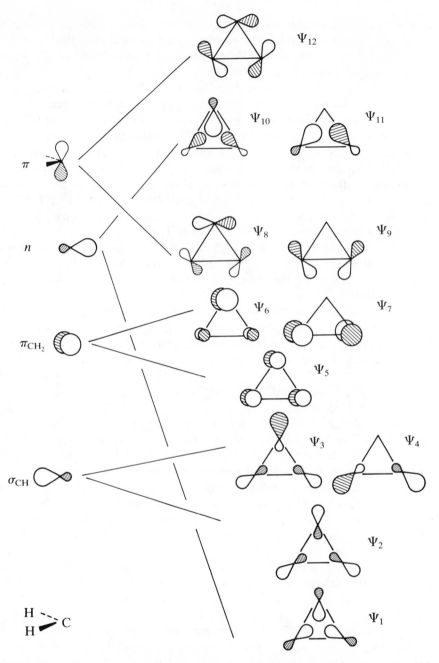

FIG. 3.6.6. The lower molecular orbitals of cyclopropane, constructed from three methylene units. The CH antibonding orbitals have been omitted for clarity.

methylene group orbital to be considered, π, cannot form a single, all-bonding molecular orbital for a three-membered ring, but instead forms a degenerate pair of molecular orbitals, Ψ_8 and Ψ_9, which are net CC-bonding in only one bond. Furthermore, because the carbon p-orbitals are directed outside the line of the CC vectors, the overlap is fairly poor and the stabilization less than that normally expected for CC sigma bonds. These two molecular orbitals are the well-known Walsh orbitals of cyclopropane, which are responsible for much of the unusual chemistry of three-membered ring systems. The strain in cyclopropane arises from the poor overlap between the carbon p-orbitals in this degenerate set, the other CC-bonding orbital (Ψ_1) being very stable. The all-antibonding combination of methylene π orbitals, Ψ_{12}, completes Fig. 3.6.6. Occupying the cyclopropane molecular orbitals with 18 valence electrons (12 for the three carbons and 1 each for the six hydrogens) gives a ground state cofiguration in which the HOMOs are the degenerate Walsh set, as shown in Fig. 3.6.6.

Using such simple group orbital approaches one can obtain a good qualitative picture of the orbitals of a complex molecule to aid the interpretation of the results of calculations. Group orbitals are naturally not limited to the one-heavy-atom units used here. The cyclopropane orbitals, for instance, are often used in considering substituent effects, and there is no reason why even larger groups should not be used. We shall now consider some typical applications of qualitative MO theory.

Walsh Diagrams

As mentioned above, the s-and-p basis set used to derive the CH_2 group orbitals applies to any main-group element A. Let us now consider the effect of removing two electrons from the AH_2 group orbitals shown in Fig. 3.6.3 (i.e., to use the same treatment for BeH_2, BH_2^+, etc.). There are two possible structures for asymmetrical AH_2 molecule, symmetrical ($D_{\infty h}$) or bent (C_{2v}). Figure 3.6.7 shows a *Walsh diagram* of the orbital changes that occur if the structure shown in Fig. 3.6.3 is made linear by bending the hydrogens away from each other. The lowest orbital, σ_{AH}, becomes slightly less stable because the effect of increasing the overlap by mixing in some p-contribution is lost. This is, however, a small effect because s-orbitals, of which the σ_{AH} is predominantly composed, are nondirectional. The π_{AH_2} orbital is strongly stabilized when the molecule is made linear because the hydrogen atoms move to the positions of maximum overlap with the heavy atom p-orbital. The n-orbital loses its s-contribution on linearization and becomes purely p, and hence degenerate with the next orbital, the perpendicular p. The n-orbital is therefore destabilized when the hydrogens move to form a linear structure. The σ^*_{AH} and $\pi^*_{AH_2}$ orbitals mirror the behavior of their bonding counterparts.

If we now consider BeH_2 or BH_2^+ with four valence electrons, only the σ_{AH} and π_{AH_2} orbitals are filled. The molecule will therefore distort in such

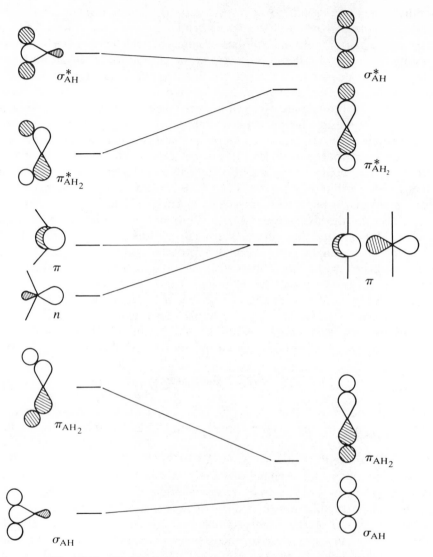

FIG. 3.6.7. Walsh diagram for bent and linear AH_2 molecules.

a way as to stabilize these two orbitals at the expense of all the others. Although the σ_{AH} is slightly destabilized when the molecule becomes linear, the π_{AH_2} is strongly stabilized, so that there is a net gain in energy. We therefore expect that four-valence-electron AH_2 molecules should be linear. The destabilization of the n- and σ_{AH} orbitals on linearization of the carbene CH_2, with two more electrons, is enough to make the bent structure more favorable. The same argument applies to water, the eight-electron case, because the π orbital is essentially independent of the H–A–H angle.

This sort of diagram and argumentation was originally used by Walsh[1] to predict the structures of heavy-atom hydrides, and has more recently been applied by Gimarc[2] to a large number of problems. Although the predictions of this simple model may fail for very polar molecules, it is nevertheless an excellent method of structure prediction.

Substituent Effects: Frontier Molecular Orbital Theory

The effect of a given substituent on the stability of a cation, anion, radical, or any other type of group can often be demonstrated using orbital interaction diagrams. Consider, for instance, the effect of substituting a methyl group for a hydrogen in the methyl cation to give the ethyl cation. The orbitals of the methyl cation are identical to those given in Fig. 3.6.4 for planar AH_3. The π orbital is unoccupied in CH_3^+, which has six valence electrons. Figure 3.6.8 shows a diagram of the interaction between a methyl cation center and a CH_3 group (one CH bond of the cation is assumed to be replaced by the CC bond). Note that in this type of diagram, which represents an interaction between chemical structures, the electrons are included for the interacting species. This is because an interaction diagram is concerned with bonding and repulsion between filled and empty orbitals, not with the construction of molecular orbitals from a basis set. This is an important distinction; to construct molecular orbitals from group orbitals one first derives the orbitals and then puts in the electrons to give the ground state, but when considering interactions between chemical moieties the electrons must be included from the start. Figure 3.6.8 introduces another useful simplification, namely that only the highest occupied and lowest unoccupied molecular orbitals (HOMO and LUMO, the frontier molecular orbitals) need be considered, because they dominate the interactions. This is a consequence of the fact that the energic effects of a two-electron interaction (i.e., that of a filled orbital with an empty one) are far larger than those of interactions involving four electrons. The frontier orbitals can indulge in the strongest two-electron interactions because the HOMO is the occupied orbital closest in energy to the unoccupied orbitals of another species, and the LUMO the empty orbital closest in energy to the filled orbitals of a donor.

Figure 3.6.8 shows the degenerate π_{CH_3} HOMOs and $\pi^*_{CH_3}$ LUMOs of a methyl group with the π_{CH_2} HOMO and π LUMO of a carbenium-ion center. Note that the orbitals that form the CC sigma bond have been omitted for clarity. The π_{CH_2} orbital of the cationic center overlaps with one of the degenerate π_{CH_3} orbitals in a four-electron (and therefore slightly destabilizing) interaction to give the bonding and antibonding combinations Ψ_1 and Ψ_3. The major interaction, however, is that between the empty π orbital of the cation and the other π_{CH_3}. This leads to a two-electron stabilization of the cation center because only the bonding combination, Ψ_2, is occupied. The $\pi^*_{CH_3}$ LUMOs of the methyl group lie high in energy and

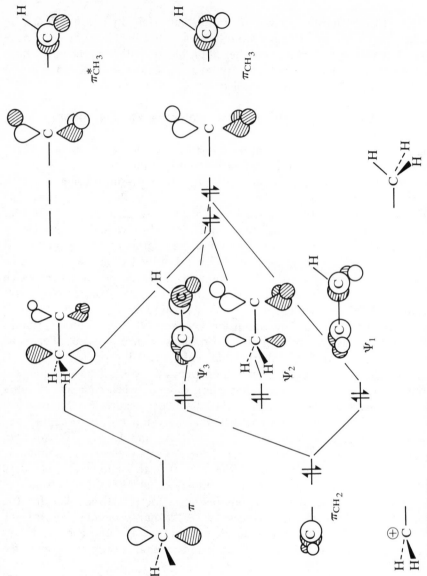

FIG. 3.6.8. Hyperconjugative stabilization of a carbenium ion center by a methyl group.

are not involved. This type of interaction between filled bonding orbitals and empty orbitals on neighboring centers is known as *hyperconjugation*. It results in a lengthening of the sigma bonds involved, because electron density is removed from the bonding orbitals. This interaction is so strong in the ethyl cation that one CH bond is partially broken and a bridged structure results:

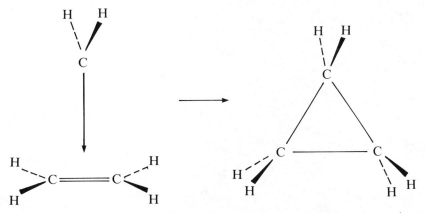

An important point about Fig. 3.6.8 is that the energy levels shown correspond to a SCF, rather than a one-electron, calculation. The π_{CH_2} orbital on the positively charged center, for instance, lies lower in energy (because of the charge) than the π_{CH_3} orbitals of the substituent. Strictly speaking, such simple interaction diagrams are applicable only to one-electron theories, but the use of more realistic energy levels improves their predictive power.

A second, and very similar, use of frontier molecular orbital (FMO) theory is to treat the interaction between two molecules. Consider the addition of singlet methylene to ethylene to give cyclopropane. The simplest possible way to effect this addition would be to allow the carbene to approach the center of the C=C bond perpendicularly, as shown in Fig. 3.6.9. This is known as the *least-motion* pathway, for obvious reasons.

FIG. 3.6.9. Least-motion (C_{2v}) attack of methylene on ethylene to give cyclopropane.

Figure 3.6.10, however, shows the interaction diagram for this least-motion approach. The HOMO of the carbene (the n-orbital) is symmetrical with respect to the ethylene σ_h plane, as is the HOMO of the olefin. The two LUMOs, the ethylene π^* and the methylene π, are both antisymmetrical with respect to this plane. The interaction diagram therefore shows a four-electron destabilization between the two HOMOs (because the anti-bonding combination is destabilized more than the bonding combination is stabilized) and an interaction between the two LUMOs that has no effect because the relevant orbitals are both unoccupied. The net result is there-fore destabilizing, and will become more so as the two fragments approach each other more closely.

An alternative reaction path, in which the carbene initially approaches with its molecular plane parallel to that of the olefin, is shown in Fig. 3.6.11. At some later stage in the reaction the carbene must swing into its final position. Figure 3.6.12 shows the interaction diagram for this case. Because the combination of the two molecules now has only C_s symmetry, all four

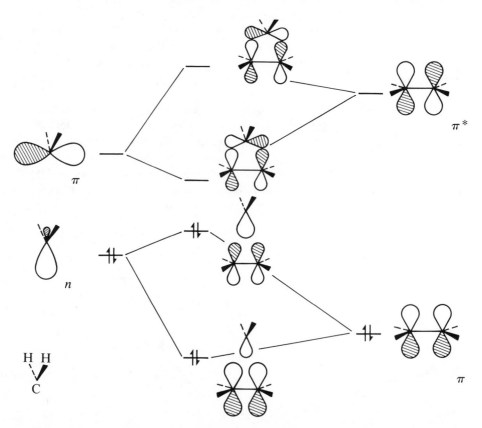

FIG. 3.6.10. Orbital interaction diagram for the least-motion reaction of singlet methylene with ethylene.

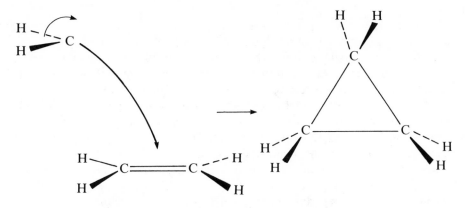

FIG. 3.6.11. C_s attack of methylene on ethylene.

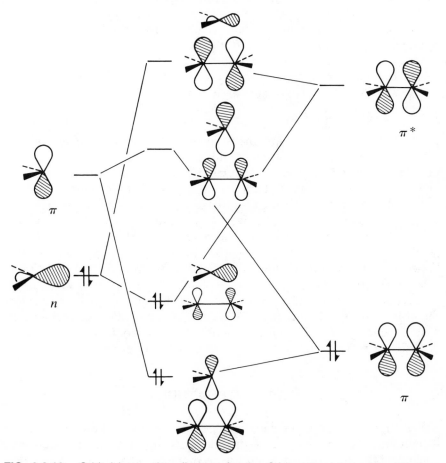

FIG. 3.6.12. Orbital interaction diagram for the C_s reaction of singlet methylene with ethylene.

frontier orbitals have the same symmetry. An analysis therefore cannot be as clear cut as for the least-motion reaction path. We can, however, use the approximate symmetry of the orbitals to estimate the most favorable overlap. The ethylene HOMO and the methylene LUMO are both approximately symmetrical with respect to a plane perpendicular to the center of the C=C bond. The ethylene LUMO is antisymmetrical with respect to this plane, and the methylene HOMO has a nodal plane that allows us to regard it as being approximately antisymmetrical. The interaction diagram obtained from these approximate symmetry assignments (Fig. 3.6.12) shows two stabilizing two-electron interactions. In reality all four orbitals can mix with each other, but Fig. 3.6.12 roughly corresponds to the first-order orbital mixing shown for methylene in Fig. 3.6.2. The clear conclusion from these qualitative considerations is that the C_s reaction path should be favored over the least-motion (C_{2v}) approach. Further analysis reveals that the least-motion approach leads to an excited state of cyclopropane, and is therefore a forbidden reaction. Both *ab initio* calculations and experiment support these conclusions, but comparison of the two reaction paths using MNDO is a useful exercise that gives good results.

The examples given above were chosen so that only the frontier molecular orbitals would be important, as is very often the case. It should be emphasized, however, that in many systems the situation is not as clear cut. Electrophilic aromatic substitution, for instance, is not controlled by the HOMO of the substrate alone because several other high-energy orbitals exist that can all interact strongly with the attacking electrophile. This is often the case for extended π systems because the separation between π energy levels is often small. These cases, for which FMO theory fails, are commonly designated as charge—rather than orbital—controlled. This distinction arises entirely from the oversimplification implicit in FMO theory, and has little physical significance. It is nevertheless often discussed at great length. Perhaps the best way to deal with the problem is to point out that the charge distribution, when it varies significantly from that in the HOMO, is controlled by lower-lying molecular orbitals, the inclusion of which in the interaction diagrams would lead to correct predictions. A general rule is that the FMO approximation is valid if the HOMO and LUMO are energetically well separated from the other orbitals, as in nonconjugated π systems, otherwise a more complete set of orbitals should be used.

3.7.   LITERATURE

This section has necessarily been only an introduction to PMO theory, and is by no means complete, although it does attempt a critical assessment of qualitative theories. It remains only to recommend other textbooks that deal with the subject at greater length and present further examples built on approaches similar to those outlined here. One of the best introductions to group orbitals is by Jorgensen and Salem,[3] from whom I have borrowed the qualitative orbital designations used above. The standard text on semiquantitative MO theory by Fukui[4] contains many examples concerning the prediction of the regioselectivity of reactions. Electrocyclic reactions and cycloadditions, which were not considered at all in the above dicussion, are well treated in the original work of Woodward and Hoffmann[5], and the excellent account by Nguyen Tron Anh.[6] As mentioned above, Gimarc[2] has described the use of Walsh diagrams for a large number of systems. Finally, one textbook that is a good deal less conventional than the others gives an impression of some of the more unusual orbital effects. Salem's *Electrons in Chemical Reactions*[7] contains a sound, nontheoretical description of a large amount of MO theory up to quite sophisticated levels, but remains thoroughly readable throughout. Apart from the classical theoretical textbook from Pople and Beveridge,[8] there are many other good introductions to quantitative MO theory. For those learning quantum chemistry, rather than seeking a reference book, however, Szabo and Ostlund[9] is possibly the best choice.

REFERENCES

1. A. D. Walsh, *J. Chem. Soc.*, **1953**, 2260, 2266, 2288, 2296, 2301, 2306, 2318, 2321, 2325, and 2330.

2. B. M. Gimarc, "Molecular Structure and Bonding," Academic Press, New York, 1979.

3. W. L. Jorgensen and L. Salem, "The Organic Chemist's Book of Orbitals," Academic Press, New York, 1973.

4. K. Fukui, "Theory of Orientation and Stereoselection," Springer, Berlin, 1975.

5. R. B. Woodward and R. Hoffmann, "The Conservation of Orbital Symmetry," Verlag Chemie, Weinheim, 1970.

6. Nguyen Trong Anh, "Les Regles de Woodward–Hoffmann," Ediscience, Paris, 1970; "Die Woodward–Hoffmann Regeln und Ihre Anwendung," Verlag Chemie, Weinheim, 1970.

7. L. Salem, "Electrons in Chemical Reactions," Wiley-Interscience, New York, 1982.

8. J. A. Pople and D. L. Beveridge, "Approximate Molecular Orbital Theory," McGraw-Hill, New York, 1970.

9. A. Szabo and N. S. Ostlund, "Modern Quantum Chemistry," Macmillan, New York, 1982.

CHAPTER **4**

SEMIEMPIRICAL METHODS

4.1. SEMIEMPIRICAL MOLECULAR ORBITAL THEORY

MINDO/3 and MNDO, the semiempirical methods to be discussed in this chapter, are members of a series of MO techniques developed by M. J. S. Dewar and his group specifically for applications in organic research. The aim of their efforts is to produce an "MO spectrometer" that should eventually be able give chemically accurate results for large molecules at a reasonable cost in computer time. The requirements of chemical accuracy and computational economy are not normally compatible, so a number of compromises must be made. The earlier semiempirical MO methods, CNDO, INDO, and NDDO, were developed by J. A. Pople and his group at a time when the available computers were able to handle *ab initio* calculations only on the smallest systems. These methods were not intended to reproduce molecular geometries and heats of formation, but rather other electronic properties, such as the dipole moment. The simplest, CNDO (Complete Neglect of Differential Overlap), assumed the atomic orbitals to be spherically symmetrical when evaluating electron repulsion integrals. The directionality of *p*-orbitals was included only *via* the the one-electron resonance integrals, the sizes of which depend on the orientations and distances of the orbitals and on a constant assigned to each type of bond. The next stage, the INDO (Intermediate Neglect of Differential Overlap) approximation, included one-center repulsion integrals between atomic orbitals on the same atom. The NDDO (Neglect of Diatomic Differential Overlap) approximation was the first in which the directionality of the atomic orbitals was considered in calculating the repulsion integrals. In this case the three- and four-center integrals in which the overlap occurs between atomic orbitals on the same atom were included. More complete descriptions are given by Pople and Beveridge[1] and by Dewar.[2]

MINDO/3 is a modified INDO method. Rather than evaluating the one-center repulsion integrals analytically, MINDO/3 uses a set of parameters to approximate them. These parameters, along with the constants used to evaluate the resonance integrals, allow the results to be fitted as closely as possible to experimental data. MINDO/3 is the last in a series of three MINDO methods and represents a milestone in the use of calculations in chemistry. It was the first easy-to-use program package with automatic geometry optimization to be made available to a wide range of nonspecialist research groups. MINDO/3 has often been heavily criticized, especially by the Pople school, which rejected parametrized INDO procedures in favor of *ab initio* calculations, but its significance in introducing the concept of structure and energy calculations to organic chemical research cannot be denied.

MNDO is not a more sophisticated version of MINDO/3, but rather an independent method based on the NDDO approximation, and was under development when MINDO/3 was published. The use of NDDO, rather than INDO, was found necessary in order to avoid some of the systematic MINDO/3 errors for molecules, such as hydrazines or polyfluoroalkanes, in which lone pair–lone pair repulsions are important. The directionality of the electron–electron repulsion terms in the NDDO approximation is particularly important in this respect.

The advantage of MINDO/3 and MNDO over *ab initio* calculations is not only that they are several orders of magnitude faster (MNDO is about 1.5 times slower than MINDO/3), but also that calculations for some very large molecules are possible only with the semiempirical methods. The neglect of large numbers of integrals not only saves computer time, but also reduces the core and disk space requirements in comparison with those for an equivalent *ab initio* job. Calculations for very much larger systems than can be managed at even the simplest *ab initio* level are therefore possible.

The approximations inherent to the neglect of differential overlap (NDO) methods naturally cause a loss of accuracy; this can, however, largely be compensated for by the parametrization. As in molecular mechanics methods the parametrization cannot be better than the available experimental data, so results for the more "exotic" elements such as beryllium and lithium cannot be expected to be as reliable as those for hydrocarbons. Another problem with the parametrization of SCF methods like MINDO/3 and MNDO to fit experimental results is that any experimental data include correlation. A more satisfactory procedure would be to parametrize so that the results of calculations with a correlation correction fit experimental data. This is the philosophy behind MNDO/C,[3] which, however, does not offer any significant improvement over MNDO for ground-state molecules, although excited states are treated considerably better than by the standard method. Configuration interaction calculations are also available in the MINDO/3 and MNDO programs, but in some ways correlation is then included twice in the calculation: once explicitly and once indirectly *via* the parametrization. The results are therefore meaningful only when compared with each other, and not as absolute values for, for instance, heats of formation.

One significant advantage of MNDO over MINDO/3 is that it needs only parameters specific to each individual element, not to combinations of elements. This means that in order to calculate compounds with nitrogen–lithium bonds, for instance, MNDO needs to be parametrized only for a set of lithium compounds and a set of nitrogen compounds. No compound containing nitrogen and lithium need be included in the parametrization set, although this would obviously be preferable. MINDO/3 could be parametrized for such molecules only if an adequate body of experimental data on N–Li compounds were available. The MOPAC program contains MINDO/3 parameters for the bonds shown in Fig. 4.1.1, and MNDO parameters for the elements shown in Fig. 4.1.2. The MNDO parameters for lithium, which do not originate from the Dewar group,[4] can easily be added to the standard program.

Figure 4.1.3 shows a plot of the number of entries in *Chemical Abstracts* listed under "MINDO/3" and "MNDO" *vs.* volume number. After an initial

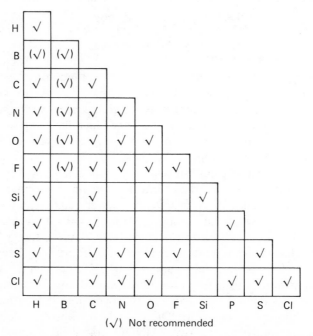

(√) Not recommended

FIG. 4.1.1. Bond parameters for MINDO/3 included in the MOPAC program.

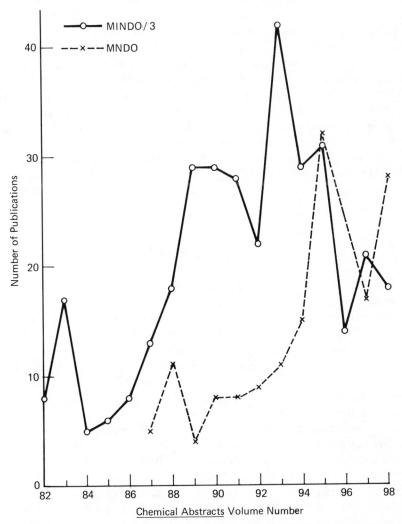

H							
	Be	B	C	N	O	F	
		Al	Si	P	S	Cl	
						Br	
						I	

FIG. 4.1.2. Elements for which MNDO parameters are included in the MOPAC program.

FIG. 4.1.3. Publications entered in Chemical Abstracts under MINDO/3 and MNDO.

peak due to the original papers of the Dewar group, there is a steady rise in the use of MINDO/3, with the method reaching a peak of popularity in 1980, 3 years after the introduction of MNDO. From 1980 onward there has been a steady decline in the use of MINDO/3. The curve for MNDO is very similar, but with about 2½ years' time lag. Surprisingly, MNDO has not yet achieved the peak popularity enjoyed by MINDO/3, although more recent figures would probably show a strong increase in the use of the newer program. The relatively rapid acceptance of the two methods is due at least partly to the fact that the programs were quickly made available through QCPE after publication of the initial papers.

The programs available through QCPE for MINDO/3 and MNDO calculations are listed in Table 1.2.1. Most are derived directly from the original Dewar programs, although the MOPN UHF/MINDO/3 program differs considerably in program structure and slightly in input format. There are essentially only three types of input for the different programs: that used by the earlier Dewar group programs, the slightly different form used by MOPN, and the free-format input used by MOPAC. The examples given in this chapter are all for the last of these, but the principles are the same for all the programs, and the format differences are only minor. There are detailed instructions for structuring the input in the relevant sections of the program listings, so there should be no problem in using the examples given with older programs. Many of the options found in MOPAC, however, are not available in the other programs. The differences are outlined in Table 1.2.1.

4.2. MINDO/3

MINDO/3 was introduced in 1975 by Dewar, Bingham, and Lo,[5] and was parametrized for hydrocarbons,[6] C–H–O–N compounds,[7] C–H–F–Cl compounds,[8] and for some types of molecules containing silicon, phosphorus, and sulfur.[9] In some MINDO/3 programs, however, a different set of parameters was included that gave wrong results for second-row elements. The parameters in early programs should therefore be checked against those in reference 9. A preliminary set of parameters was also reported for boron,[5] but these proved to be unsatisfactory and no further work in this area appeared. Olah et al.[10] later published parameters for the sulfur–oxygen bond in sulfoxides, and Frenking et al.[11] redetermined some of the phosphorus parameters using better experimental data. The same authors later published parameters for PO, PF, and PCl bonds.[12] Figure 4.1.1 shows the types of bonds for which MINDO/3 parameters are available in the MOPAC program.

The original MINDO/3 treatment for open-shell systems used the half-electron (HE) method. Bischof[13] published an unrestricted Hartree–Fock (UHF) version in 1976, as did Dewar and Olivella[14] somewhat later. A generalized coupling operator approach was also tried by the Austin group,[15] but this gave results very similar to those of the HE method at a greater cost in computer time. The MOPAC program performs both HE and UHF MINDO/3 calculations for open-shell systems.

Among the wide variety of applications of MINDO/3 to problems other than structure and energy calculations are the calculation of ESCA chemical shifts,[16] nmr coupling constants[17–19] and chemical shifts,[20–22] ^{14}N nuclear quadrupole resonance (nqr) coupling constants,[23] polarizabilities,[24] non-linear optical coefficients,[25] vibrational frequencies[26–28] (including isotope effects[29–32]), and even partition coefficients.[33] In general the coupling constant calculations were no improvement on those obtained using INDO, and in some cases they were worse. Except for the use of MINDO/3 for vibrational problems, such as isotope effects, these applications were mostly one-off experiments which did not lead to general use of the method in these areas. The success or failure of MINDO/3 in these applications is generally critically assessed in the original publications. Among the variations of MINDO/3 theory itself are a π-only version for very large molecules[34] and its use in crystal orbital calculations for TTF and TCNQ.[35]

Almost immediately after its publication, MINDO/3 was the subject of heavy criticism[36,37] because of its relatively poor performance on some problems in comparison with *ab initio* programs. A reply to these criticisms appeared promptly.[38] One of the main purposes of this section is to review the performance of MINDO/3 critically and to assess the strengths and weaknesses of the method on the basis of the published literature. The original papers introducing the method[5–9] contain assessments of its performance and point out several weak points often later rediscovered by other authors. These five papers are worth reading not only in order to avoid the obvious shortcomings of MINDO/3, but also because they contain a vast amount of calculated data that often prove useful in other contexts.

One of the strong points of MINDO/3 is its performance in carbocation calculations. It is often the method of choice for systems that are too large for good *ab initio* calculations, and is usually preferable to minimal-basis-set Hartree–Fock calculations. Koehler and Lischka[39] have compared MINDO/3 results for a variety of carbocations with large-basis-set *ab initio* calculations performed with a CEPA (correlated electron pair approximation) correction for electron correlation. The performance of MINDO/3 is impressive, rather more so than that of the intermediate-level *ab initio* calculations. There are several other papers in that MINDO/3 predictions for carbocations compare well with more expensive calculations.[40–48] Naturally, there are exceptions to the rule, especially when other unusual features occur in the cation. Two examples appear to be protonated ethane, $C_2H_7^+$,[49] and cations containing small rings, the MINDO/3 results for which are too stable.[50] MINDO/3 has, however, often been used successfully to predict and interpret mass spectrometric fragmentation processes.[51–56]

Although Dewar and Thiel[57] have specifically warned about spurious high-lying sigma orbitals in MINDO/3 calculations on aromatic systems, the method has often been used with great success for the interpretation of photoelectron (PE) spectra. There are also indications that MINDO/3 severely underestimates the through-space effect,[58] although the through-bond effect is reproduced well.[59–62] Many of the published papers in this area have been devoted to hydrocarbons, [58,59,63–67] although diones[68–71] are also well treated by MINDO/3, and the method has been applied to conjugated carbonyl compounds,[72] ethers and thioethers,[73] and diamines.[74]

Other areas in which MINDO/3 has been noted to give good results are in studies of homoaromaticity,[45,75–77] proton affinities of carbonyl compounds,[78] and the inversion barriers of ammonia[79] and simple primary amines,[80] although calculated inversion barriers for aziridines are too low.[80] Excellent results were also obtained in studies of the protonation of alkyl chlorides[81,82] and deuterium isotope effects.[29,30] MINDO/3 has been to be found preferable to MNDO for calculating the structures of fluoropyridines,[83] and also performs well for oxohydropyridines, pyrimidines,[84] and dihydropyridines,[85] in accord with the fact that errors are small for six-membered rings.[86]

There are, however, several areas in which MINDO/3 has been shown to give poor results. Dewar et al.[5–9] list compounds for which the errors are particularly large, and have pointed out some systematic errors. Two of MINDO/3's more important deficiencies for organic calculations are its overestimation of the stability of triple bonds and the underestimation of the stability of aromatic compounds. The former can lead to false conclusions when isomeric unsaturated systems are compared; those with triple bonds are always found to be relatively too stable. Another weakness, which is common to many NDO methods, is the inability of MINDO/3 to describe rotation barriers in conjugated systems correctly. This has been pointed out in studies on benzylidene aniline, stilbene, and azobenzene;[87] nitrobenzene;[88] butadiene and acrolein;[89] and biacetyl.[90] Good results were, however, obtained for vinyl azide.[91] The problem has often been avoided by constraining the molecule to planarity during the optimization, so MINDO/3 results for conjugated molecules should be examined carefully.

A further serious defect of MINDO/3, which it shares with MNDO, is its failure to reproduce hydrogen bonds.[92,93] This is particularly unfortunate because it limits the usefulness of the method for modeling biological systems, for which its relatively small requirements for computer time would otherwise recommend it. This weakness has been noted in contexts other than static hydrogen bonding, such as in the study of diotropic 1,4 proton transfers.[94] A variation of MINDO/3, MINDO/3(H), was developed to overcome this problem.[95]

Another problem area for MINDO/3 is its handling of lone-pair repulsions, the strength of which it seriously underestimates. Boyd[96] carried out a model study in which two dimethyl sulfide molecules were forced to approach each other in such a way that the lone pairs would interact most strongly. Rather than the expected repulsive curve, MINDO/3 found a shallow minimum corresponding to an S...S dimer. This underestimation of lone-pair repulsions also manifests itself in the poor performance of MINDO/3 for *gem*-difluoro and similar compounds and for hydrazines, as pointed out by Dewar et al.[7,8]

Bantle and Ahlrichs[97] have discussed other serious MINDO/3 errors. It is, for instance, not suited to the calculation of transition states for bimolecular substitution reactions. MINDO/3 predicts a minimum, rather than a maximum, for the well known D_{3h} transition state for a degenerate S_N2 reaction. This quirk, however, is shared by minimal-basis-set *ab initio* calculations. Bantle and Ahlrichs also pointed out MINDO/3's weaknesses in describing the silicon–carbon double bond and the donor–acceptor bond in $BH_3:NH_3$. The latter example probably does not represent a general deficiency of MINDO/3 in this respect, but rather highlights the unsuitability of the method for boron, and presumably for other electropositive elements for which no MINDO/3 parameters have been published.

There have been a few attempts to correct for systematic MINDO/3 errors, notably its underestimation of the stability of highly branched alkane systems, by applying correction factors to the results. Because branching errors are closely related to the structure of the molecule, they can be predicted, and hence corrected for, *via* comparison with experimental data for a variety of molecules. This approach has been used several times with good results,[86,98,99] as have scaling factors for use in calculating the frequencies of normal vibrations.[100] In the most detailed study, the average error in the calculated heats of formation was reduced from 17.3 kcal mol^{-1} before correction to 1.3 kcal mol^{-1}.[99] This was, however, for a set of molecules chosen to illustrate this type of error. The errors expected from uncorrected MINDO/3 are normally much smaller. McManus and Smith[86] analyzed MINDO/3 errors for alkanes and concluded that they are generally smaller for cyclic molecules than for acyclic ones, and are particularly small for six-membered rings. The performance of MINDO/3 for three- and five-membered rings is average (although small-ring carbocations[50] and spiro-fused cyclopropanes[101] give consistently too stable results), but poorer for four- and seven-membered rings. Very large errors were noted for cyclic sulfides. It has also been found that ring structures are systematically calculated to be too flat.[102]

One point of controversy surrounding MINDO/3 was its treatment of cycloaddition transition states, which are usually found to be strongly unsymmetrical,[103,104] in contrast to results found using *ab initio* calculations.[105] This difference was originally attributed to improper location of the transition state in the *ab initio* work,[103] although this was not the case. The *ab initio* transition state was fully characterized by diagonalization of the force-constant matrix.[105] Dewar has, however, recently suggested that all reactions involving the making or breaking of two or more bonds must proceed asynchronously.[106,107]

In general MINDO/3 should always be considered for large carbocations as long as branching errors are taken into account. It also seems to give useful results for the interpretation of PE spectra, although the artifacts pointed out by Dewar and Thiel[57] should be considered. It should not be used for problems in which hydrogen bonds, lone-pair repulsions or the rotation barriers of conjugated systems are important. One may obtain misleading results when comparing isomeric systems with different numbers of triple bonds (including cyanides) or in comparing benzenoid with nonbenzenoid molecules. *Gem*-difluoro or similar compounds in which lone-pair repulsions are important should be avoided. MINDO/3 appears to perform well for six-membered rings in general and to deal well with nitrogen inversion barriers in most cases. One important exception is trimethylamine, which it predicts to be planar, probably as a consequence of overestimation of the methyl–methyl repulsions. It should not be used for hydrazines or similar compounds. Although it gives moderately good results for esr (UHF/MINDO/3) and nmr coupling constants (that cannot be calculated with the QCPE programs), it is not as good as INDO for these applications. A useful compromise would be to use MINDO/3 or MNDO optimized geometries for single-point INDO calculations in order to calculate the coupling constants. Good results have been obtained with MINDO/3 for the activation energies of a variety of reactions, but bimolecular substitutions are an important exception. The tendency of MINDO/3 to give unsymmetrical transition states for cycloadditions does not appear to have serious energetic consequences, so such reactions can be studied. The symmetry of such transition states is in any case still uncertain. Finally, any calculation involving boron should not be carried out using MINDO/3.

4.3. MNDO

MNDO (Modified Neglect of Diatomic Differential Overlap) was introduced by Dewar and Thiel[57] in 1977. The motivation for developing this method was the realization that calculations such as MINDO/3 that are based on the INDO formalism cannot properly reproduce effects due to lone-pair repulsions. By basing MNDO on NDDO these authors hoped to avoid many of the weaknesses of MINDO/3. Their attempt was in many respects successful, and MNDO performs significantly better on many problems than its predecessor. Some of the characteristic failings of MINDO/3 are, however, shared by MNDO.

One significant advantage of MNDO over MINDO/3 is the range of compounds to which it is applicable. Figure 4.1.2 shows the available MNDO parameters. The original paper[57] gives parameters for H, B, C, O, N, and F, with subsequent publications reporting the performance for C, H, O, and N compounds[108] and for molecules containing boron.[109] Parameters were later published for beryllium;[110] aluminum;[111] silicon, phosphorus, sulfur, and chlorine;[112] and bromine.[113] Parameters for iodine† are included in the MOPAC program, and a preliminary set of lithium parameters[4] has been used with success in predicting the structures of large organolithium compounds.

The major improvements over MINDO/3 noted by Dewar and Thiel[57] were with respect to unsaturated molecules, compounds with adjacent lone-pairs, the calculation of bond angles, and the ordering of molecular orbitals. The relative stabilities of doubly and triply bonded isomers are reproduced reliably by MNDO, in contrast to MINDO/3, which has problems with such systems. Compounds such as hydrazines are treated adequately by MNDO because of the improved description of the lone-pair repulsions. The rotational-potential surfaces of H_2O_2 and H_4N_2 are, however, not reproduced well,[108] although many *ab initio* methods also fail at these tasks. Dewar and Thiel also noted a considerable improvement in calculated bond angles compared with MINDO/3. This is probably also due to the improved description of directional effects in NDDO. The poor performance of MINDO/3 in this respect was the subject of much of the early criticism.[37] MNDO was also found to be more reliable for the prediction of molecular orbital ordering. It gives, for instance, no spurious high-lying sigma orbitals in aromatic systems, and treats the problem molecules N_2 and F_2 correctly. Branching errors, which are particularly severe in MINDO/3, are lower in MNDO, but still substantial.[99] Nevertheless, MNDO performs significantly better for globular molecules, such as adamantane, than does MINDO/3.[57]

MNDO was not applied to as wide a range of problems as MINDO/3 in the initial stages, but was successfully used for the calculation of polymer vibrational frequencies,[114,115] and a general MNDO treatment for linear polymers was developed.[116] In general the frequencies calculated by MNDO are comparable with those given by MINDO/3 and *ab initio* methods,[117] and several studies of this topic,[117,118] including some on isotope effects,[29,119–121] have been published. Although Dewar and Thiel[57] originally pointed out that MNDO gives better orbital ordering than MINDO/3, the earlier method has been used more extensively for the interpretation of PE spectra, especially of hydrocarbons. MNDO has been used for poly-ynes[115] and paracyclophanes,[122] but its main advantages over MINDO/3 are in its treatment of more polar molecules.[74,123–129] The performance of MNDO in this respect is generally satisfactory, the major exception being perfluorotricyclo [4.2.0.0^{2,5}] octa-3,7-diene,[125] in which the fused four-membered rings may be the problem. That the splitting between the nitrogen lone-pair orbitals in 3,7-diazabicyclo [3.3.1] nonane is calculated to be too small[74] may reflect an underestimation of the through-bond effect, as in MINDO/3.

---

† Dewar *et al.*[513] have recently published the MNDO parameters for iodine with an assessment of the results obtained for iodine-containing molecules.

MNDO performs exceptionally well in predicting the conformations and rotation barriers of aminophosphines[130] and in treating the thermolyses of 1,1-diethyl diazene and 3,4-diaza-hex-2-ene.[131] In general there are considerable improvements over MINDO/3 for all polar molecules, but not necessarily for nonglobular hydrocarbons. Repulsion between adjacent methyl groups is overestimated in MNDO, but not as severely as in MINDO/3.

Calculation of carbocations, a particular strength of MINDO/3, is treated less well by MNDO. The methyl cation is predicted to be too stable, but the hyperconjugative stabilization due to alkyl substituents is severely underestimated,[108] so that the *t*-butyl cation is calculated to be too unstable. One significant difference between MNDO and MINDO/3 is the underestimation of the strength of three-center bonds by the former. This was noted in connection with carboranes[109,132] and is reflected in the fact that MINDO/3 calculates a nonclassical structure for the 2-norbornyl cation,[39,133] whereas MNDO favors the classical ion.[108] The most recent *ab initio* calculations[134] give a nonclassical geometry, so that MINDO/3 may be preferable in this respect.

Dewar and Rzepa[135] noted that MNDO yielded major improvements over MINDO/3 for fluorine compounds, especially those with FO and FN bonds. MINDO/3 was, however, found to be slightly superior for compounds containing only C, H, and F, although *gem*-difluoroalkanes are an obvious exception. MNDO is consistently superior to MINDO/3 for in handling carbenes, cations, anions, and radicals containing fluorine, although it generally trends to make radicals too stable.[108,135] In the same paper,[135] these authors noted a characteristic weakness common to MINDO/3 and MNDO, namely, the inability to treat hydrogen-bonded systems correctly. This was later confirmed for a wider variety of systems[136,137] and, as for MINDO/3,[95] a modified version of MNDO for hydrogen bonds has recently been published.[138] A further common weakness of the two methods is their inability to treat rotational profiles of conjugated molecules. This has been noted for benzylideneaniline, stilbene, and azobenzene,[139] and is probably caused by the approximations common to INDO and NDDO. In one case, however, namely the ethylene radical cation,[140] MNDO has been found to perform extremely well, reproducing the twisted structure obtained only at the highest levels of *ab initio* theory. This result should be treated with some caution, because the general preference of MNDO for twisted π systems and its underestimation of hyperconjugative stabilization of cationic centers may make results for substituted alkene radical cations unreliable.

MNDO has been remarkably successful in the calculation of electron affinities, normally a particularly difficult problem. In a comprehensive study,[141] Dewar and Rzepa found that although MNDO calculated electron affinities for atoms, diatomic molecules, and localized anions were systematically too negative, the average error for delocalized anions was only 0.43

eV (10 kcal mol^{-1}). In that paper open-shell species were optimized using UHF/MNDO for computational efficiency and single-point calculations using the HE method were performed in order to calculate energies on the same scale as that for the closed-shell molecules. This technique is recommended for use whenever comparison between open- and closed-shell systems is important. Bischof and Friedrich[142] have found that although UHF/MNDO results for radicals are generally good (with a tendency to make them too stable), the correlation between MNDO calculated spin densities and esr hyperfine coupling constants is poor, much more so than with MINDO/3. Once again, performing INDO calculations on MNDO geometries appears to be an acceptable way to solve this problem.

Some systematic errors in MNDO have been noted for specific elements. Carbon–sulfur bonds in thiocyanates are consistently too short,[143] and sulfur ionization potentials (IPs) are too high by about 1 eV.[144] Verwoerd[145] has found MNDO results for silanes to be consistently worse than those given by MINDO/3. Like those for sulfur, chlorine IPs are consistently 1 eV too high,[146] and IPs for second-row elements for orbitals involving large 2s contributions may be 2–6 eV too high, probably as a consequence of the frozen-core approximation. Results for bromine-containing molecules are generally good,[147] although positive ions show a systematic error. Ionization potentials are once again too high, but Dewar and Healy[147] have suggested a simple correction that reduces the average error to 0.3 eV.

Because MNDO only uses s- and p-orbitals it cannot be expected to give reliable results for hypervalent molecules. Errors may be very large in such cases, up to 200 kcal mol^{-1}. This has been noted for F_3NO,[135] ClF_3[146] and especially for sulfones.[144] This deficiency, which is not an error in the MNDO method but rather a consequence of the omission of d-orbitals, severely limits the use of MNDO for compounds containing –SO and –SO_2 groups.

In conclusion, MNDO is generally superior to MINDO/3, although the latter performs better for carbocations and for silanes and comparably for C–H–F compounds without *gem*-difluoro groups. Both methods have problems treating four-membered rings, which they calculate as too flat and too stable,[108] with rotation barriers in conjugated π systems, and with hydrogen bonds. MNDO results for diatomic molecules are often unreliable, and hyperconjugative stabilization of cationic centers is not well reproduced. The systematic overestimation of IPs for S, Cl, and Br compounds is predictable, and can be corrected for. The performance of MNDO for delocalized anions is impressive, although small or localized negative ions give problems. Vibrational frequencies and isotope effects can be treated well. MNDO calculations for boron compounds are fairly reliable, although three-center bonds are calculated to be too unstable. Some systematic errors occur for beryllium compounds,[110] but the method still gives a useful degree of accuracy, as is also the case for lithium.[4]

4.4. MOPAC INPUT AND OUTPUT

The MOPAC program can perform both MINDO/3 and MNDO calculations. The latter will be used here for illustration, but using the keyword
MINDO3 (without the "/") with the input examples given here will give
MINDO/3 jobs. This section is not intended to reproduce the lists of options
and other features from the MOPAC manual, which is available from
QCPE (separately from the program itself if necessary), but to illustrate the
use of the program itself. Figure 4.4.1 shows the MOPAC input for a simple
job, the optimization of singlet methylene using C_{2v} symmetry, as defined in
the Z-matrix shown in Fig. 3.3.2.

The first line of the input is the keyword card, on which the options to be
used by the program are defined. If no keywords are specified the program
performs a RHF calculation (HE for open-shell molecules) with geometry
optimization, but without symmetry constraints, on the ground state of the
molecule. The time limit for the job is 60 minutes. A minimum of output
information about the job (geometry, energy, IP, dipole moment, eigenvalues, and atomic orbital charges) will be printed. In Fig. 4.4.1 the three
keywords (the first line or card) indicate that the geometry optimization is to
be performed within the given symmetry constraints (SYMMETRY) and
that the molecular orbitals (VECTORS) and bond-order matrix (BONDS)
will also be printed. It is important that the keywords be written in capital
letters and spelled correctly; otherwise the computer will not be recognize
them. This can be a dangerous type of error; when, for instance, the
keywords "CHARGE" or "TRIPLET" are written incorrectly, the program
will calculate the entire job, but not for the species intended. The next two
lines of the input are simply title information—in this case the number of the
example and the name of the molecule. It is prettier to leave an empty space
at the beginning of the line, as the first character is cut off in some parts of
the output.

The Z-matrix then follows, one line per atom as outlined in Section 3.3.
Note that the symbol for a dummy atom in MOPAC is "XX", and that the
order of the elements in each line of the Z-matrix is different from that used
by GAUSSIAN82. The two hydrogen atoms, for instance, are defined in the

```
VECTORS SYMMETRY BONDS
Example 4.1
Singlet methylene (C2v)
XX 0.0 0     0. 0      0. 0     0  0  0
C  1.0 0     0. 0      0. 0     1  0  0
H  1.1 1   122. 1      0. 0     2  1  0
H  1.1 0   122. 0    180. 0     2  1  3
0
3,1,4,
3,2,4,
```

FIG. 4.4.1. One possible MOPAC input for methylene in C_{2v} symmetry.

following way:

```
H  1.1    1  122.0    1    0.0  0    2  1  0
H  1.1    0  122.0    0  180.0  0    2  1  3
```

Using the format given in Section 3.3 these two lines would be as follows:

$$
\left.\begin{array}{ll} \text{H} \quad 2 \quad 1.1 \\ \text{H} \quad 2 \quad 1.1 \end{array}\right\}* \quad
\left.\begin{array}{ll} 1 \quad 122.0 \\ 1 \quad 122.0 \end{array}\right\}* \quad 3 \quad 180.0
$$

The reference atoms to which the current atom is bound and with respect to which the bond and dihedral angles are defined are given at the end of the line in MOPAC ("2 1" and "2 1 3" in the two cards shown above). A "1" or a "0" behind each geometrical parameter (bond length, bond angle, or dihedral angle) indicates that the parameter will or will not be optimized respectively.

The Z-matrix is terminated by a zero as the symbol for the next atom. The next two cards are used to define the symmetry constraints, exactly as illustrated in Section 3.3. The symmetry cards simply contain a series of numbers that define the reference atom, the type of symmetry to be used, and the atoms to be held symmetrical. Thus the two symmetry cards shown in Fig. 4.4.1 state that for reference atom **3** the bond length (function **1**) will be used to define atom **4** and (second line) for reference atom **3** the bond angle (function **2**) will also be used to define atom **4**. The bond length is optimized for the reference atom, but the new values obtained during the optimization are used to define the atom being held symmetrical. Thus

```
XX
C    1  1.0
H    2  1.1}*    1  122.0}*
H    2  1.1      1  122.0      3  180.0
```

translates to

```
XX    0.0  0      0.0  0      0.0  0    0  0  0
C     1.0  0      0.0  0      0.0  0    1  0  0
H     1.1  1    122.0  1      0.0  0    2  1  3
H     1.1  0    122.0  0    180.0  0    2  1  3
0
3,1,4
3,2,4
```

in the MOPAC input format. Further examples will clarify this point. The MOPAC Z-matrices shown in Appendix A include the full symmetry cards for the optimizations outlined in Section 3.3. Of the symmetry functions available in MOPAC (given in a table in the manual), only functions 1 (bond length held equal to that of the reference atom), 2 (bond angle held equal to that of the reference atom), 3 (dihedral angle held equal to that of the reference atom) and 14, (dihedral angle is set equal to minus that of the reference atom) are commonly used. Function 14 can be used to define AH_2 groups in C_s symmetry, although the use of a dummy atom on the H–A–H bisector, as shown in Section 3.3, is preferable.

Figure 4.4.2 shows the output produced by MOPAC from the input given in Fig. 4.4.1. The program first lists and defines the keywords used, in this case "VECTORS," "BONDS," and "SYMMETRY." Note that even if symmetry cards were included in the input, they would not be read unless the keyword "SYMMETRY" were included in the first line. The "parameter dependance" (not all computers can spell correctly) data are then given. These simply summarize the symmetry data defined in the input and define the functions used. If this table does not appear in a job that should use symmetry, the keyword is either missing or has not been recognized by the program (usually because of misspelling or the use of lower-case letters).

```
*****************************************************************************
                          MNDO CALCULATION RESULTS

*****************************************************************************
*                        VERSION  1.25
*   VECTORS  - FINAL EIGENVECTORS TO BE PRINTED
*   BONDS    - FINAL BOND-ORDER MATRIX TO BE PRINTED
*   SYMMETRY - SYMMETRY CONDITIONS TO BE IMPOSED
*****************************************************************************


        PARAMETER DEPENDANCE DATA

            REFERENCE ATOM        FUNCTION NO.      DEPENDANT ATOM(S)
                 3                    1                   4
                 3                    2                   4

               DESCRIPTIONS OF THE FUNCTIONS USED

     1        BOND LENGTH    IS SET EQUAL TO THE REFERENCE BOND LENGTH
     2        BOND ANGLE     IS SET EQUAL TO THE REFERENCE BOND ANGLE
  VECTORS SYMMETRY BONDS
  Example 4.1
  Singlet methylene (C2v)

     ATOM    CHEMICAL   BOND LENGTH     BOND ANGLE      TWIST ANGLE
    NUMBER    SYMBOL    (ANGSTROMS)     (DEGREES)        (DEGREES)
     (I)                  NA:I           NB:NA:I         NC:NB:NA:I    NA  NB  NC

      1       XX
      2       C        1.00000                                         1
      3       H        1.10000 *       122.00000 *                     2   1
      4       H        1.10000         122.00000        180.00000      2   1   3
```

FIG. 4.4.2. Output produced by MOPAC from the input shown in Fig. 4.4.1.

CARTESIAN COORDINATES

NO.	ATOM	X	Y	Z
1	6	1.0000	0.0000	0.0000
2	1	1.5829	0.9329	0.0000
3	1	1.5829	-0.9329	0.0000

RHF CALCULATION, NO. OF DOUBLY OCCUPIED LEVELS = 3

INTERATOMIC DISTANCES

		C 1	H 2	H 3
C	1	0.000000		
H	2	1.100000	0.000000	
H	3	1.100000	1.865706	0.000000

```
CYCLE: 1 TIME:    2.79 TIME LEFT:  3595.84 GRADIENT:    4.927 HEAT: 107.3803
CYCLE: 2 TIME:    0.98 TIME LEFT:  3594.86 GRADIENT:    3.159 HEAT: 107.3677
HEAT OF FORMATION TEST SATISFIED
PETERS TEST SATISFIED
VECTORS SYMMETRY BONDS
Example 4.1
Singlet methylene (C2v)
```

PETERS TEST WAS SATISFIED IN FLETCHER-POWELL OPTIMISATION
SCF FIELD WAS ACHIEVED

MNDO   CALCULATION

VERSION 1.25

FINAL HEAT OF FORMATION = 107.366493 KCAL

ELECTRONIC ENERGY   = -240.256190 EV
CORE-CORE REPULSION   =   88.669822 EV

IONISATION POTENTIAL =   9.136801

29-FEB-84

NO. OF FILLED LEVELS =   3

SCF CALCULATIONS =   16
COMPUTATION TIME =   6.97 SECONDS

ATOM NUMBER (I)	CHEMICAL SYMBOL	BOND LENGTH (ANGSTROMS) NA:I	BOND ANGLE (DEGREES) NB:NA:I	TWIST ANGLE (DEGREES) NC:NB:NA:I	NA	NB	NC
1	XX						
2	C	1.00000			1		
3	H	1.09114 *	124.42857 *		2	1	
4	H	1.09114	124.42857	180.00000	2	1	3

FIG. 4.4.2. (*continued*)

INTERATOMIC DISTANCES

```
             C   1       H   2       H   3
------------------------------------------------
  C   1   0.000000
  H   2   1.091141    0.000000
  H   3   1.091141    1.800015    0.000000
```

EIGENVECTORS

```
ROOT NO.      1            2            3            4            5            6

          -26.860812  -14.205754   -9.136801   -0.008549    4.349087    4.48139

S  C   1   0.806473    0.000000   -0.402047    0.000000    0.000000   -0.43354
PX C   1   0.140397    0.000000    0.842480    0.000000    0.000000   -0.52011
PY C   1   0.000000   -0.637589    0.000000    0.000000    0.770376    0.00000
PZ C   1   0.000000    0.000000    0.000000   -1.000000    0.000000    0.00000

S  H   2   0.406134   -0.544738    0.253560    0.000000   -0.450844    0.52034

S  H   3   0.406134    0.544738    0.253560    0.000000    0.450844    0.52034
```

NET ATOMIC CHARGES AND DIPOLE CONTRIBUTIONS

```
      ATOM NO.    TYPE        CHARGE       ATOM  ELECTRON DENSITY
         1         C         0.1039            3.8961
         2         H        -0.0520            1.0520
         3         H        -0.0520            1.0520
```

```
DIPOLE          X          Y          Z       TOTAL

POINT-CHG.   -0.308      0.000      0.000     0.308

HYBRID        1.851      0.000      0.000     1.851

SUM           1.543      0.000      0.000     1.543
```

CARTESIAN COORDINATES

```
 NO.     ATOM        X          Y          Z

  1        C       1.0000     0.0000     0.0000
  2        H       1.6169     0.9000     0.0000
  3        H       1.6169    -0.9000     0.0000
```

ATOMIC ORBITAL ELECTRON POPULATIONS

```
  1.62408    1.45897    0.81304    0.00000    1.05195    1.05195
```

BOND ORDERS AND VALENCIES

```
             C   1       H   2       H   3
------------------------------------------------
  C   1   1.958149
  H   2   0.979074    0.997301
  H   3   0.979074    0.018226    0.997301
```

FIG. 4.4.2. (*continued*)

The keywords and the two title lines are then printed, followed by the initial Z-matrix (i.e., that given in the input). The parameters to be optimized are marked with asterisks, but the symmetry-dependent parameters are not marked in any way. In this case the bond length and bond angle of atom 3 are marked for optimization, but not those of atom 4. If, however, other values were input for atom 4, the program would print those defined by the symmetry data (in this case 1.1 Å and 122.0°).

The program then prints the initial Cartesian coordinates calculated from the input Z-matrix. The dummy atom has been eliminated, and the atom numbering given in this table is used for the remainder of the output. The MOPAC program has placed the dummy (atom 1 of the Z-matrix) at the origin and the second atom (the carbon) along the x-axis. The position of this atom is therefore 1.0, 0.0, 0.0. The two hydrogen atoms are then placed in the xy-plane symmetrically, as discussed in Section 3.3. The orientation indicated by the Cartesian coordinates will be important when we consider the molecular orbitals later in the output.

The program then informs the user that it will perform a RHF (closed-shell) calculation with three doubly occupied orbitals (i.e., six electrons). Possible errors are often revealed in this message. The omission of a hydrogen atom or of the charge for an ion will result in calculation of a radical, in which case the program would indicate that there is one singly occupied orbital.

The table of interatomic distances that follows is very simple in this case, but can be very informative for larger molecules, especially those involving rings, for which the last bond distance is not usually defined explicitly in the Z-matrix. The program performs a check on the interatomic distances to ensure that no two atoms are unreasonably close to another. If one or more interatomic distances are too short, the job is aborted. This check can be turned off by use of the keyword "GEO-OK," but this should be used only in cases where two atoms are genuinely too close to each other. In all other calculations GEO-OK is a dangerous option that allows an erroneous or sloppily defined starting geometry to be optimized. This unnecessarily wastes computer time or leads to a job being run on a nonsense geometry.

The program now proceeds to the geometry optimization itself. Information about the time in seconds needed for each cycle, the time left before the limit (in this case the 60-minute default) is reached, the size of the current gradients (atomic forces), and the current heat of formation is printed for each cycle. The times per cycle given in this example are for a VAX 11/780 without a floating point accelerator (FPA, a device that speeds up calculations involving nonintegers) and would be considerably shorter for a machine in the more usual configuration with FPA. If the remaining time is too short for the program to be sure that it can finish the next optimization cycle, the job is terminated and the density matrix and current geometry optimization information are written onto two disk files. The calculation can then be continued by adding the keyword "RESTART" to the original input. In a normal optimization the gradient should diminish to a value close to zero and the heat of formation ("HEAT:" in the output) should become more negative, finally converging on its optimum value. MOPAC uses several different types of test on the gradients and heat of formation to determine whether the optimization is complete. The heat of formation test (which checks that the change in HEAT between consecutive cycles is less than a threshhold value) alone is not enough to terminate the job, but in this case Peter's test was also passed, so the geometry is considered to be optimized. The various tests for a completed optimization are defined in the MOPAC manual.

MOPAC then prints the keywords and the two title lines once more before summarizing the results of the calculation. The first message indicates that the geometry optimization was successfully completed, and that the SCF calculations converged satisfactorily. The final heat of formation is then given to an excessive degree of accuracy (the normal threshold value for the change in the heat of formation is 0.002 kcal mol^{-1}). This energy should normally be quoted to at most the second place after the decimal point, but the full value may be useful for checking the accuracy of the calculation when converting from one machine to another.

The electronic energy is the potential energy of the electrons in the molecule, and is therefore negative. The core–core repulsion energy is simply the electrostatic interaction between the positively charged atomic cores, which are calculated as point charges, and is thus positive. The sum of the electronic energy and the core–core repulsion energy is the total energy of the system (not printed), in this case -151.586368 eV. In order to calculate the heat of formation the program now calculates the atomization energy using the MNDO energies (total and electronic are identical for atoms) for a carbon and two hydrogen atoms:

$$E_{atom.} = -E_{tot(CH_2)} + E_C + 2E_H ,$$

where

$$E_C = -120.500606 ,$$
$$E_H = -11.906276 ;$$

thus

$$E_{atom.} = 151.586368 - 120.500606 - 23.812552$$
$$= 7.27321 \text{ eV}$$
$$= 167.72 \text{ kcal mol}^{-1}.$$

The heat of formation is then calculated using the experimental heats of formation of the carbon (170.89 kcal mol^{-1}) and hydrogen (52.102 kcal

mol^{-1}) atoms:

$$\Delta H°_{f(CH_2)} = \Delta H°_{f(C)} + 2 \times \Delta H°_{f(H)} - E_{atom.}$$
$$= 170.89 + 104.20 - 167.72 = 107.37 \text{ kcal mol}^{-1}.$$

The ionization potential given in this section is that obtained from Koopman's theorem. The value given is not calculated directly, but is minus the eigenvalue (energy) of molecular orbital number 3, the HOMO. Koopman's theorem simply states that this energy is equal to that required to remove an electron from the relevant molecular orbital (i.e., that no change in the molecular orbitals occurs on ionization). This is illustrated diagrammatically below:

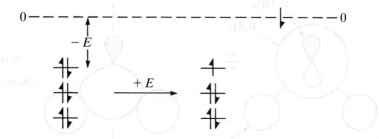

This simple approximation works remarkably well, not only for the first ionization potential, but also for the higher-energy ionizations from lower-lying orbitals. Koopman's theorem is therefore used to correlate the results of MO calculations with PE spectra, the peak energies corresponding to minus the eigenvalues of the individual orbitals. (This will be illustrated below for ethylene.)

After some information about the number of SCF calculations and the computer time required for the job, the optimized geometry is printed in Z-matrix form. In this case the optimized CH bond length is 1.091 Å and the dummy–C–H angles are 124.4°, giving an H–C–H angle of 111.1°:

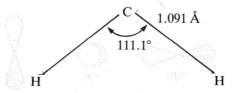

Note that the bond length and angle for hydrogen 4 have been held equal to those for atom 3, as defined in the symmetry cards. The interatomic-distance matrix for the optimized geometry is then printed.

The next section of the output, the molecular orbitals, is printed because the keyword "VECTORS" was given in the first line of the input. The orbitals are listed in order of increasing energy, with their eigenvalues (energy levels) in electron volts. Each vertical column represents one molecular orbital. The first number in the column is the eigenvalue and the remainder are the coefficients of the individual atomic orbitals in the LCAO equations, as shown in Section 3.1. Molecular orbital 1, for instance, has an energy of -26.860812 eV. The coefficient of the carbon $1s$-orbital is 0.806, of the carbon p_x 0.140, and of each of the hydrogen $1s$-orbitals 0.406. The orbital phases are shown in Fig. 4.4.3. The unshaded lobes correspond to a positive sign for an s-orbital, and the shaded to a negative sign. The convention for p-orbitals is simply that a positive coefficient means that the positive lobe of the orbital extends along the axis in the positive direction. Adding the contributions of molecular orbital 1 together gives the σ_{CH} orbital derived qualitatively in Section 3.6:

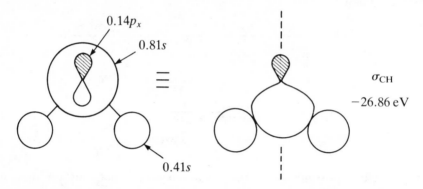

The charge distribution in this molecular orbital may now be assigned to the individual atoms. Note that the sum of the squares of the coefficients is 1 (because of the normalization condition), so that the charge assignable to a

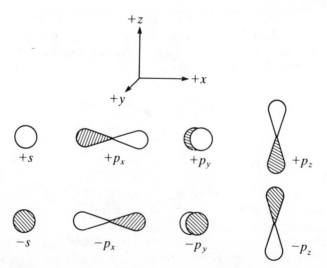

FIG. 4.4.3. The relationship between the Cartesian axes and orbital phases. (The Cartesian axes are abritrary in this and other examples.)

given atom is simply the sum of the squares of the coefficients of the orbitals centered on that atom times the number of electrons in the molecular orbital (in this case two). Thus the carbon atom can be assigned $(0.81^2 + 0.14^2) \times 2 = 1.35$ electrons and the hydrogen atoms $(0.41^2) \times 2 = 0.33$ electrons each. This is the charge distribution in molecular orbital 1, not in the whole molecule.

Molecular orbital 2, which has an energy of -14.21 eV, has zero coefficients on carbon except for the p_y-orbital, which has a coefficient of -0.64. The hydrogen that extends in the positive y-direction (see the Cartesian coordinates table) has a negative coefficient and the other a positive one, both with absolute values of 0.54. Using the orbital phases given in Fig. 4.4.3 we thus obtain the π_{CH_2} orbital shown in Section 3.6:

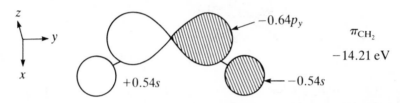

The charge distribution in this orbital, which is doubly occupied, is 0.82 electrons on carbon and 0.58 electrons on each hydrogen. There is more negative charge on hydrogen than in the first molecular orbital because the carbon atom uses only a p-orbital, which is higher in energy than the predominantly s-contribution in the σ_{CH}. The electronegativity effect demonstrated in Section 3.1 for lithium hydride therefore places more charge on hydrogen than in the lower-lying molecular orbital.

Molecular orbital 3 is the last of the doubly occupied orbitals (see "NO. OF FILLED LEVELS" in the output). This HOMO has an energy of -9.14 eV and is composed of s- and p_x-orbitals from carbon and of hydrogen s-contributions:

$$0.84p_x$$
$$-0.40s$$
$$0.25s$$

The charge distribution in this orbital, which is the carbon in-plane lone-pair, is 1.73 electrons on carbon and 0.13 electrons on each hydrogen. The electronic charges for the three occupied orbitals can now be added to give the total electron densities for each atom. The carbon thus has $1.35 + 0.82 + 1.73 = 3.90$ electrons, and the hydrogens each have $0.33 + 0.58 + 0.13 = 1.04$ electrons. The more exact atomic electron densities are printed in the table of net atomic charges and dipole distributions. The charges are calculated by simply subtracting the atomic electron densities from the nuclear charges. The nuclear charge for carbon in MNDO is 4 because the nonvalence electrons (and hence the equivalent number of nuclear charges) are omitted. The resulting net atomic charges are positive (0.10) for carbon and slightly negative (-0.05) for hydrogen.

Molecular orbital 4, the LUMO, is a pure carbon p_z orbital:

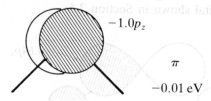

The two remaining molecular orbitals are the antibonding equivalents of molecular orbitals 1 and 2:

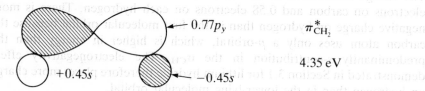

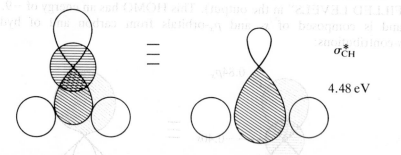

Note that there is a jump in energy between the HOMO and the LUMO because of the inherent instability of virtual orbitals in the SCF treatment, as discussed in Section 3.1. Two final points to note about the methylene molecular orbitals are that if there were any charge in the virtual orbitals the charge distributions would be reversed between the π_{CH_2} (molecular orbital 2) and its antibonding equivalent, the $\pi^*_{CH_2}$(molecular orbital 5), and that the sum of the squares of the coefficients in a horizontal row is also 1, as discussed in Section 3.1. Figure 4.4.4 shows pseudo-three-dimensional electron-density

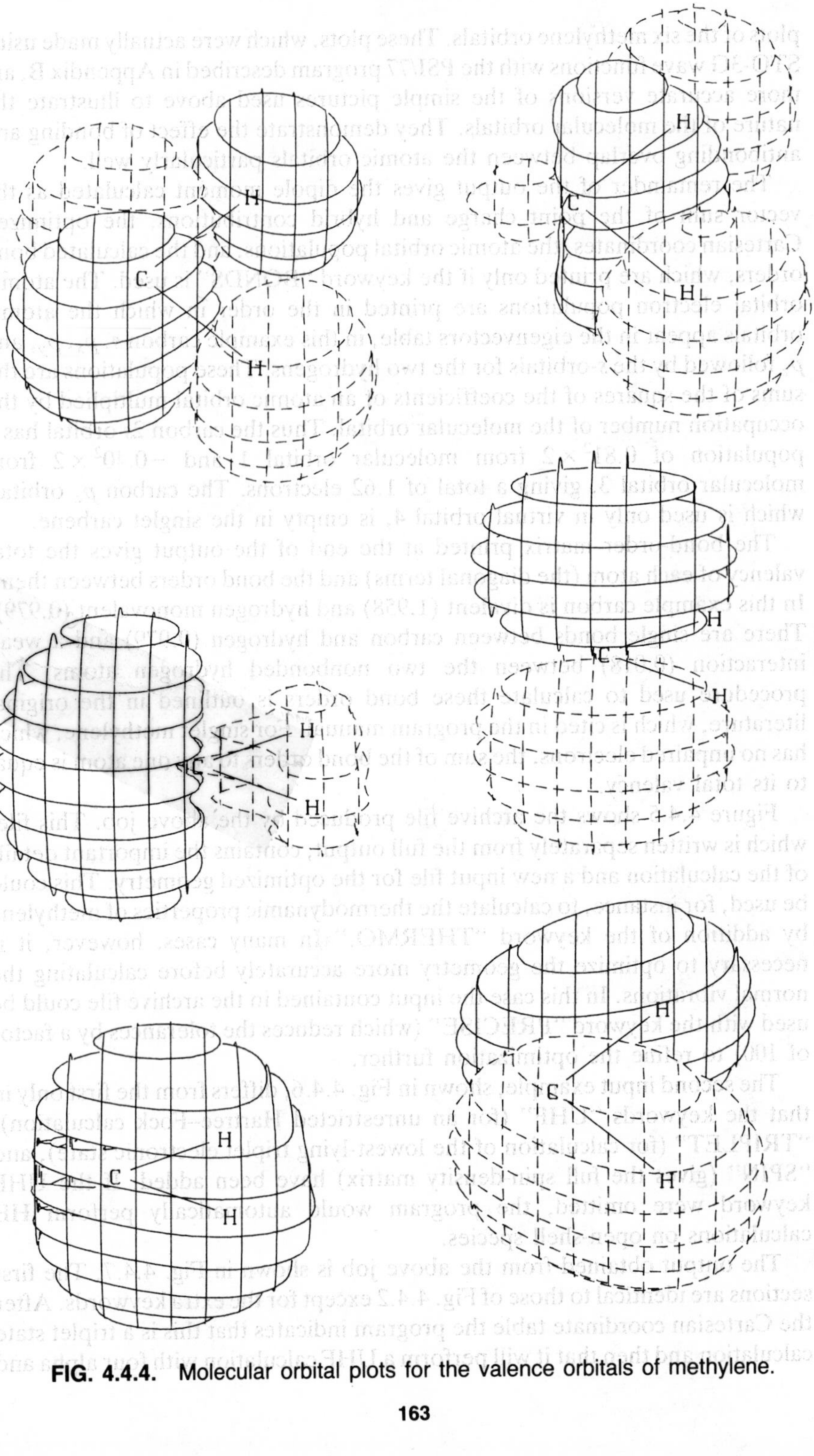

FIG. 4.4.4. Molecular orbital plots for the valence orbitals of methylene.

plots of the six methylene orbitals. These plots, which were actually made using STO-3G wave functions with the PSI/77 program described in Appendix B, are more accurate versions of the simple pictures used above to illustrate the nature of the molecular orbitals. They demonstrate the effect of bonding and antibonding overlap between the atomic orbitals particularly well.

The remainder of the output gives the dipole moment calculated as the vector sum of the point charge and hybrid contributions, the optimized Cartesian coordinates, the atomic orbital populations, and the calculated bond orders, which are printed only if the keyword "BONDS" is used. The atomic orbital electron populations are printed in the order in which the atomic orbitals appear in the eigenvectors table, in this example carbon s, p_x, p_y, and p_z followed by the s-orbitals for the two hydrogens. These populations are the sums of the squares of the coefficients of an atomic orbital multiplied by the occupation number of the molecular orbital. Thus the carbon $2s$ orbital has a population of $0.81^2 \times 2$ from molecular orbital 1 and $-0.40^2 \times 2$ from molecular orbital 3, giving a total of 1.62 electrons. The carbon p_z orbital, which is used only in virtual orbital 4, is empty in the singlet carbene.

The bond-order matrix printed at the end of the output gives the total valency of each atom (the diagonal terms) and the bond orders between them. In this example carbon is divalent (1.958) and hydrogen monovalent (0.979). There are single bonds between carbon and hydrogen (0.979) and a weak interaction (0.018) between the two nonbonded hydrogen atoms. The procedure used to calculate these bond orders is outlined in the original literature, which is cited in the program manual. For singlet methylene, which has no unpaired electrons, the sum of the bond orders to any one atom is equal to its total valency.

Figure 4.4.5 shows the archive file produced by the above job. This file, which is written separately from the full output, contains the important details of the calculation and a new input file for the optimized geometry. This could be used, for instance, to calculate the thermodynamic properties of methylene by addition of the keyword "THERMO." In many cases, however, it is necessary to optimize the geometry more accurately before calculating the normal vibrations. In this case the input contained in the archive file could be used with the keyword "PRECISE" (which reduces the tolerances by a factor of 100) to refine the optimization further.

The second input example, shown in Fig. 4.4.6, differs from the first only in that the keywords "UHF" (for an unrestricted Hartree–Fock calculation), "TRIPLET" (for calculation of the lowest-lying triplet electronic state), and "SPIN" (gives the full spin-density matrix) have been added. If the UHF keyword were omitted, the program would automatically perform HE calculations on open-shell species.

The output obtained from the above job is shown in Fig. 4.4.7. The first sections are identical to those of Fig. 4.4.2 except for the extra keywords. After the Cartesian coordinate table the program indicates that this is a triplet state calculation and then that it will perform a UHF calculation with four alpha and

```
                 SUMMARY OF  MNDO   CALCULATION

                                                 VERSION  1.25

   C    H2
                                                 29-FEB-84
   Example 4.1
   Singlet methylene (C2v)


         PETERS TEST WAS SATISFIED IN FLETCHER-POWELL OPTIMISATION
         SCF FIELD WAS ACHIEVED


              HEAT OF FORMATION     =    107.366493 KCAL
              ELECTRONIC ENERGY     =   -240.256190 EV
              CORE-CORE REPULSION   =    88.669822 EV
              DIPOLE                =    1.54316 DEBYE
              NO. OF FILLED LEVELS  =    3
              IONISATION POTENTIAL  =    9.136801 EV
              SCF CALCULATIONS      =    16
              COMPUTATION TIME      =    7.21 SECONDS


         CARTESIAN COORDINATES

   NO.       ATOM            X              Y              Z


    1         C        1.0000000000   0.0000000000   0.0000000000
    2         H        1.6169072042   0.9000072751   0.0000000000
    3         H        1.6169072042  -0.9000072751   0.0000000000


         FINAL GEOMETRY OBTAINED
   VECTORS SYMMETRY BONDS
   Example 4.1
   Singlet methylene (C2v)
   XX   0.000000 0   0.000000 0    0.000000 0   0  0  0
   C    1.000000 0   0.000000 0    0.000000 0   1  0  0
   H    1.091141 1 124.428568 1    0.000000 0   2  1  0
   H    1.091141 0 124.428568 0  180.000000 0   2  1  3
   0    0.000000 0   0.000000 0    0.000000 0   0  0  0
   3,   1,   4,
   3,   2,   4,
```

FIG. 4.4.5. The archive file produced by MOPAC from the input shown in Fig. 4.4.1.

```
         UHF TRIPLET VECTORS SYMMETRY BONDS SPIN
          Example 4.2
          Triplet methylene (C2v)
          XX 0.0 0    0. 0      0. 0    0  0  0
          C  1.0 0    0. 0      0. 0    1  0  0
          H  1.1 1  122. 1      0. 0    2  1  0
          H  1.1 0  122. 0    180. 0    2  1  3
          0
          3,1,4,
          3,2,4,
```

FIG. 4.4.6. One possible MOPAC input for a UHF calculation on triplet methylene.

```
******************************************************************************
                        MNDO CALCULATION RESULTS


******************************************************************************
*                       VERSION  1.25
*   VECTORS  - FINAL EIGENVECTORS TO BE PRINTED
*   SPIN     - FINAL UHF SPIN MATRIX TO BE PRINTED
*   BONDS    - FINAL BOND-ORDER MATRIX TO BE PRINTED
*   UHF      - UNRESTRICTED HARTREE-FOCK CALCULATION
*   TRIPLET  - TRIPLET
*   SYMMETRY - SYMMETRY CONDITIONS TO BE IMPOSED
******************************************************************************

          PARAMETER DEPENDANCE DATA

              REFERENCE ATOM       FUNCTION NO.     DEPENDANT ATOM(S)
                    3                   1                 4
                    3                   2                 4

                 DESCRIPTIONS OF THE FUNCTIONS USED

      1       BOND LENGTH     IS SET EQUAL TO THE REFERENCE BOND LENGTH
      2       BOND ANGLE      IS SET EQUAL TO THE REFERENCE BOND ANGLE
UHF TRIPLET VECTORS SYMMETRY BONDS SPIN
Example 4.2
Triplet methylene (C2v)

       ATOM    CHEMICAL    BOND LENGTH     BOND ANGLE      TWIST ANGLE
       NUMBER  SYMBOL      (ANGSTROMS)     (DEGREES)       (DEGREES)
       (I)                 NA:I            NB:NA:I         NC:NB:NA:I      NA  NB  NC


        1      XX
        2      C           1.00000                                        1
        3      H           1.10000 *       122.00000 *                    2   1
        4      H           1.10000         122.00000       180.00000      2   1   3


           CARTESIAN COORDINATES

     NO.      ATOM         X          Y          Z

      1        6        1.0000     0.0000     0.0000
      2        1        1.5829     0.9329     0.0000
      3        1        1.5829    -0.9329     0.0000

TRIPLET STATE CALCULATION


              UHF CALCULATION, NO. OF ALPHA ELECTRONS =  4
                              NO. OF BETA  ELECTRONS =  2


           INTERATOMIC DISTANCES

                  C   1      H   2      H   3
       ------------------------------------------------
     C   1    0.000000
     H   2    1.100000   0.000000
     H   3    1.100000   1.865706   0.000000
```

FIG. 4.4.7. The output produced by MOPAC from the input shown in Fig. 4.4.6.

```
CYCLE:  1 TIME:    7.90 TIME LEFT:   3589.25 GRADIENT:   36.944 HEAT:  75.1053
CYCLE:  2 TIME:    2.38 TIME LEFT:   3586.87 GRADIENT:   25.764 HEAT:  74.0996
CYCLE:  3 TIME:    6.50 TIME LEFT:   3580.37 GRADIENT:    4.248 HEAT:  73.9429
CYCLE:  4 TIME:    2.18 TIME LEFT:   3578.19 GRADIENT:    2.400 HEAT:  73.9368
HEAT OF FORMATION TEST SATISFIED
PETERS TEST SATISFIED
 UHF TRIPLET VECTORS SYMMETRY BONDS SPIN
 Example 4.2
 Triplet methylene (C2v)


     PETERS TEST WAS SATISFIED IN FLETCHER-POWELL OPTIMISATION
     SCF FIELD WAS ACHIEVED


                        MNDO    CALCULATION
                                              VERSION  1.25



          FINAL HEAT OF FORMATION =    73.935199 KCAL



          ELECTRONIC ENERGY        =  -243.399193 EV
          CORE-CORE REPULSION      =    90.363135 EV

          IONISATION POTENTIAL     =     9.918751
                                                   29-FEB-84

          NO. OF ALPHA ELECTRONS   =     4
          NO. OF BETA  ELECTRONS   =     2


          SCF CALCULATIONS    =    26
          COMPUTATION TIME    =   26.59 SECONDS
```

ATOM NUMBER (I)	CHEMICAL SYMBOL	BOND LENGTH (ANGSTROMS) NA:I	BOND ANGLE (DEGREES) NB:NA:I	TWIST ANGLE (DEGREES) NC:NB:NA:I	NA	NB	NC
1	XX						
2	C	1.00000			1		
3	H	1.05173 *	103.53378 *		2	1	
4	H	1.05173	103.53378	180.00000	2	1	3

```
     INTERATOMIC DISTANCES


             C   1      H   2      H   3
------------------------------------------------
  C   1   0.000000
  H   2   1.051733   0.000000
  H   3   1.051733   2.045058   0.000000
```

FIG. 4.4.7. (*continued*)

ALPHA EIGENVECTORS

ROOT NO.	1	2	3	4	5	6
	-28.227035	-15.485033	-10.245397	-9.918751	3.898854	5.33706

			1	2	3	4	5	6
S	C	1	0.856793	0.000000	-0.159454	0.000000	-0.490387	0.00000
PX	C	1	0.028982	0.000000	0.964377	0.000000	-0.262940	0.00000
PY	C	1	0.000000	0.700755	0.000000	0.000000	0.000000	0.71340
PZ	C	1	0.000000	0.000000	0.000000	-1.000000	0.000000	0.00000
S	H	2	0.364050	0.504452	0.149251	0.000000	0.587530	-0.49550
S	H	3	0.364050	-0.504452	0.149251	0.000000	0.587530	0.49550

BETA EIGENVECTORS

ROOT NO.	1	2	3	4	5	6
	-24.716383	-14.975892	1.985845	2.128741	5.121025	6.27123

			1	2	3	4	5	6
S	C	1	0.785808	0.000000	-0.207161	0.000000	0.582743	0.00000
PX	C	1	0.083397	0.000000	0.969121	0.000000	0.232057	0.00000
PY	C	1	0.000000	-0.641205	0.000000	0.000000	0.000000	-0.76737
PZ	C	1	0.000000	0.000000	0.000000	-1.000000	0.000000	0.00000
S	H	2	0.433330	-0.542612	0.094578	0.000000	-0.550709	0.45340
S	H	3	0.433330	0.542612	0.094578	0.000000	-0.550709	-0.45340

NET ATOMIC CHARGES AND DIPOLE CONTRIBUTIONS

ATOM NO.	TYPE	CHARGE	ATOM ELECTRON DENSITY
1	C	-0.2170	4.2170
2	H	0.1085	0.8915
3	H	0.1085	0.8915

DIPOLE	X	Y	Z	TOTAL
POINT-CHG.	0.257	0.000	0.000	0.257
HYBRID	0.260	0.000	0.000	0.260
SUM	0.517	0.000	0.000	0.517

FIG. 4.4.7. (*continued*)

CARTESIAN COORDINATES

NO.	ATOM	X	Y	Z
1	C	1.0000	0.0000	0.0000
2	H	1.2461	1.0225	0.0000
3	H	1.2461	-1.0225	0.0000

ATOMIC ORBITAL ELECTRON POPULATIONS

1.37702   0.93782   0.90220   1.00000   0.89148   0.89148

$$\langle SZ \rangle \quad = \quad 1.000000$$
$$\langle S**2 \rangle \quad = \quad 2.016834$$

SPIN DENSITY MATRIX

			S C 1	PX C 1	PY C 1	PZ C 1	S H 2	S H 3
S C	1		-0.142026					
PX C	1		0.194476	-0.923908				
PY C	1		0.000000	0.000000	-0.079914			
PZ C	1		0.000000	0.000000	0.000000	-1.000000		
S H	2		0.052397	-0.118347	-0.005571	0.000000	0.072924	
S H	3		0.052397	-0.118347	0.005571	0.000000	-0.006990	0.072924

BOND ORDERS AND VALENCIES

		C 1	H 2	H 3
C	1	3.836387		
H	2	0.923509	0.988224	
H	3	0.923509	0.042566	0.988224

FIG. 4.4.7. *(continued)*

two beta electrons. This corresponds to two doubly occupied and two singly occupied molecular orbitals:

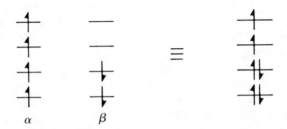

The optimization is then performed as for the singlet carbene. Note, however, that more cycles are required in the optimization and that more computer time is used than for the singlet. This arises partly because the starting geometry is closer to that of the singlet than to that of the triplet and partly because open-shell calculations are inherently slower than closed-shell ones. The heat of formation, electronic energy, and so forth are then printed. The triplet carbene is calculated to be 33.5 kcal mol^{-1} more stable than the singlet and to have a wider H–C–H angle:

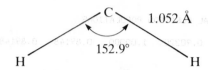

The energetic comparison is unbalanced because the triplet was calculated using the UHF formalism and the singlet with RHF. In this case, however, the difference is small, as shown by the next example. Generally differences between UHF and RHF (HE or ROHF) energies are smaller for MNDO and MINDO/3 than for *ab initio* calculations. This is because the electron–electron repulsions that are largely responsible for the differences are generally ignored in the semiempirical methods. Triplets are generally calculated to be too stable relative to singlets by MNDO.

The alpha and beta molecular orbitals are now printed exactly as for a closed-shell molecule except that each orbital is singly occupied or empty, as discussed above. The four occupied alpha orbitals (1–4) and the two occupied beta orbitals (1 and 2) all have negative energies, but the two beta orbitals (3 and 4) that correspond to the singly occupied molecular orbitals are unbound (i.e., their eigenvalues are positive). This is usually the case for neutral molecules, and results from the same conditions as for the unoccupied orbitals in closed-shell calculations (an electron in, for instance, alpha orbital 4 is repelled by one less electron than an extra electron in beta orbital 4 would be). The alpha and beta occupied molecular orbitals are shown in Fig. 4.4.8. Note that the alpha and beta molecular orbitals for the two doubly occupied levels are not identical. This results from a process known as spin polarization. If we consider the two singly occupied orbitals (3 and 4) we find most of the electron density on carbon. The electrons in the doubly occupied orbitals are affected differently by the alpha (conventionally spin up) electrons in these two orbitals. Electrons with common spin repel each other normally in all regions common to a singly and a doubly occupied molecular orbital. The Pauli exclusion principle, however, does not allow two electrons of the same spin to occupy the same space at the same time. This means that electrons of the same spin automatically avoid each other in areas common to the two orbitals. This, in turn, results in a lower repulsion between the electrons of the same spin than between those of opposite spin. The net result looks as if electrons of the same spin attract each other, but a better way to describe the phenomenon is that electrons of the same spin repel each other less than those of opposite spin do. The molecular orbitals reflect these trends. The high alpha-electron density on carbon in orbitals 3 and 4 polarizes the beta orbitals 1 and 2 toward hydrogen

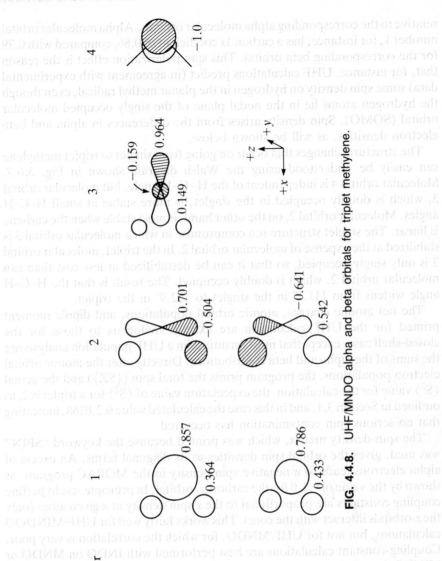

FIG. 4.4.8. UHF/MNDO alpha and beta orbitals for triplet methylene.

relative to the corresponding alpha molecular orbitals. Alpha molecular orbital number 1, for instance, has a carbon $1s$ coefficient of 0.86, compared with 0.79 for the corresponding beta orbital. This spin polarization effect is the reason that, for instance, UHF calculations predict (in agreement with experimental data) some spin density on hydrogen in the planar methyl radical, even though the hydrogen atoms lie in the nodal plane of the singly occupied molecular orbital (SOMO). Spin density arises from the differences in alpha and beta electron densities, as will be shown below.

The structural changes that occur on going from singlet to triplet methylene can easily be understood using the Walsh diagram shown in Fig. 3.6.7. Molecular orbital 4 is independent of the H–C–H angle, but molecular orbital 3, which is doubly occupied in the singlet, is more stable at small H–C–H angles. Molecular orbital 2, on the other hand, is most stable when the carbene is linear. The singlet structure is a compromise in which molecular orbital 3 is stabilized at the expense of molecular orbital 2. In the triplet, molecular orbital 3 is only singly occupied, so that it can be destabilized at less cost than can molecular orbital 2, which is doubly occupied. The result is that the H–C–H angle widens from 111.1° in the singlet to 152.9° in the triplet.

The net atomic charges, atomic orbital populations, and dipole moment printed for the UHF calculation are exactly analogous to those for the closed-shell case, except that most quantities in a UHF population analysis are the sums of the alpha and beta contributions. Directly after the atomic orbital electron populations, the program prints the total spin ($\langle SZ \rangle$) and the actual $\langle S^2 \rangle$ value for the calculation. the expectation value of $\langle S^2 \rangle$ for a triplet is 2, as outlined in Section 3.1, and in this case the calculated value is 2.0168, indicating that no serious spin contamination has occurred.

The spin-density matrix, which was printed because the keyword "SPIN" was used, gives the orbital spin densities as its diagonal terms. An excess of alpha electrons leads to a negative spin density in the MOPAC program, as shown by the value of -1.0 for the carbon p_z orbital. In principle, esr hyperfine coupling constants are proportional to the s-spin density at a given atom (only the s-orbitals interact with the core). This works fairly well for UHF/MINDO/3 calculations, but not for UHF/MNDO, for which the correlation is very poor. Coupling-constant calculations are best performed with INDO on MNDO or MINDO/3 optimized geometries.

The final matrix contains the bond orders and valencies, exactly as for the singlet carbene. The carbon atom now has a valency of almost 4, however, reflecting the free valencies of the two unpaired electrons. These two valencies do not appear anywhere else in the matrix (i.e., the sum of the CH bonds is almost 2 less than the carbon total valency).

Figure 4.4.9 shows the output obtained from the same input (Fig. 4.4.6) without the keyword "UHF." In this case the program performs a HE calculation and the keyword "SPIN" is ignored because it is effective only in conjunction with "UHF." After "TRIPLET STATE CALCULATION" the program reports that it will perform an RHF calculation with two doubly

```
***************************************************************************
                      MNDO CALCULATION RESULTS


***************************************************************************
*                    VERSION  1.25
*  VECTORS  - FINAL EIGENVECTORS TO BE PRINTED
*  SPIN     - FINAL UHF SPIN MATRIX TO BE PRINTED
*  BONDS    - FINAL BOND-ORDER MATRIX TO BE PRINTED
*  TRIPLET  - TRIPLET
*  SYMMETRY - SYMMETRY CONDITIONS TO BE IMPOSED
***************************************************************************



        PARAMETER DEPENDANCE DATA

            REFERENCE ATOM      FUNCTION NO.    DEPENDANT ATOM(S)
                  3                  1                4
                  3                  2                4


              DESCRIPTIONS OF THE FUNCTIONS USED

       1      BOND LENGTH    IS SET EQUAL TO THE REFERENCE BOND LENGTH
       2      BOND ANGLE     IS SET EQUAL TO THE REFERENCE BOND ANGLE
    TRIPLET VECTORS SYMMETRY BONDS SPIN
    Example 4.3
    Triplet methylene (C2v)


        ATOM      CHEMICAL    BOND LENGTH    BOND ANGLE    TWIST ANGLE
        NUMBER    SYMBOL      (ANGSTROMS)    (DEGREES)      (DEGREES)
        (I)                     NA:I          NB:NA:I      NC:NB:NA:I    NA  NB  NC


          1       XX
          2       C          1.00000                                    1
          3       H          1.10000 *      122.00000 *                 2   1
          4       H          1.10000        122.00000     180.00000     2   1   3



            CARTESIAN COORDINATES

     NO.      ATOM       X         Y         Z

      1        6      1.0000    0.0000    0.0000
      2        1      1.5829    0.9329    0.0000
      3        1      1.5829   -0.9329    0.0000



    SYSTEM IS A BIRADICAL



    TRIPLET STATE CALCULATION


            RHF CALCULATION, NO. OF DOUBLY OCCUPIED LEVELS =   2

                      NO. OF SINGLY OCCUPIED LEVELS =   2
```

FIG. 4.4.9. MOPAC output for a HE calculation on triplet methylene.

```
           INTERATOMIC DISTANCES

             C    1      H   2      H   3
-------------------------------------------------
  C   1   0.000000
  H   2   1.100000    0.000000
  H   3   1.100000    1.865706    0.000000

CYCLE:  1 TIME:     8.14 TIME LEFT:  3588.01 GRADIENT:    38.175 HEAT:  78.6105
CYCLE:  2 TIME:     4.09 TIME LEFT:  3583.92 GRADIENT:    27.768 HEAT:  77.3723
CYCLE:  3 TIME:     3.57 TIME LEFT:  3580.35 GRADIENT:     5.454 HEAT:  77.3006
HEAT OF FORMATION TEST SATISFIED
PETERS TEST SATISFIED
 TRIPLET VECTORS SYMMETRY BONDS SPIN
 Example 4.3
 Triplet methylene (C2v)


      PETERS TEST WAS SATISFIED IN FLETCHER-POWELL OPTIMISATION
      SCF FIELD WAS ACHIEVED


                      MNDO    CALCULATION

                                                      VERSION  1.25

          FINAL HEAT OF FORMATION =     77.300058 KCAL


          ELECTRONIC ENERGY      =   -243.320930 EV
          CORE-CORE REPULSION    =     90.430783 EV

          IONISATION POTENTIAL   =      3.876891

                                                    29-FEB-84
          NO. OF FILLED LEVELS   =      2
          AND NO. OF OPEN LEVELS =      2

          HALF-ELECTRON CORRECTION FOR TRIPLET STATE

          -0.25*(<II|II>+<JJ|JJ>)-0.5*<IJ|JI>

                          <II|II>=    11.326612
                          <JJ|JJ>=    11.080000
                          <IJ|JI>=     0.678315


          SCF CALCULATIONS  =       26
          COMPUTATION TIME  =    23.49 SECONDS



    ATOM    CHEMICAL   BOND LENGTH    BOND ANGLE    TWIST ANGLE
    NUMBER  SYMBOL     (ANGSTROMS)    (DEGREES)     (DEGREES)
    (I)                   NA:I          NB:NA:I       NC:NB:NA:I   NA  NB  NC

       1    XX
       2    C        1.00000                                       1
       3    H        1.05132 *      105.24162 *                    2   1
       4    H        1.05132        105.24162     180.00000        2   1   3
```

FIG. 4.4.9. (*continued*)

174

```
                 INTERATOMIC DISTANCES

                     C   1      H   2      H   3
        ----------------------------------------------
        C   1   0.000000
        H   2   1.051315   0.000000
        H   3   1.051315   2.028672   0.000000


                         EIGENVECTORS

  ROOT NO.        1           2           3           4           5           6

             -26.423578  -15.166531   -4.185380   -3.876891    4.473943    5.74022


  S   C   1   0.822611   0.000000   -0.213840   0.000000   -0.526862   0.00000
  PX  C   1   0.066669   0.000000    0.956470   0.000000   -0.284113   0.00000
  PY  C   1   0.000000  -0.669724    0.000000   0.000000    0.000000  -0.74261
  PZ  C   1   0.000000   0.000000    0.000000  -1.000000    0.000000   0.00000

  S   H   2   0.399291  -0.525105    0.140424   0.000000    0.566434   0.47356

  S   H   3   0.399291   0.525105    0.140424   0.000000    0.566434  -0.47356


            NET ATOMIC CHARGES AND DIPOLE CONTRIBUTIONS

    ATOM NO.    TYPE        CHARGE       ATOM ELECTRON DENSITY
       1         C         -0.2199            4.2199
       2         H          0.1099            0.8901
       3         H          0.1099            0.8901

  DIPOLE          X           Y           Z         TOTAL

  POINT-CHG.    0.292       0.000       0.000       0.292

  HYBRID        0.389       0.000       0.000       0.389

  SUM           0.681       0.000       0.000       0.681


               CARTESIAN COORDINATES
     NO.    ATOM        X           Y           Z

      1       C       1.0000      0.0000      0.0000
      2       H       1.2764      1.0143      0.0000
      3       H       1.2764     -1.0143      0.0000


          ATOMIC ORBITAL ELECTRON POPULATIONS

   1.39910   0.92372   0.89706   1.00000   0.89006   0.89006


               BOND ORDERS AND VALENCIES

                  C   1      H   2      H   3
        ----------------------------------------------
        C   1   3.806309
        H   2   0.922873   0.987912
        H   3   0.922873   0.045320   0.987912
```

FIG. 4.4.9. (*continued*)

occupied and two singly occupied orbitals. Although the optimization requires three cycles, rather than the four needed for the UHF job, the computer time needed is very similar. Generally UHF optimizations are faster than the corresponding HE jobs. The calculated heat of formation, which is on the same energy scale as the RHF singlet, is 77.3 kcal mol^{-1}, compared with the 73.9 kcal mol^{-1} given by the UHF calculation. Note that the Koopman's theorem ionization potential in a HE calculation is meaningless because the eigenvalues of the SOMOs are not corrected for the pairing energy of the two half electrons. This correction is, however, applied to the total energy, as printed in the output. The geometry is similar to that given by the UHF calculation, although the H–C–H angle is 149.5°, compared with the 152.9° from the UHF approach. Only one set of orbitals is calculated, 1 and 2 being doubly and 2 and 3 singly occupied. The main uses of HE calculations are to make possible direct comparison of the energies of open- and closed-shell species and in cases where spin contamination is very severe in UHF calculations. For energy comparisons the most economical procedure is often to perform a UHF optimization followed by a HE single-point calculation on the UHF geometry.

In the following examples only the relevant parts of the output will be reproduced to avoid repetition. Figure 4.4.10 shows the input for D_{2h} ethylene. The symmetry definition corresponds to that given in Fig. 3.3.3. The output (Fig. 4.4.11) shows a double bond length of 1.334 Å, C–C–H angles of 123.2°, and a heat of formation of 15.4 kcal mol^{-1}. The molecular orbitals are shown in Fig. 4.4.12(a), and correspond to those derived qualitatively in section 3.6. The ionization potentials derived from the MO eigenvalues using Koopman's theorem can be compared with the experimental PE spectrum, as shown in Fig. 4.4.13. This procedure has been used very widely, especially with MINDO/3, and although the use of Koopman's theorem for such applications, has been criticized, it appears to be reliable for the first three or four peaks in the spectrum. Configuration interaction (CI) calculations show the lowest four radical cation states of ethylene to be essentially pure Koopman's configurations, so the simple correlation used here is applicable. High-energy (>15 eV) ionizations should, however, be treated carefully.

```
LOCALISE VECTORS SYMMETRY BONDS
Example 4.5
Ethylene
C    0.0  0    0.0    0    0.0  0    0   0   0
C    1.34 1    0.0    0    0.0  0    1   0   0
H    1.08 1  122.0    1    0.0  0    2   1   0
H    1.08 0  122.0    0  180.0  0    2   1   3
H    1.08 0  122.0    0    0.0  0    1   2   3
H    1.08 0  122.0    0  180.0  0    1   2   3
0
3,1,4,5,6,
3,2,4,5,6,
```

FIG. 4.4.10. One possible MOPAC input for ethylene in D_{2h} symmetry with localization of the molecular orbitals.

```
++++++++++++++++++++++++++++++++++++++++++++++++++++++++++++++++++++++++++++++++

      PETERS TEST WAS SATISFIED IN FLETCHER-POWELL OPTIMISATION
      SCF FIELD WAS ACHIEVED


                           MNDO    CALCULATION
                                                    VERSION  1.25


         FINAL HEAT OF FORMATION =    15.350072 KCAL


         ELECTRONIC ENERGY      =  -742.357328 EV
         CORE-CORE REPULSION    =   430.538699 EV

         IONISATION POTENTIAL   =    10.177637
                                                    29-FEB-84
         NO. OF FILLED LEVELS   =     6


         SCF CALCULATIONS   =     11
         COMPUTATION TIME   =   21.81 SECONDS
```

ATOM NUMBER (I)	CHEMICAL SYMBOL	BOND LENGTH (ANGSTROMS) NA:I	BOND ANGLE (DEGREES) NB:NA:I	TWIST ANGLE (DEGREES) NC:NB:NA:I	NA	NB	NC
1	C						
2	C	1.33448 *			1		
3	H	1.08917 *	123.19955 *		2	1	
4	H	1.08917	123.19955	180.00000	2	1	3
5	H	1.08917	123.19955	0.00000	1	2	3
6	H	1.08917	123.19955	180.00000	1	2	3

```
         INTERATOMIC DISTANCES
```

		C 1	C 2	H 3	H 4	H 5	H 6
C	1	0.000000					
C	2	1.334482	0.000000				
H	3	2.135149	1.089171	0.000000			
H	4	2.135149	1.089171	1.822768	0.000000		
H	5	1.089171	2.135149	2.527248	3.116001	0.000000	
H	6	1.089171	2.135149	3.116001	2.527248	1.822768	0.000000

```
         EIGENVECTORS


 ROOT NO.    1           2           3           4           5           6

          -35.312267  -22.653780  -15.831952  -14.607457  -12.640527  -10.177637


 S  C  1 -0.625343  -0.472897   0.000000   0.015222   0.000000   0.000000
 PX C  1 -0.165701   0.278645   0.000000   0.596418   0.000000   0.000000
 PY C  1  0.000000   0.000000  -0.519116   0.000000  -0.441257   0.000000
 PZ C  1  0.000000   0.000000   0.000000   0.000000   0.000000   0.707107
```

FIG. 4.4.11. Extract from the MOPAC output for ethylene with localized orbitals.

```
  S C   2 -0.625343   0.472897    0.000000    0.015222    0.000000   0.000000
  PX C  2  0.165701   0.278645    0.000000   -0.596418    0.000000   0.000000
  PY C  2  0.000000   0.000000   -0.519116    0.000000    0.441257   0.000000
  PZ C  2  0.000000   0.000000    0.000000    0.000000    0.000000   0.707107

  S H   3 -0.201853   0.315219   -0.339499   -0.268379    0.390700   0.000000

  S H   4 -0.201853   0.315219    0.339499   -0.268379   -0.390700   0.000000

  S H   5 -0.201853  -0.315219   -0.339499   -0.268379   -0.390700   0.000000

  S H   6 -0.201853  -0.315219    0.339499   -0.268379    0.390700   0.000000


  ROOT NO.    7          8           9          10          11         12


            1.320191   3.786734    4.795120    5.420344    5.576594   6.530834

  S C   1  0.000000   0.000000   -0.279726    0.000000   -0.329719  -0.445109
  PX C  1  0.000000   0.000000    0.374308    0.000000    0.341803  -0.531272
  PY C  1  0.000000  -0.480124    0.000000    0.552533    0.000000   0.000000
  PZ C  1  0.707107   0.000000    0.000000    0.000000    0.000000   0.000000

  S C   2  0.000000   0.000000    0.279726    0.000000   -0.329719   0.445109
  PX C  2  0.000000   0.000000    0.374308    0.000000   -0.341803  -0.531272
  PY C  2  0.000000  -0.480124    0.000000   -0.552533    0.000000   0.000000
  PZ C  2 -0.707107   0.000000    0.000000    0.000000    0.000000   0.000000

  S H   3  0.000000   0.367070   -0.375265    0.312016    0.370443  -0.099065

  S H   4  0.000000  -0.367070   -0.375265   -0.312016    0.370443  -0.099065

  S H   5  0.000000   0.367070    0.375265   -0.312016    0.370443   0.099065

  S H   6  0.000000  -0.367070    0.375265    0.312016    0.370443   0.099065
```

NET ATOMIC CHARGES AND DIPOLE CONTRIBUTIONS

ATOM NO.	TYPE	CHARGE	ATOM ELECTRON DENSITY
1	C	-0.0798	4.0798
2	C	-0.0798	4.0798
3	H	0.0399	0.9601
4	H	0.0399	0.9601
5	H	0.0399	0.9601
6	H	0.0399	0.9601

DIPOLE	X	Y	Z	TOTAL
POINT-CHG.	0.000	0.000	0.000	0.000
HYBRID	0.000	0.000	0.000	0.000
SUM	0.000	0.000	0.000	0.000

CARTESIAN COORDINATES

NO.	ATOM	X	Y	Z
1	C	0.0000	0.0000	0.0000
2	C	1.3345	0.0000	0.0000
3	H	1.9309	0.9114	0.0000
4	H	1.9309	-0.9114	0.0000
5	H	-0.5964	0.9114	0.0000
6	H	-0.5964	-0.9114	0.0000

FIG. 4.4.11. (*continued*)

178

ATOMIC ORBITAL ELECTRON POPULATIONS

```
1.22983    0.92163    0.92838    1.00000    1.22983    0.92163    0.92838    1.00000
0.96008    0.96008    0.96008    0.96008
```

BOND ORDERS AND VALENCIES

```
                C   1      C   2      H   3      H   4      H   5      H   6
-------------------------------------------------------------------------------------
C   1   3.932995
C   2   1.986307   3.932995
H   3   0.008975   0.964369   0.998406
H   4   0.008975   0.964369   0.012442   0.998406
H   5   0.964369   0.008975   0.002300   0.010321   0.998406
H   6   0.964369   0.008975   0.010321   0.002300   0.012442   0.998406
```

NUMBER OF ITERATIONS = 10   CONVERGANCE PARAMETER = 1.250000

LOCALISATION VALUE =   2.974967634

NUMBER OF CENTERS   % COMPOSITION OF ORBITALS

```
2.0230    C   2 49.72    C   1 49.71
2.0194    C   1 51.88    H   6 47.54
2.0194    C   2 51.88    H   3 47.54
2.0194    C   1 51.88    H   5 47.55
2.0195    C   2 51.88    H   4 47.55
2.0005    C   2 49.99    C   1 49.99
```

LOCALISED ORBITALS

```
ROOT NO.      1          2          3          4          5          6

        -25.186466  -18.885329  -18.884968  -18.883173  -18.882059  -10.501625

S  C   1 -0.459291  -0.446130   0.026825   0.446028   0.026883   0.067480
PX C   1 -0.525001   0.298718   0.020091  -0.298670   0.020109   0.077096
PY C   1 -0.000007   0.480137  -0.038935   0.480232   0.038973   0.000001
PZ C   1  0.102757   0.000017   0.000024  -0.000017   0.000031   0.699601

S  C   2 -0.459373   0.026835  -0.446105  -0.026813  -0.445968   0.067507
PX C   2  0.524953  -0.020033  -0.298756   0.020058  -0.298728  -0.077080
PY C   2  0.000036   0.038916  -0.480132   0.038894   0.480245  -0.000010
PZ C   2  0.102757   0.000017   0.000024  -0.000017   0.000031   0.699601

S  H   3  0.037778  -0.034799  -0.689523  -0.016420   0.040745  -0.005526

S  H   4  0.037721   0.016410   0.040591   0.034846  -0.689533  -0.005511

S  H   5  0.037794   0.040604   0.016364   0.689532  -0.034780  -0.005534

S  H   6  0.037813  -0.689521  -0.034815  -0.040744   0.016368  -0.005538
```

FIG. 4.4.11. (*continued*)

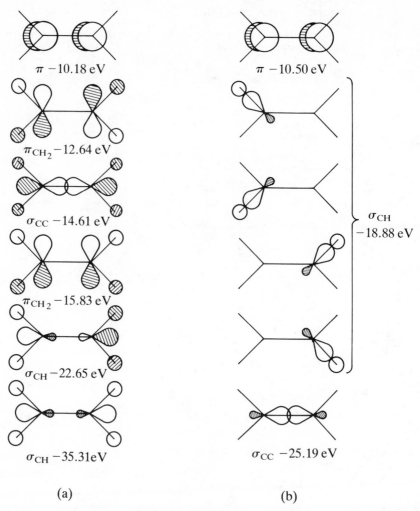

FIG. 4.4.12. The (a) delocalized and (b) localized orbitals of ethylene.

The molecular orbitals of ethylene are shown in Fig. 4.4.12(a) along with the localized orbitals [Fig. 4.4.12(b)] produced by the keyword "LOCALISE" (with "s"). Note that the four delocalized carbon–hydrogen bonding orbitals localize to the four equivalent CH bonds, which are practically degenerate. The CC sigma and π bonds are easily recognizable as the two remaining orbitals. There are three main types of localization scheme: the Perkins–Stewart method used in MOPAC, the Ruedenberg localization scheme, and Boys localization, which is available in some versions of GAUSSIAN70 and GAUSSIAN76. The major difference between the three methods is that Boys localization gives two "banana" sigma bonds for a double

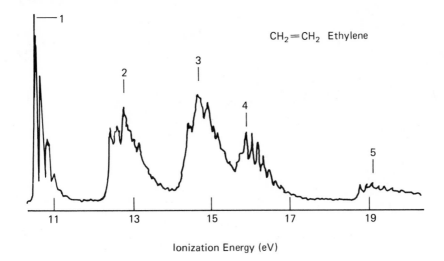

$CH_2{=}CH_2$ Ethylene

Ionization Energy (eV)

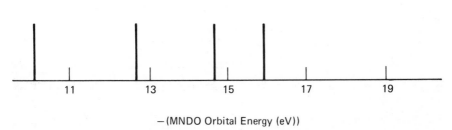

−(MNDO Orbital Energy (eV))

FIG. 4.4.13. Correlation between the observed photoelectron spectrum of ethylene and the MNDO energy levels. The experimental spectrum is taken from K.Kimura, S.Katsumata, Y.Achiba, T.Yamazaki, and S.Iwata, "Handbook of HeI Photoelectron Spectra of Fundamental Organic Molecules", Japan Scientific Societies Press, Tokyo/Halsted, New York, 1981, and is reproduced by permission of Japan Scientific Societies Press.

bond, rather than the separate sigma and π contributions shown above. One of the main uses of localized orbitals is in the interpretation of the bonding in electron-deficient molecules. Figure 4.4.14, for instance, shows the localized orbitals for diborane. The three-center two-electron bonds to the two bridging hydrogens are clearly discernible, far more so than they would be from the delocalized molecular orbitals.

The MOPAC manual provides other examples and full definitions of all the available keywords, so further input and output examples need not be described here. A few extra features and possible problems should, however, be discussed.

CARTESIAN COORDINATES

NO.	ATOM	X	Y	Z
1	H	1.0263	0.0000	0.0000
2	B	0.0000	0.8767	0.0000
3	B	0.0000	-0.8767	0.0000
4	H	-1.0263	0.0000	0.0000
5	H	0.0000	1.4492	-1.0133
6	H	0.0000	1.4492	1.0133
7	H	0.0000	-1.4492	1.0133
8	H	0.0000	-1.4492	-1.0133

EIGENVECTORS

ROOT NO.		1	2	3	4	5	6
		-27.592632	-19.173455	-14.873384	-14.842508	-13.967567	-12.785307
S H	1	0.322232	0.000000	-0.543661	0.000000	0.276483	0.000000
S B	2	0.554569	0.513938	0.000000	0.000000	-0.128082	0.000000
PX B	2	0.000000	0.000000	-0.452143	0.000000	0.000000	0.000000
PY B	2	-0.176511	0.208817	0.000000	0.000000	-0.475517	0.000000
PZ B	2	0.000000	0.000000	0.000000	-0.477445	0.000000	0.422830
S B	3	0.554569	-0.513938	0.000000	0.000000	-0.128082	0.000000
PX B	3	0.000000	0.000000	-0.452143	0.000000	0.000000	0.000000
PY B	3	0.176511	0.208817	0.000000	0.000000	0.475517	0.000000
PZ B	3	0.000000	0.000000	0.000000	-0.477445	0.000000	-0.422830
S H	4	0.322232	0.000000	0.543661	0.000000	0.276483	0.000000
S H	5	0.169505	0.310051	0.000000	0.368813	-0.300862	-0.400759
S H	6	0.169505	0.310051	0.000000	-0.368813	-0.300862	0.400759
S H	7	0.169505	-0.310051	0.000000	-0.368813	-0.300862	-0.400759
S H	8	0.169505	-0.310051	0.000000	0.368813	-0.300862	0.400759

NUMBER OF CENTERS   % COMPOSITION OF ORBITALS

2.8313	H	4 46.87	B	3 25.83	B	2 25.83
2.8313	H	1 46.87	B	2 25.83	B	3 25.83
2.0267	H	6 50.37	B	2 48.96		
2.0268	H	7 50.37	B	3 48.96		
2.0268	H	8 50.37	B	3 48.96		
2.0268	H	5 50.37	B	2 48.95		

(a)

FIG. 4.4.14. (a) Extract from the MOPAC output for diborane with localized orbitals.

ROOT NO.		1	2	3	4	5	6
		−18.457409	−18.457232	−16.581097	−16.580122	−16.579936	−16.579057
S H	1	0.084235	0.684614	−0.003561	−0.003565	0.003575	0.003585
S B	2	−0.244053	0.244049	0.483336	−0.030698	0.030698	−0.483180
PX B	2	0.319712	0.319715	−0.000001	0.000000	−0.000001	−0.000003
PY B	2	0.310704	−0.310703	0.231116	0.022266	−0.022272	−0.231079
PZ B	2	0.000024	−0.000020	0.450075	0.027275	0.027325	0.450201
S B	3	−0.244056	0.244046	−0.030625	0.483272	−0.483249	0.030692
PX B	3	0.319712	0.319715	−0.000001	0.000005	−0.000001	−0.000003
PY B	3	−0.310703	0.310704	−0.022289	−0.231097	0.231093	0.022288
PZ B	3	0.000003	−0.000008	0.027313	0.450126	0.450147	0.027318
S H	4	−0.684613	−0.084242	−0.003559	−0.003577	0.003577	0.003591
S H	5	0.043509	−0.043515	−0.059764	0.030857	0.001100	−0.709719
S H	6	0.043551	−0.043548	0.709699	−0.001143	−0.030819	0.059961
S H	7	0.043528	−0.043538	−0.001060	0.709708	0.059877	−0.030837
S H	8	0.043528	−0.043529	0.030865	−0.059851	−0.709709	0.001102

FIG. 4.4.14.a. (*continued*)

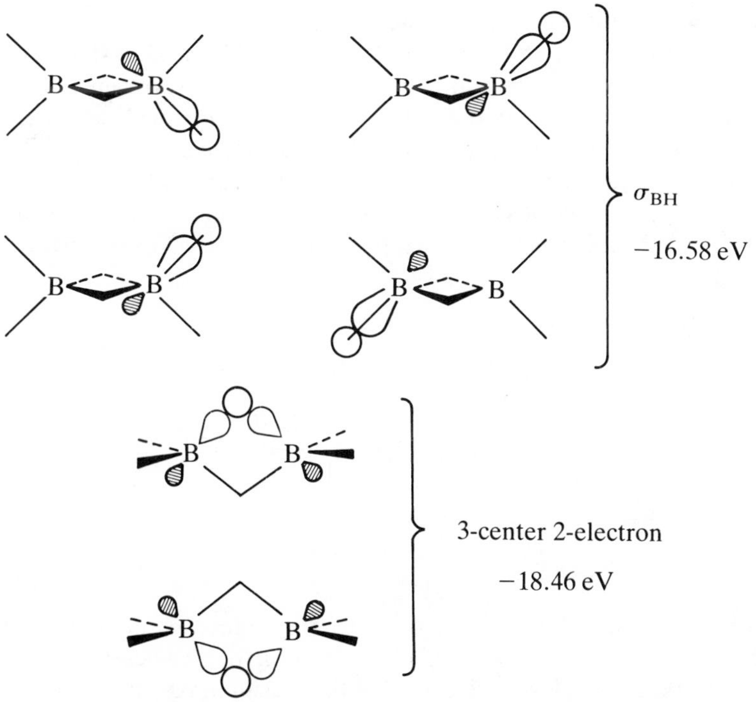

σ_{BH}

-16.58 eV

3-center 2-electron

-18.46 eV

FIG. 4.4.14. (b) Localized orbitals of diborane.

183

Degenerate Orbitals

Degenerate orbitals may be represented in an infinite number of ways. Any proper combinations of the orbitals of the degenerate set may be printed by the program. This means, for instance, that the degenerate Walsh HOMOs of cyclopropane or the triply degenerate HOMOs of methane may not be recognizable as the p_x, p_y, and p_z orientations used in Section 3.5. If easily recognizable orientations are required, such as for MO plots, they can be obtained by a slight reduction in the symmetry to a nondegenerate point group. This symmetry reduction need not be large enough to be visible in the molecular geometry.

Geometry Optimizations

A job may fail in the middle of a geometry optimization for two reasons that are not immediately obvious. The first, which usually results in an arithmetic error message being given from the computer operating system, rather than from the MOPAC program, is that three atoms used to define the dihedral angle for a fourth become linear, or very nearly so, during the optimization. The position of the fourth atom is then no longer definable. This type of problem can be prevented by the use of extra dummy atoms or by defining dihedrals relative to atoms that cannot possibly become linear.

The second type of problem may also be caused by an almost linear linkage. If, for instance, the substituent A in the vinyl cation shown below moves from one side of the molecule to the other, the two substituents X and Y also swap places if their dihedral angles are defined relative to A. This has two consequences; first the geometry cannot be optimized, and second the program detects very large changes in the Cartesian coordinates when small changes are made in the Z-matrix. The optimization is then aborted and an error message printed.

This message may also be printed if a ring is defined sequentially, especially with symmetry. In this case small changes in the ring angles and bond lengths are magnified by the lever effect so that they result in very large changes in the distance between the first and last ring atoms. This bond length is not explicitly defined, as shown below:

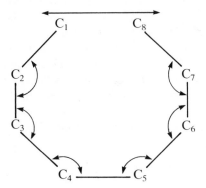

The solution to these problems is usually to use extra dummy atoms to define the structure uniquely. For ring molecules a strategy similar to that used for cyclopentane (Fig. 3.3.9) is usually the most effective.

UNABLE TO ACHIEVE
SELF CONSISTENCE

This error is relatively rare for correctly defined geometries in MNDO and MINDO/3. It means that a satisfactory solution has not been found by the SCF procedure within the given number of cycles. Very often this results from errors in the geometry definition that give such an outlandish structure that the SCF procedure can find no reasonable solution. If the geometry *is* correct, three techniques may help. The first is simply to shorten the bond lengths between heavy atoms. This is remarkably effective, and was the standard method before the advent of the MOPAC program, which has two options to improve convergence. The keyword "SHIFT=n" introduces damping into the SCF by artificially raising the energy of the virtual orbitals by n eV, so that they do not mix as strongly into the occupied molecular orbitals and oscillation is avoided. A reasonable value for n is 10. The PULAY procedure, the third alternative, may work if all else fails, but can be somewhat unpredictable.

Symmetry of the Wave Function

A molecule calculated within a given symmetry should have a total wave function of that symmetry. This means that the charges and spin densities of equivalent atoms should be identical, and that the coefficients of atomic orbitals related by symmetry should either be equal or opposite. In 90% of all cases where this is not true the geometry has been defined incorrectly, so that the molecule does not have the intended symmetry. Checking the symmetry of the electronic properties is therefore a good test for correct geometry. There may be cases, however, in which the wave function has a lower symmetry than the molecular geometry. This is the case, for instance, for Jahn–Teller molecules, in which a degenerate set of molecular orbitals is not fully occupied. Such problems, which are more common for radicals than for closed-shell molecules, indicate that the assumed molecular symmetry is too high. An example is the methane radical cation. An UHF/MNDO calculation within T_d symmetry gives the hydrogen spin densities shown in (1) below, and indicates that a distortion to C_{3v} symmetry is necessary, in which case structure (2), which is 17.7 kcal mol^{-1} more stable than (1), is obtained:

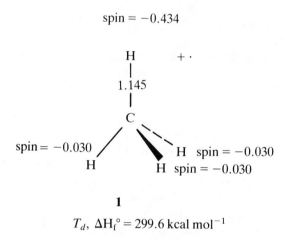

1

$$T_d, \ \Delta H_f^\circ = 299.6 \text{ kcal mol}^{-1}$$

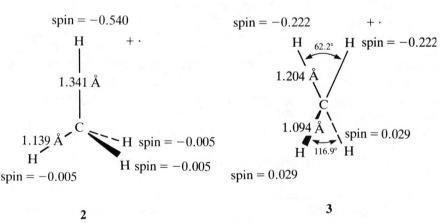

2   **3**

$$C_{3v}, \ \Delta H_f^\circ = 281.9 \text{ kcal mol}^{-1} \qquad C_{2v}, \ \Delta H_f^\circ = 274.5 \text{ kcal mol}^{-1}$$

A symmetrical wave function, however, is no guarantee that a given structure is a minimum, as demonstrated by the fact that geometry (3), with C_{2v} symmetry, is more stable than either (1) or (2).

Reaction Paths

A series of calculations along a reaction path do not necessarily give the lowest-energy reaction profile, as shown in Fig. 4.4.15. If the CC distance is used as the reaction coordinate in the symmetrical opening of a three-membered ring, there may be a discontinuity in the curve (indicated by the dotted line) because the reaction valley is not uniquely defined for the distance between r_1 and r_2 (i.e., the lowest energy pathway doubles back on itself). This could result from a significant shortening of the CX bonds after the transition state. In such a situation the transition state would be missed if one used the CC distance as a reaction coordinate, but could easily be located with respect to the C–X–C angle. It is therefore always necessary to define a transition state completely by diagonalization of the force-constant matrix. The "SADDLE" keyword in MOPAC can be used to locate transition states given two structures, one on the reactant and one on the product side. Because the transition-state search requires a great deal of computer time the two structures should be as close to the transition state as possible. A prior reaction-path calculation can therefore often save time. Details of this and other options are given in the MOPAC manual.

Forbidden Reactions

Forbidden reactions involve a discontinuity in the electronic state at some stage along the reaction path (i.e., the occupancy of the molecular orbitals must

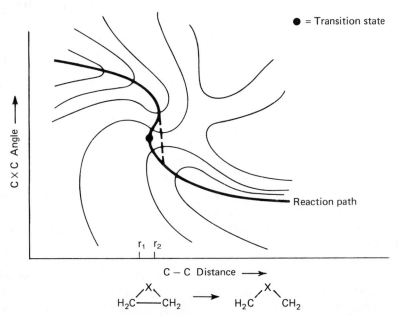

FIG. 4.4.15. An energy surface in which the minimum cannot be found by using one parameter (the CC distance) as a reaction coordinate.

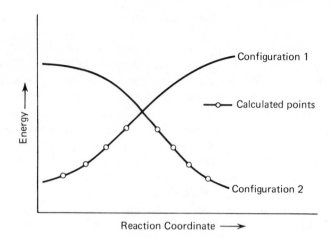

FIG. 4.4.16. Schematic energy profile for a forbidden reaction at the single determinant level.

change). If such a reaction is calculated at the single-determinant level, it may appear to give a normal potential-energy surface, but the program will not be able to locate a transition state. This situation is shown in Fig. 4.4.16. What appears to be a normal potential curve is in fact two curves that cross each other. This crossing is, however, not real if the electronic states of the two curves have the same symmetry. In that case the crossing is avoided to give the true potential energy curve, which, however, can be correctly calculated only by taking electron correlation into account, because the wave function near the transition state is a combination of the two states involved. Methods that explicitly include correlation, such as MNDO/C or even MNDO with CI, are necessary for correct treatment of forbidden reactions. In any case, a careful check that the wave function is continuous should be made for all reaction-path calculations.

4.5 MINDO/3 AND MNDO SUBJECT INDEX

The following subject index contains references to publications in which MINDO/3 and MNDO have been used. It covers literature cited in *Chemical Abstracts* up to the end of volume 98, but is probably relatively incomplete because the *Chemical Abstracts* citations for the two methods do not include all relevant papers. The choice of subjects indexed is strongly influenced by the author's own interpretations of the work, so that one subject may appear under a variety of headings. I have made no attempt to standardize the nomenclature of chemical compounds; these are indexed according to the author's designations.

REFERENCES

1. J. A. Pople and D. L. Beveridge, "Approximate Molecular Orbital Theory," McGraw-Hill, New York, 1970.

2. M. J. S. Dewar, "The Molecular Orbital Theory of Organic Chemistry," McGraw-Hill, New York, 1969; see also J. N. Murrell and A. J. Harget, "Semiempirical Self-Consistent-Field Molecular Orbital Theory of Molecules," Wiley-Interscience, London, 1972.

3. W. Thiel, *J. Am. Chem. Soc..*, 1981, **103**, 1413.

4. W. Thiel and T. Clark, unpublished.

5. R. C. Bingham, M. J. S. Dewar, and D. H. Lo, *J. Am. Chem. Soc.*, 1975, **97**, 1285.

6. R. C. Bingham, M. J. S. Dewar, and D. H. Lo, *J. Am. Chem. Soc.*, 1975, **97**, 1294.

7. R. C. Bingham, M. J. S. Dewar, and D. H. Lo, *J. Am. Chem. Soc.*, 1975, **97**, 1302.

8. R. C. Bingham, M. J. S. Dewar, and D. H. Lo, *J. Am. Chem. Soc.*, 1975, **97**, 1307.

9. M. J. S. Dewar, D. H. Lo, and C. A. Ramsden, *J. Am. Chem. Soc.*, 1975, **97**, 1311.

10. G. A. Olah, D. J. Donovan, H. C. Lin, H. Mayr, and P. Andreozzi, *J. Org. Chem.*, 1978, **43**, 2268.

11. G. Frenking, H. Goetz, and F. Marschner, *J. Am. Chem. Soc.*, 1978, **100**, 5295.

12. G. Frenking, F. Marschner, and H. Goetz, *Phosphorus Sulfur*, 1980, **8**, 337.

13. P. Bischof, *J. Am. Chem. Soc.*, 1976, **98**, 6844.

14. M. J. S. Dewar and S. Olivella, *J. Am. Chem. Soc.*, 1978, **100**, 5290.

15. M. J. S. Dewar and S. Olivella, *J. Chem. Soc. Faraday II*, 1979, **75**, 829.

16. M. J. S. Dewar and D. H. Lo, *Chem. Phys. Lett.*, 1975, **97**, 2933.

17. M. J. S. Dewar, D. Landman, S. H. Suck, and P. K. Weiner, *J. Am. Chem. Soc.*, 1977, **99**, 3951.

18. J. M. Schulman, *J. Magn. Res.*, 1977, **28**, 137.

19. P. K. K. Pandey and P. Chandra, *Theor. Chim. Acta*, 1978, **50**, 211.

20. M. E. Van Dommelen, H. M. Buck, and J. W. DeHaan, *Chem. Phys. Lett.*, 1978, **57**, 80.

21. M. Jallali-Heravi and G. A. Webb, *Org. Magn. Res.*, 1979, **12**, 174.

22. M. Jallali-Heravi and G. A. Webb, *J. Mol. Struct.*, 1979, **55**, 113.

23. M. J. S. Dewar, H. Kollmar, and S. H. Suck, *J. Am. Chem. Soc.*, 1975, **97**, 5590.

24. M. J. S. Dewar, S. H. Suck, P. K. Weiner, and J. G. Bergman, Jr., *Chem. Phys. Lett.*, 1976, **38**, 226.

25. M. J. S. Dewar, S. H. Suck, P. K. Weiner and J. G. Bergman, Jr., *Chem. Phys. Lett.*, 1976, **38**, 228.

26. M. J. S. Dewar and G. P. Ford, *J. Am. Chem. Soc.*, 1977, **99**, 7822.

27. B. Silvi, *J. Chim. Phys. Phys.-Biol. Chim. Biol.*, 1979, **76**, 21.

28. A.-L. K. Al-Jibari, K. H. Al-Niami, and M. Shansal, *Theor. Chim. Acta*, 1979, **53**, 327.

29. H. J. Koehler, *Z. Chem.*, 1979, **19**, 235.

30. G. Engel, H. J. Koehler, and C. Weiss, *Tetrahedron Lett.*, 1979, 2975.

31. T. Ando, H. Yamataka, S. Yabushita, K. Yamaguchi, and T. Fueno, *Bull. Chem. Soc. Japan*, 1981, **54**, 3613.

32. T. Clark, J. Chandrasekhar, P. v. R. Schleyer, and M. Saunders, *J. Chem.Soc. Chem. Comm.*, 1980, 265.

33. G. Klopman and L. D. Iroff, *J. Comput. Chem.*, 1981, **2**, 157.

34. O. Kikuchi, A. J. Hopfinger, and G. Klopman, *J. Theor. Biol.*, 1979, **77**, 129.

35. R. D. Singh and J. Ladik, *Chem. Phys. Lett.*, 1978, **65A**, 264.

36. J. A. Pople, *J. Am. Chem. Soc.*, 1975, **97**, 5306.

37. W. J. Hehre, *J. Am. Chem. Soc.*, 1975, **97**, 5308.

38. M. J. S. Dewar, *J. Am. Chem. Soc.*, 1975, **97**, 6591.

39. H. J. Koehler and H. Lischka, *J. Am. Chem. Soc.*, 1979, **101**, 3479.

40. H. J. Koehler, D. Heidrich and H. Lischka, *Z. Chem.*, 1977, **17**, 67.

41. J. Chandrasekhar, P. v. R. Schleyer, and H. B. Schlegel, *Tetrahedron Lett.*, 1978, 3393.

42. J. Chandrasekhar and P. v. R. Schleyer, *Tetrahedron Lett.*, 1979, 4057.

43. H. Schwarz, W. Franke, J. Chandrasekhar, and P. v. R. Schleyer, *Tetrahedron*, 1979, **35**, 1969.

44. Y. Apeloig, J. B. Collins, D. Cremer, T. Bally, E. Haselbach, J. A. Pople, J. Chandrasekhar, and P. v. R. Schleyer, *J. Org. Chem.*, 1980, **45**, 3496.

45. G. A. Olah, G. K. S. Prakash, G. Liang, P. W. Westerman, K. Kunde, J. Chandrasekhar, and P. v. R. Schleyer, *J. Am. Chem. Soc.*, 1980, **102**, 4485.

46. W. Francke, H. Schwarz, H. Thies, J. Chandrasekhar, P. v. R. Schleyer, W. J. Hehre, M. Saunders, and G. Walker, *Angew. Chem.*, 1980, **92**, 488.

47. D. Wirth and N. L. Bauld, *J. Comput. Chem.*, 1980, **1**, 189.

48. P. v. R. Schleyer and J. Chandrasekhar, *J. Org. Chem.*, 1981, **46**, 225.

49. H. J. Koehler and H. Lischka, *Chem. Phys. Lett.*, 1978, **58**, 175.

50. D. A. Krause, R. J. Day, W. L. Jorgensen, and R. G. Cooks, *Int. J. Mass. Spectrom. Ion. Phys.*, 1978, **27**, 227.

51. C. A. Wellington and S. H. Khowaiter, *Tetrahedron*, 1978, **34**, 2183.

52. R. J. Day, D. A. Krause, W. L. Jorgensen, and R. G. Cooks, *Int. J. Mass. Spectrom. Ion. Phys.*, 1979, **30**, 83.

53. J. R. Bews and C. Glidewell, *J. Mol. Struct.*, 1980, **64**, 75.

54. R. J. Day and R. G. Cooks, *Int. J. Mass. Spectrom. Ion. Phys.*, 1980, **35**, 293.

55. J. R. Bews and C. Glidewell, *J. Mol. Struct.*, 1980, **67**, 151.

56. M. Tasaka, M. Ogata, and H. Ichikawa, *J. Am Chem. Soc.*, 1981, **103**, 1885.

57. M. J. S. Dewar and W. Thiel, *J. Am. Chem. Soc.*, 1977, **99**, 4899.

58. P. Bischof, M. Boehm, R. Gleiter, R. A. Snow, C. W. Doecke, and L. A. Paquette, *J. Org. Chem.*, 1978, **43**, 2387.

59. P. Bischof, R. Gleiter, and R. Haider, *Angew. Chem.*, 1977, **89**, 122.

60. P. Hemmersbach, M. Klessinger, and P. Brackmann, *J. Am. Chem. Soc*, 1978, **100**, 6344.

61. D. A. Dougherty, H. B. Schlegel, and K. Mislow, *Tetrahedron*, 1978, **34**, 1441.

62. R. Sarneel, C. W. Worell, P. Pasman, J. W. Verhoeven, and G. F. Mes, *Tetrahedron*, 1980, **36**, 3241.

63. P. Bischof, R. Gleiter, H. Durr, B. Ruge, and P. Herbst, *Chem. Ber.*, 1976, **109**, 1412.

64. M. J. S. Dewar, G. J. Fonken, T. B. Jones, and D. E. Minter, *J. Chem. Soc. Perkin II*, 1976, 764.

65. P. Bischof, R. Gleiter, R. T. Taylor, A. R. Browne, and L. A. Paquette, *J. Org. Chem.*, 1978, **43**, 2391.

66. M. C. Boehm and R. Gleiter, *Chem. Ber.*, 1978, **111**, 3516.

67. J. Spanget-Larsen, R. Gleiter, G. Klein, C. W. Doecke, and L. A. Paquette, *Chem. Ber.*, 1980, **113**, 2120.

68. R. Gleiter, P. Schang, and G. Seitz, *Chem. Phys. Lett.*, 1978, **55**, 144.

69. R. Bartetzko, R. Gleiter, J. L. Muthard, and L. A. Paquette, *J. Am. Chem. Soc.*, 1978, **100**, 5589.

70. S. D. Worley, S. H. Gerson, N. Bodor, and J. J. Kaminski, *Chem. Phys. Lett.*, 1978, **60**, 104.

71. R. Gleiter, P. Hofmann, P. Schang, and A. Sieber, *Tetrahedron*, 1980, **36**, 655.

72. P. Masclet and G. Mouvier, *J. Electron. Spectrosc. Relat. Phenom.*, 1978, **14**, 77.

73. P. C. Bargers, C. W. Worell, and M. P. Groenewege, *Spectroscop. Lett.*, 1980, **13**, 391.

74. P. Livant, K. A. Roberts, M. D. Eggers, and S. D. Worley, *Tetrahedron*, 1981, **37**, 1853.

75. W. L. Jorgensen, *J. Am. Chem. Soc.*, 1976, **98**, 6784.

76. W. L. Jorgensen, *Tetrahedron Lett.*, 1976, 3029.

77. R. C. Haddon, *Tetrahedron Lett.*, 1975, 863.

78. H. Dits, N. M. M. Nibbering, and J. W. Verhoeven, *Chem. Phys. Lett.*, 1977, **51**, 95.

79. D. B. Boyd, *Int. J. Quant. Chem. Symp.*, 1977, **4**, 161.

80. W. B. Jennings and S. D. Worley, *J. Chem. Soc. Perkin II*, 1980, 1512.

81. W. L. Jorgensen, *J. Am. Chem. Soc.*, 1978, **100**, 1049.

82. W. L. Jorgensen, *J. Am. Chem. Soc.*, 1978, **100**, 1511.

83. M. J. S. Dewar, Y. Yamaguchi, S. Doraiswamy, S. Sharma, and S. H. Suck, *Chem. Phys.*, 1979, **41**, 21.

84. R. Czerminski, B. Lesyng, and A. Pohorille, *Int. J. Quant. Chem.*, 1979, **16**, 1141.

85. N. Bodor and R. Pearlman, *J. Am. Chem. Soc.*, 1978, **100**, 4946.

86. S. P. Mcmanus and M. R. Smith, *Tetrahedron Lett.*, 1978, 1897.

87. T. Bally, E. Haselbach, S. Lanyiova, F. Marschner, and M. Rossi, *Helv. Chem. Acta*, 1976, **59**, 486.

88. L. P. Davis and R. M. Guidry, *Aust. J. Chem.*, 1979, **32**, 1369.

89. H. Dodziuk, *J. Mol. Struct.*, 1979, **55**, 107.

90. J. Tyrell, *J. Am. Chem. Soc.*, 1979, **101**, 3766.

91. G. Favini and R. Todecshini, *J. Mol. Struct.*, 1978, **59**, 191.

92. T. J. Zielinski, D. L. Breen, and R. Rein, *J. Am. Chem. Soc.*, 1978, **100**, 6266.

93. G. Klopman, P. Andreozzi, A. J. Hopfinger, O. Kikuchi, and M. J. S. Dewar, *J. Am. Chem. Soc.*, 1978, **100**, 6267.

94. B. Y. Simkin, B. V. Golyanskii, and V. I. Minkin, *Zh. Org. Khim.*, 1981, **17**, 3.

95. S. N. Mohammed and A. J. Hopfinger, *Int. J. Quant. Chem.*, 1982, **22**, 1189.

96. D. B. Boyd, *J. Phys. Chem.*, 1978, **82**, 1407.

97. S. Bantle and R. Ahlrichs, *Chem. Phys. Lett.*, 1978, **53**, 148.

98. S. D. Beatty, S. D. Worley, and S. P. McManus, *J. Am. Chem. Soc.*, 1978, **100**, 4254.

99. S. P. McManus, R. M. Smith, M. B. Smith, and S. G. Schafer, *J. Comput. Chem.*, 1980, **1**, 233.

100. D. Van Hemelrijk, W. Versichel, C. Van Alsenoy, and J. H. Geise, *J. Comput. Chem.*, 1981, **2**, 63.

101. S. P. Zil'berg, A. I. Ioffe, and O. M. Nefedov, *Izv. Akad. Nauk. SSSR, Ser. Khim.*, 1983, 261.

102. P. Felker, D. M. Hayes, and L. A. Hull, *Theor. Chim. Acta*, 1980, **55**, 293.

103. M. J. S. Dewar, S. Olivella, and H. S. Rzepa, *J. Am. Chem. Soc.*, 1978, **100**, 5650.

104. P. R. Andrews and R. C. Haddon, *Aust. J. Chem.*, 1979, **32**, 1921.

105. D. Poppinger, *J. Am. Chem. Soc.*, 1975, **97**, 7486.

106. M. J. S. Dewar, *J. Am. Chem. Soc.*, 1984, **106**, 209.

107. A. Komornicki, J. D. Goddard, and H. F. Schaeffer III, *J. Am. Chem. Soc.*, 1980, **102**, 1763.

108. M. J. S. Dewar and W. Thiel, *J. Am. Chem. Soc.*, 1977, **99**, 4907.

109. M. J. S. Dewar and M. L. McKee, *J. Am. Chem. Soc.*, 1977, **99**, 5231.

110. M. J. S. Dewar and H. S. Rzepa, *J. Am. Chem. Soc.*, 1978, **100**, 777.

111. L. P. Davis, R. M. Guidry, J. R. Williams, M. J. S. Dewar, and H. S. Rzepa, *J. Comput. Chem.*, 1981, **2**, 433.

112. M. J. S. Dewar, M. L. McKee, and H. S. Rzepa, *J. Am. Chem. Soc.*, 1978, **100**, 3607.

113. M. J. S. Dewar and E. Healy, *J. Comput. Chem.*, 1983, **4**, 542.

114. M. J. S. Dewar, Y. Yamaguchi, and S. H. Suck, *Chem. Phys. Lett.*, 1977, **51**, 175.

115. M. J. S. Dewar, G. P. Ford, and H. S. Rzepa, *Chem. Phys. Lett.*, 1977, **50**, 262.

116. M. J. S. Dewar, Y. Yamaguchi, and S. H. Suck, *Chem. Phys.*, 1979, **55**, 145.

117. M. J. S. Dewar, G. P. Ford, M. L. McKee, H. S. Rzepa, W. Thiel, and Y. Yamaguchi, *J. Mol. Struct.*, 1978, **43**, 135.

118. M. J. S. Dewar, G. P. Ford, and H. S. Rzepa, *J. Mol. Struct.*, 1979, **51**, 275.

119. W. G. Bowman and T. G. Spiro, *J. Chem. Phys.*, 1980, **73**, 5482.

120. H. S. Rzepa, *J. Chem. Soc. Chem. Comm.*, 1981, 939.

121. S. Gabbay and H. S. Rzepa, *J. Chem. Soc. Faraday II*, 1982, **78**, 671.

122. H. Schmidt, A. Schweig, W. Thiel, and M. Jones, Jr., *Chem. Ber.*, 1978, **111**, 1958.

123. E. Veit, A. Schweig, and H. Vermeer, *Tetrahedron Lett.*, 1978, 2433.

124. D. Nelson, M. J. S. Dewar, J. M. Buschek, and E. McCarthy, *J. Org. Chem.*, 1979, **44**, 4109.

125. H. J. Chiang and S. D. Worley, *J. Electron. Spectrosc. Relat. Phenom.*, 1980, **21**, 121.

126. F. S. Joergensen, L. Carlsen, and F. Duus, *Acta Chem. Scand. [B]*, 1980, **34**, 695.

127. R. Gleiter and R. Bartetzko, *Z. Naturforsch. [B]*, 1981, **36**, 492.

128. F. S. Joergensen, *J. Chem. Res. (S)*, 1981, 212.

129. W. B. Jennings, D. Randall, S. D. Worley, and J. Hargis, *J. Chem. Soc. Perkin II*, 1981, 1411.

130. W. B. Jennings, J. H. Hargis, and S. Worley, *J. Chem. Soc. Chem. Comm.*, 1980, 30.

131. J. J. Dannanberg and D. Rocklin, *J. Org. Chem.*, 1982, **47**, 4529.

132. M. J. S. Dewar and M. L. McKee, *Inorg. Chem.*, 1978, **17**, 1569.

133. G. Wenke and D. Lenoir, *Tetrahedron*, 1979, **35**, 489.

134. K. Raghavachari, R. C. Haddon, P. v. R. Schleyer, and H. Schaefer III, *J. Am. Chem. Soc.*, 1983, **105**, 5915.

135. M. J. S. Dewar and H. S. Rzepa, *J. Am. Chem. Soc.*, 1978, **100**, 58.

136. M. J. S. Dewar and G. P. Ford, *J. Am. Chem. Soc.*, 1979, **101**, 5558.

137. S. Schneider, *Theor. Chim. Acta*, 1980, **57**, 71.

138. K. Y. Burstein and A. N. Isaev, *Theor. Chim. Acta*, 1984, **64**, 397.

139. H. Perrin and G. Berges, *THEOCHEM*, 1981, **1**, 299.

140. D. J. Belville and N. L. Bauld, *J. Am. Chem. Soc.*, 1982, **104**, 294.

141. M. J. S. Dewar and H. S. Rzepa, *J. Am. Chem. Soc.*, 1978, **100**, 784.

142. P. Bischof and G. Friedrich, *J. Comput. Chem.*, 1982, **3**, 486.

143. V. Barone, P. Cristinziano, F. Lelj, and N. Russo, *THEOCHEM*, 1982, **3**, 239.

144. M. J. S. Dewar and M. L. McKee, *J. Comput. Chem.*, 1983, **4**, 84.

145. W. S. Verwoerd, *J. Comput. Chem.*, 1982, **3**, 445.

146. M. J. S. Dewar and H. S. Rzepa, *J. Comput. Chem.*, 1983, **4**, 158.

147. M. J. S. Dewar and E. Healy, *J. Comput. Chem.*, 1983, **4**, 542.

148. R. Gleiter and K. Gubernator, *Chem. Ber.*, 1982, **115**, 3811.

149. V. I. Faustov and S. S. Yufit, *Zh. Fiz. Khim.*, 1982, **56**, 2226.

150. A. Shigihara, H. Ichikawa, and M. Tsuchiya, *Shitsuryo Buñseki*, 1982, **39**, 145.

151. J. G. Andrade, T. Clark, J. Chandrasekhar, and P. v. R. Schleyer, *Tetrahedron Lett.*, 1981, **22**, 2957.

152. M. N. Glukhotsev, B. Ya. Simkin, and I. A. Yudilevich, *Teor. Eksp. Khim.*, 1982, **18**, 726.

153. L. M. Loew and E. Sacher, *J. Macromol. Phys.*, 1978, *B*15, 619.

154. M. J. S. Dewar, A. H. Pakiari, and A. B. Pierini, *J. Am. Chem. Soc.*, 1982, **104**, 3242.

155. B. Ya. Simkin, M. E. Kletskii, R. M. Minyaev, and V. I. Minkin, *Zh. Org. Khim.*, 1983, **19**, 3.

156. H. S. Rzepa, *Tetrahedron*, 1981, **37**, 3107.

157. A. Sygula and A. Buda, *THEOCHEM*, 1983, **9**, 267.

158. J. R. Bews and C. Glidewell, *Inorg. Chim. Acta*, 1980, **39**, 217.

159. P. F. Dicks, S. A. Glover, A. Goosen, and C. W. McCleland, *S. Afr. J. Chem.*, 1984, **34**, 101.

160. W. S. Verwoerd, *Int. J. Quant. Chem.*, 1980, **18**, 1449.

161. D. J. Belville and N. L. Bauld, *J. Am. Chem. Soc.*, 1982, **104**, 5700.

162. P. Bischof, *Croat. Chem. Acta*, 1980, **53**, 51.

163. P. Bischof, *Helv. Chim. Acta*, 1980, **63**, 1434.

164. T. Sordo, M. Campillo, A. Oliva, and J. Bertran, *Chem. Phys. Lett.*, 1982, **85**, 225.

165. Y. Shinagawa, *Int. J. Quant. Chem. Quant. Biol. Symp.*, 1978, **5**, 269.

166. O. Kikuchi, A. J. Hopfinger, and G. Klopman, *Biopolymers*, 1980, **19**, 325.

167. S. N. Mohammed and A. J. Hopfinger, *J. Theor. Biol.*, 1980, **87**, 401.

168. G. P. Ford and J. D. Scribner, *J. Am. Chem. Soc.*, 1983, **105**, 349.

169. J. I. Seeman, R. Galzerano, K. Curtis, J. C. Schug, and J. M. Viers, *J. Am. Chem. Soc.*, 1981, **103**, 5982.

170. H. Bock, T. Hirabayashi, and S. Mohammed, *Chem. Ber.*, 1981, **114**, 2595.

171. J. A. Mosbo, R. K. Atkins, P. L. Bock, and B. N. Storhoff, *Phosphorus Sulfur*, 1981, **11**, 11.

172. R. Weiss, H. Wolf, U. Schubert, and T. Clark, *J. Am. Chem. Soc.*, 1981, **103**, 6142.

173. J. A. Defina and P. R. Andrews, *Int. J. Quant. Chem.*, 1980, **18**, 797.

174. B. Hoesterey, W. C. Neely, and S. D. Worley, *Chem. Phys. Lett.*, 1983, **94**, 311.

175. B. Maouche and J. Gayuso, *Int. J. Quant. Chem.*, 1983, **23**, 891.

176. M. Yanaka, S. Enomoto, Y. Inoue, and R. Chujo, *Bull. Chem. Soc. Japan*, 1981, **54**, 2420.

177. W. B. Jennings, J. M. Hargis, and S. D. Worley, *J. Chem. Soc. Chem. Comm.*, 1980, 30.

178. J. Mirek and A. Sygula, *THEOCHEM*, 1981, **3**, 85.

179. S. M. Adams, M. J. Murphy, and L. Kaminsky, *Mol. Pharmacol.*, 1981, **20**, 423.

180. M. J. S. Dewar and Y. Yamaguchi, *J. Comput. Chem.*, 1978, **2**, 25.

181. G. Buerni, A. Raudino, and F. Zuccarello, *THEOCHEM*, 1981, **1**, 113.

182. R. H. Conteras, D. G. DeKowaleski, and J. C. Facelli, *J. Mol. Struct.*, 1982, **81**, 147.

183. M. J. S. Dewar and D. J. Nelson, *J. Org. Chem.*, 1982, **47**, 2614.

184. S. Amirkhalili, R. Boese, U. Hoehner, D. Kampmann, G. Schmid, and P. Rademacher, *Chem. Ber.*, 1982, **115**, 732.

185. R. D. Mitchell and N. L. Bauld, *Isr. J. Chem.*, 1980, **29**, 319.

186. J. Tyrell, V. M. Kolb, and C. Y. Meyers, *J. Am. Chem. Soc.*, 1979, **101**, 3497.

187. A. W. Zwaard, A. M. Brouwer, and J. J. C. Mulder, *Recl. J. R. Neth. Chem. Soc.*, 1982, **101**, 137.

188. V. I. Faustov and S. S. Yufit, *Izv. Akad. Nauk SSSR Ser. Khim.*, 1982, 50.

189. M. Shansal, *Z. Naturforsch. [A]*, 1978, **33**, 1069.

190. C. Glidewell, *THEOCHEM*, 1981, **2**, 365.

191. W. Kuehnel, E. Gey, and H. J. Spangenberg, *Z. Phys. Chem. (Leipzig)*, 1982, **263**, 641.

192. C. Sourisseau and J. Hervieu, *J. Mol. Struct.*, 1977, **40**, 167.

193. M. Mehnert and J. K. Dohrmann, *Ber. Bunsenges. Phys. Chem.*, 1979, **83**, 825.

194. O. Kikuchi, S. Tanaka, T. Kanekiyo, K. Naruchi, and K. Yamada, *Kogakubu Kenkyu Hokuku (Chiba Daigaku)*, 1981, **32**, 53.

195. L. A. Paquette, H. C. Berk, C. R. Degenhardt, and G. D. Ewing, *J. Am. Chem. Soc.*, 1977, **99**, 4764.

196. M. J. S. Dewar, M. A. Fox, and D. J. Nelson, *J. Organomet. Chem.*, 1980, **185**, 157.

197. R. Czerminski, B. Lesyng, and A. Pohorille, *Int. J. Quant. Chem.*, 1979, **16**, 605.

198. O. Hofer, *Monatsschr. Chem.*, 1978, **109**, 405.

199. M. J. S. Dewar, R. C. Haddon, and P. J. Student, *J. Chem. Soc. Chem.. Comm.* 1974, 569.

200. R. Gandar, *Tetrahedron*, 1980, **36**, 1001.

201. W. I. Parsons, *J. Fluorine Chem.*, 1982, **21**, 445.

202. G. P. Ford and J. D. Scribner, *J. Am. Chem. Soc.*, 1981, **103**, 4281.

203. R. Arnaud, R. Subra, and V. Barone, *Nouv. J. Chim.*, 1982, **6**, 91.

204. E.-U. Wuerthwein, H. Halim, H. Schwarz, and N. M. M. Nibbering, *Chem. Ber.*, 1982, **115**, 2626.

205. M. J. S. Dewar, G. P. Ford, J. P. Ritchie, and H. S. Rzepa, *J. Chem. Res.(S)*, 1976, 26.

206. S. Olivella and J. Vilarrasa, *J. Heterocycl. Chem.*, 1979, **16**, 685.

207. N. A. Tausenko, V. G. Avakyan, and A. V. Belik, *Zh. Strukt. Khim.*, 1978, 541.

208. J. S. Yadav, H. Labischinski, G. Barnickel, and H. Bradaczek, *J. Theor. Biol.*, 1981, **88**, 441.

209. P. R. Andrews and G. P. Jones, *Int. J. Quant. Chem. Quant. Biol. Symp.*, 1979, **6**, 439.

210. P. R. Andrews and G. P. Jones, *Eur. J. Med. Chem.-Chim. Ther.*, 1981, **16**, 139.

211. T. W. Bentley and C. A. Wellington, *Org. Mass Spec.*, 1981, **16**, 523.

212. K. Lammertsma and P. v. R. Schleyer, *J. Am. Chem. Soc.*, 1983, **105**, 1049.

213. J. E. Ferrell, Jr., and G. H. Loew, *J. Am. Chem. Soc.*, 1979, **101**, 1385.

214. D. M. Hayes, S. D. Nelson, W. A. Garland, and P. A. Kollman, *J. Am. Chem. Soc.*, 1980, **102**, 1255.

215. T. Sordo, M. Arumi, and J. Bertran, *J. Chem. Soc. Perkin II*, 1981, 708.

216. J. T. Kleghorn, F. W. McKonkey, and K. Lundy, *J. Chem. Res. (S)*, 1978, 418.

217. T. Sordo, J. Bertran, and E. Canadell, *J. Chem. Soc. Perkin II*, 1979, 1486.

218. L. A. Paquette, F. Bellamy, G. J. Wells, M. C. Boehm, and R. Gleiter, *J. Am. Chem. Soc.*, 1981, **103**, 7122.

219. N. Z. Huang, T. C. W. Mak, and W.-K. Li, *Tetrahedron Lett.*, 1981, **22**, 3765.

220. G. Favini, C. Rubino, and R. Todeschini, *J. Mol. Struct.*, 1979, **53**, 267.

221. M. J. S. Dewar and S. Kirschner, *J. Am. Chem. Soc.*, 1975, **97**, 2932.

222. C. Cone, M. J. S. Dewar, and D. Landman, *J. Am. Chem. Soc.*, 1977, **99**, 372.

223. C. Decoret, J. Roger, and J. J. Dannenberg, *J. Org. Chem.*, 1981, **46**, 4074.

224. M. J. S. Dewar and D. Landman, *J. Am. Chem. Soc.*, 1977, **99**, 7439.

225. O. Kikuchi, A. Hiyama, H. Yoshida, and K. Suzuki, *Bull. Chem. Soc. Japan*, 1978, **51**, 11.

226. O. Kikuchi, K. Suzuki, and K. Tokumaru, *Bull. Chem. Soc. Japan*, 1979, **52**, 1086.

227. K. Yagamuchi, A. Nishio, S. Yabushita, and T. Fueno, *Chem. Phys. Lett.*, 1978, **53**, 109.

228. J. O. Noell and M. D. Newton, *J. Am. Chem. Soc.*, 1979, **101**, 51.

229. C. Glidewell, *J. Organomet. Chem.*, 1981, **217**, 273.

230. J. G. Andrade, J. Chandrasekhar, and P. v. R. Schleyer, *J. Comput. Chem.*, 1981, **2**, 207.

231. R. Gleiter, K. Gubernator, M. Eckert-Maksic, J. Spanget-Larsen, B. Bianco, G. Gardillon, and U. Burger, *Helv. Chim. Acta*, 1981, **64**, 1312.

232. S.-W. Chiu and W.-K. Li, *Croat. Chem. Acta*, 1981, **54**, 183.

233. R. Todeschini, D. Pitea, and G. Favini, *J. Mol. Struct.*, 1981, **71**, 279.

234. M. Rubio, A. Garcia-Hernandez, J. P. Daudy, R. Cetina, and A. Diaz, *J. Org. Chem.*, 1980, **45**, 150.

235. R. Todeschini and G. Favini, *J. Mol. Struct.*, 1980, **64**, 47.

236. T. C. W. Mak and W.-K. Li, *THEOCHEM*, 1982, **6**, 281.

237. G. Wencke and D. Lenoir, *Tetrahedron Lett.*, 1979, 2823.

238. M. J. S. Dewar and S. Kirschner, *J. Chem. Soc. Chem. Comm.*, 1975, 461.

239. M. J. S. Dewar and G. P. Ford, *J. Chem. Soc. Chem. Comm.*, 1977, 539.

240. A. H. Cowley, M. C. Cusher, M. Lattman, M. L. McKee, J. S. Szobota, and J. C. Wilburn, *Pure Appl. Chem.*, 1980, **52**, 789.

241. J. Janssen and W. Luettke, *J. Mol. Struct.*, 1982, **81**, 73.

242. B. Maouche, J. Gayoso, and O. Ouamerali, *C. R. Seances Acad. Sci.*, Ser.2, 1981, **293**, 141.

243. M. J. S. Dewar and M. L. McKee, *J. Am. Chem. Soc.*, 1978, **100**, 7499.

244. J. J. Rafelko, H. S. Rzepa, and B. I. Swanson, *J. Mol. Spectrosc.*, 1979, **75**, 363.

245. B. I. Swanson, J. J. Rafalko, H. S. Rzepa, and M. J. S. Dewar, *J. Am. Chem. Soc.*, 1977, **99**, 7829.

246. J. Martelli, F. Tonnard, R. Carrie, and R. Sustmann, *Nouv. J. Chim.*, 1978, **2**, 609.

247. W. W. Schoeller and E. Yurtsever, *J. Am. Chem. Soc.*, 1978, **100**, 7548.

248. W. W. Schoeller and N. Aktekin, *J. Chem. Soc. Chem. Comm.*, 1982, 20.

249. D. J. Belville, R. Chelsky, and N. L. Bauld, *J. Comput. Chem.*, 1982, **3**, 548.

250. M. J. S. Dewar and C. Doubleday, *J. Am. Chem. Soc.*, 1978, **100**, 4935.

251. F. Zuccarello, G. Buerni, and A. Raudino, *J. Mol. Struct.*, 1978, **50**, 183.

252. H. Iwamura, M. Iwai, and H. Kihara, *Chem. Lett.*, 1977, 881.

253. G. Deleris, J. P. Pillot, and J. C. Rayez, *Tetrahedron*, 1980, **36**, 2215.

254. G. Bouchoux and Y. Hoppilliard, *Int. J. Mass Spectrom. Ion Phys.*, 1983, **47**, 105.

255. R. Wolfschuetz, H. Halim and H. Schwarz, *Z. Naturforsch. [B]*, 1982, **37**, 734.

256. G. H. Loew, E. Kukjian, and M. Rabagliat, *Chem.-Biol. Interact.*, 1983, **43**, 33.

257. R. Nayori and H. Yamakawa, *Tetrahedron Lett.*, 1980, 2851.

258. H. Kollmar, F. Carrion, M. J. S. Dewar, and R. C. Bingham, *J. Am. Chem. Soc.*, 1981, **103**, 5292.

259. W.-K. Li and S. M. Rothstein, *Chem. Phys. Lett.*, 1978, **57**, 211.

260. W. W. Schoeller, *J. Am. Chem. Soc.*, 1979, **101**, 4811.

261. E. E. Waali, *J. Am. Chem. Soc.*, 1981, **103**, 3604.

262. A. F. Cuthbertson and C. Glidewell, *THEOCHEM*, 1982, **4**, 71.

263. U. Burger, G. Cardillon, and J. Mareda, *Helv. Chem. Acta*, 1981, **64**, 844.

264. M. J. S. Dewar, R. C. Haddon, W.-K. Li, W. Thiel, and P. K. Wiener, *J. Am. Chem. Soc.*, 1975, **97**, 4540.

265. C. S. Chung, *J. Chem. Soc. Faraday II*, 1976, **72**, 456.

266. V. I. Minkin, *Izv. Sib. Otd. Akad. Nauk, SSSR, Ser. Khim. Nauk*, 1980, 87.

267. E. P. Kyba, *J. Am. Chem. Soc.*, 1977, **99**, 8330.

268. R. Noyori, M. Yamakawa, and W. Ando, *Bull. Chem. Soc. Japan*, 1978, **51**, 811.

269. S.-W. Chiu and W.-K. Li, *J. Mol. Struct.*, 1980, **66**, 221.

270. M. J. S. Dewar, D. J. Nelson, P. B. Shevlin, and K. A. Biesiada, *J. Am. Chem. Soc.*, 1981, **103**, 2802

271. J. M. Figuera, P. B. Shevlin, and S. D. Worley, *J. Am. Chem. Soc.*, 1976, **98**, 3820.

272. M. C. Boehm, R. Gleiter, and P. Schang, *Tetrahedron Lett.*, 1979, 2575.

273. E.-U. Wuerthwein, J. Chandrasekhar, E. D. Jemmis, and P. v. R. Schleyer, *Tetrahedron Lett.*, 1981, 843.

274. D. C. Crans and J. P. Snyder, *J. Am. Chem. Soc.*, 1980, **102**, 7152.

275. J. Chandrasekhar and P. v. R. Schleyer, *J. Chem. Soc. Chem. Comm.*, 1981, 260.

276. J. Chandrasekhar, E.-U. Wuerthwein, and P. v. R. Schleyer, *Tetrahedron*, 1981, **37**, 921.

277. Y. Sawaki, H. Kato, and Y. Ogata, *J. Am. Chem. Soc.*, 1981, **103**, 3832.

278. L. Carlsen, J. P. Snyder, A. Holm, and E. Pedersen, *Tetrahedron*, 1981, **37**, 1257.

279. G. Frenking, J. Schmidt, and H. Schwarz, *Z. Naturforsch. [B]*, 1982, **37**, 355.

280. H. Lischka and H. J. Koehler, *J. Am. Chem. Soc.*, 1978, **100**, 5297.

281. W. A. M. Castenmiller and H. M. Buck, *Rec. Trav. Chim. Pays Bas*, 1977, **96**, 207.

282. M. Tasaka, M. Ogata, and H. Ichikawa, *J. Am. Chem. Soc.*, 1981, **103**, 1885.

283. C. W. Jefford, J. Mareda, J. C. Perberger, and U. Burger, *J. Am. Chem. Soc.*, 1979, **101**, 1370.

284. W. L. Jorgensen, *J. Am. Chem. Soc.*, 1977, **99**, 4272.

285. G. Frenking and H. Schwarz, *J. Comput. Chem.*, 1982, **3**, 251.

286. G. Bouchoux and Y. Hoppilliard, *Int. J. Mass Spectrom. Ion Phys.*, 1982, **43**, 63.

287. S. P. McManus and S. D. Worley, *Tetrahedron Lett.*, 1977, 555.

288. S. P. McManus, *J. Org. Chem.*, 1982, **47**, 3070.

289. J. M. Harris, S. G. Schafer, and S. D. Worley, *J. Comput. Chem.*, 1982, **3**, 208.

290. G. Frenking, J. Schmidt, and H. Schwarz, *Z. Naturforsch. [B].*, 1980, **35**, 1031.

291. N. L. Bauld, J. Cessac, and R. L. Holloway, *J. Am. Chem. Soc.*, 1977, **99**, 8140.

292. R. C. Haddon, F. Wudl, M. L. Kaplan, J. H. Marshall, R. E. Cais, and F. B. Bramwell, *J. Am. Chem. Soc.*, 1978, **100**, 7629.

293. C. F. Wilcox, Jr., I. Szele, and D. E. Sunko, *Tetrahedron Lett.*, 1975, 4457.

294. M. J. S. Dewar and H. S. Rzepa, *J. Am. Chem. Soc.*, 1979, **99**, 7432.

295. M. J. S. Dewar and G. P. Ford, *J. Mol. Struct.*, 1979, **51**, 281.

296. P. v. R. Schleyer and J. Chandrasekhar, *J. Org. Chem.*, 1981, **46**, 225.

297. F. Lelj, G. Sindona, and N. Uccella, *Ann. Chim. (Rome)*, 1977, **67**, 773.

298. A. Corvers, P. C. H. Scheers, W. A. M. Castenmiller, and H. M. Buck, *Tetrahedron*, 1978, **34**, 457.

299. K. B. Astin, *Tetrahedron Lett.*, 1980, **21**, 3713.

300. S. H. Al-Khowaiter and C. A. Wellington, *Tetrahedron*, 1977, **33**, 2843.

301. J. A. Hashmall, V. Horak, L. E. Khoo, C. O. Quicksall, and M. K. Sun, *J. Am. Chem. Soc.*, 1981, **103**, 289.

302. W. L. Jorgensen, *J. Am. Chem. Soc.*, 1977, **99**, 280.

303. J. L. Ginsberg and R. F. Langler, *Can. J. Chem.*, 1983, **61**, 589.

304. P. Bischof, *Angew. Chem.* 1976, **88**, 609.

305. D. M. Storch, C. J. Dymek, J. Chester, Jr., and L. P. Davis, *J. Am. Chem. Soc.*, 1983, **105**, 1765.

306. J. R. Bews and C. Glidewell, *J. Mol. Struct.*, 1980, **64**, 87.

307. I. Tvaroska, *Coll. Czech. Chem. Comm.*, 1982, **47**, 3199.

308. M. J. S. Dewar, G. P. Ford, M. L. McKee, H. S. Rzepa, and L. E. Wade, *J. Am. Chem. Soc.*, 1977, **99**, 5069.

309. P. Brant, J. A. Hashmall, F. L. Carter, R. A. DeMarco, and W. B. Fox, *J. Am. Chem. Soc.*, 1981, **103**, 329.

310. P. Brant, A. D. Berry, R. A. Demarco, F. L. Carter, and W. B. Fox, *J. Electron. Spectrosc. Rel. Phenom.*, 1981, **22**, 119.

311. M. J. S. Dewar and I. J. Turchi, *J. Chem. Soc. Perkin II*, 1977, 724.

312. D. R. Arnold, P. C. Wong, A. J. Maroulis, and T. S. Cameron, *Pure Appl. Chem.*, 1980, **52**, 2609.

313. M. Duran and J. Bertran, *J. Chem. Soc. Perkin II*, 1982, 681.

314. M. J. S. Dewar and H. W. Kollmar, *J. Am. Chem. Soc.*, 1975, **97**, 2933.

315. R. C. Haddon and G. R. J. Williams, *J. Am. Chem. Soc.*, 1975, **97**, 6582.

316. A. Schweig and W. Thiel, *Tetrahedron Lett.*, 1978, 1841.

317. A. Schweig and W. Thiel, *J. Am. Chem. Soc.*, 1979, **101**, 4742.

318. O. Kikuchi, *Bull. Chem. Soc. Japan*, 1982, **55**, 1669.

319. K. U. Ingold and J. C. Walton, *J. Chem. Soc. Chem. Comm.*, 1980, 604.

320. Y. Ogata, K. Tomizawa, K. Furuta, and H. Kato, *J. Chem. Soc. Perkin II*, 1981, 110.

321. N. Heinrich, R. Wolfschuetz, G. Frenking, and H. Schwarz, *Int. J. Mass Spectrom. Ion Phys.*, 1982, **44**, 81.

322. J. Spanget-Larsen and R. Gleiter, *Angew. Chem.*, 1978, **90**, 471.

323. T. Clark and P. v. R. Schleyer, *Nouv. J. Chim.*, 1978, **2**, 665.

324. G. W. Klumpp, J. Fleischhauer, and W. Schleker, *J. R. Neth. Chem. Soc.*, 1982, **101**, 208.

325. R. Houriet, H. Schwarz, W. Zummack, J. G. Andrade, and P. v. R. Schleyer, *Nouv. J. Chim.*, 1981, **5**, 505.

326. R. W. Alder and W. Grimme, *Tetrahedron*, 1981, **37**, 1809.

327. R. L. Al-Nia'mi and M. Shansal, *Z. Naturforsch. [A]*, 1980, **35**, 129.

328. S. P. Zil'berg, A. I. Ioffe, and O. M. Nefedov, *Izv. Akad. Nauk, SSSR, Ser. Khim.*, 1982, 2481.

329. T. Clark, G. W. Spitznagel, R. Klose, and P. v. R. Schleyer, *J. Am. Chem. Soc.*, 1984, **106**, 4412.

330. J. A. Pincock and R. J. Boyd, *Can. J. Chem.*, 1977, **55**, 2482.

331. G. Frenking, L. Huelskaemper, and P. Weyerstahl, *Chem. Ber.*, 1982, **115**, 2826.

332. M. A. Fox, C. C. Cheng, and K. A. Campbell, *J. Org. Chem.*, 1983, **48**, 321.

333. A. T. Pudzianowski and G. H. Loew, *J. Am. Chem. Soc.*, 1980, **102**, 5443.

334. A. T. Pudzianowski and G. H. Loew, *J. Mol. Catal.*, 1982, **17**, 1.

335. A. T. Pudzianowski and G. H. Loew, *Int. J. Quant. Chem.*, 1983, **23**, 1257.

336. A. T. Pudzianowski, G. H. Loew, B. A. Mico, R. V. Branchflower, and L. R. Pohl, *J. Am. Chem. Soc.*, 1983, **105**, 3434.

337. M. H. Palmer, W. J. Ross, J. S. Kwiatowski, and B. Lesyng, *THEOCHEM*, 1983, **9**, 283.

338. A. Buda and A. Sygula, *THEOCHEM*, 1983, **9**, 255.

339. H. E. Audier, G. Bouchoux, M. Fetizon, P. Jaudon, and J. C. Tabet, *Tetrahedron*, 1981, **37**, 3121.

340. T. Minato and S. Yamabe, *J. Org. Chem.*, 1983, **48**, 1479.

341. V. I. Faustov and S. S. Yufit, *Izv. Akad. Nauk, SSSR, Ser. Khim.*, 1979, 2797.

342. J. M. Ribo, M. D. Masip, and A. Valles, *Monatsschr. Chem.*, 1981, **112**, 359.

343. D. Wilhelm, T. Clark, and P. v. R. Schleyer, *J. Chem. Soc. Chem. Comm.*, 1983, 211.

344. M. J. S. Dewar, G. P. Ford, and H. S. Rzepa, *J. Chem. Soc. Chem. Comm.*, 1977, 728.

345. M. J. S. Dewar and S. Kirschner, *J. Chem. Soc. Chem. Comm.*, 1975, 463.

346. W. S. Verwoerd and F. J. Kok, *Surf. Sci.*, 1979, **80**, 89.

347. D. Wilhelm, T. Clark, and P. v. R. Schleyer, *J. Chem. Soc. Perkin II*, 1984, 915.

348. D. Wilhelm, T. Clark, P. v. R. Schleyer, K. Buckl, and G. Boche, *Chem. Ber.*, 1983, **116**, 1669.

349. D. Burkholder, W. E. Jones, K. W. Ling, and J. S. Wasson, *Theor. Chim. Acta*, 1980, **55**, 325.

350. H. Prinzbach, G. Sedelmaier, C. Krueger, R. Goddard, H. D. Martin, and R. Gleiter, *Angew. Chem.*, 1978, **90**, 297.

351. R. Chadha and N. K. Ray, *Theor. Chim. Acta*, 1982, **60**, 573.

352. E. L. Andersen, R. L. DeKock, and T. P. Fehlner, *J. Am. Chem. Soc.*, 1980, **102**, 2644.

353. Y. Apeloig and D. Arad, *J. Am. Chem. Soc.*, 1981, **103**, 4258.

354. A. G. Turner, *Inorg. Chim. Acta*, 1982, **65**, L201.

355. K. J. Miller, K. F. Moschner, and K. T. Potts, *J. Am. Chem. Soc.*, 1983, **105**, 1705.

356. K. Riemenschneider, H. Bartels, W. Eichel, and P. Boldt, *Tetrahedron Lett.*, 1979, 189.

357. H. D. Scharf, H. Plum, J. Fleischhauer, and W. Schleker, *Chem. Ber.*, 1979, **112**, 862.

358. C. H. Bushweller, B. J. Laurenzi, J. G. Brennan, M. J. Goldberg, and R. P. Marcantonio, in "*Stereodyn. Mol. Syst. Proc. Symp.*," R. H. Sarma, Ed., Pergamon, Elmsford, NY, 1979, p.113.

359. S. Bohm and J. Kuthan, *Coll. Czech. Chem. Comm.*, 1981, **46**, 2068.

360. J. R. Bews and C. Glidewell, *THEOCHEM*, 1982, **7**, 151.

361. B. L. Bublev and L. S. Chuiko, *Vysokomol. Soedin Ser. A*, 1982, **22**, 1481.

362. P. Letchken, *Chem. Ber.*, 1978, **111**, 1413.

363. J. C. Barnes, J. D. Paton, J. R. Damewood, Jr., and K. Mislow, *J. Org. Chem.*, 1981, **46**, 4975.

364. G. Marconi, G. Orlandi, and G. Poggi, *J. Photochem.*, 1982, **19**, 329.

365. K. Yamaguchi, S. Yabushita, T. Fueno, S. Kato, and K. Morokuma, *Chem. Phys.Lett.*, 1980, **79**, 27.

366. V. Galasso, E. Montoneri, and G. C. Pappalardo, *THEOCHEM*, 1981, **1**, 43.

367. J. M. Schulman and R. L. Disch, *J. Am. Chem. Soc.*, 1978, **100**, 5677.

368. M. C. A. Dankersloot and H. M. Buck, *J. Am. Chem. Soc.*, 1981, **103**, 6554.

369. M. C. A. Dankersloot and H. M. Buck, *J. Am. Chem. Soc.*, 1981, **103**, 6549.

370. P. R. Andrews, *A. C. S. Symp. Ser.*, 1979, **112** (Computer assisted drug design), 149.

371. V. N. Kokorev, N. N. Vyshinskii, V. P. Maslenikov, I. A. Abronin, G. M. Zhidomirov, and Yu. A. Aleksandrov, *Zh. Strukt. Khim.*, 1982, **23**, 13.

372. L. L. Combs and M. Rossi, *Spectroscop. Lett.*, 1976, **9**, 495.

373. K. Ishida and S. Mayama, *Theor. Chim. Acta*, 1983, **62**, 245.

374. M. Martin Munoz, *An. Quim.*, 1978, **74**, 1456.

375. S. Olivella and R. Caballol, *THEOCHEM*, 1982, **5**, 45.

376. M. Arai, T. Ando, M. Tanura, and Y. Tadao, *Kogyo Kayaku*, 1980, **41**, 8.

377. R. M. Guidry and L. P. Davis, *Model Simul.*, 1978, **9**, 331.

378. W. R. Carper, L. P. Davis, and M. W. Extine, *J. Phys. Chem.*, 1982, **86**, 459.

379. R. M. Minyaev and V. I. Natanzon, *Izv. Sev. Nauk. Nauchn. Tsentra Vyssh. Shk. Estestv. Nauk.*, 1980, 55.

380. G. E. Scuseria, A. R. Engelman, and R. H. Contreras, *Theor. Chim. Acta*, 1982, **61**, 49.

381. M. J. S. Dewar and H. S. Rzepa, *Inorg. Chem.*, 1979, **18**, 602.

382. M. A. Masson, A. Bouchy, G. Roussy, G. Serratrice, and J. J. Delpuech, *J. Mol. Struct.*, 1980, **68**, 307.

383. A. H. Cowley, R. A. Kemp, M. Lattman, and M. L. McKee, *Inorg. Chem.*, 1982, **21**, 85.

384. R. M. Minyaev and M. E. Kletskii, *Teor. Eksp. Khim.*, 1980, **16**, 368.

385. J. Capdevila and E. Canadell, *J. Heterocycl. Chem.*, 1981, **18**, 1055.

386. P. R. Andrews, M. N. Iskander, C. P. Jones, and D. A. Winkler, *Int. J. Quant. Chem. Quant. Biol. Symp.*, 1982, **9**, 345.

387. A. M. Butkus, *Synth. Meth.*, 1980, **2**, 215.

388. G. Boeck, H. J. Koehler, and C. Weiss, *Z. Chem.*, 1980, **20**, 455.

389. H. Bock, B. Solouki, G. Fritz, and W. Hoeldrich, *Z. Anorg. Allg. Chem.*, 1979, **458**, 53.

390. E. Osawa, Y. Onaki, and K. Mislow, *J. Am. Chem. Soc.*, 1981, **103**, 7475.

391. A. Ferse, K. Lunkwitz, J. Morgenstern, E. Gey, and R. Vetter, *Z. Phys. Chem. (Leipzig)*, 1983, **264**, 181.

392. M. J. S. Dewar and M. L. McKee, *Inorg. Chem.*, 1978, **16**, 1075.

393. N. K. Ray and R. Chadra, *Proc. Indian Acad. Sci. [Ser.]: Chem. Sci.*, 1982, **91**, 129.

394. R. Chadra and N. K. Ray, *J. Indian Chem. Soc.*, 1982, **59**, 204.

395. N. K. Ray, *Indian J. Chem. Sect. A*, 1981, **20**, 492.

396. J. A. Fleischhauer, *Z. Naturforsch. [A]*, 1977, **32**, 1564.

397. J. A. Fleischhauer, G. Raabe, and H. Thiele, *Z. Naturforsch. [A]*, 1978, **33**, 1230.

398. S. O. Paul and T. A. Ford, *J. Mol. Struct.*, 1980, **61**, 373.

399. M. Kanoda, H. Shinoda, K. Kobashi, J. Hase, and S. Nagahara, *J. Pharmacabiodyn.*, 1982, **5**, S-49.

400. O. Kikuchi, T. Kanekiyo, S. Tanaka, K. Naruchi, and K. Yamada, *Bull. Chem. Soc. Japan*, 1982, **55**, 1509.

401. A. M. Lobo, S. Prabhakar, M. R. Tavares, and H. S. Rzepa, *Tetrahedron Lett.*, 1981, **22**, 3007.

402. K. Kirste and P. Rademacher, *J. Mol. Struct.*, 1981, **73**, 171.

403. Y. Nakajima, M. Ogata, and H. Ichikawa, *Shitsuryo Bunseki*, 1980, **28**, 243.

404. M. F. Teitell, S. H. Suck, and J. L. Fox, *Theor. Chim. Acta*, 1981, **60**, 127.

405. C. Glidewell, *J. Organomet. Chem.*, 1981, **217**, 11.

406. J. R. Bews and C. Glidewell, *THEOCHEM*, 1982, **3**, 217.

407. P. Bischof, *J. Am. Chem. Soc.*, 1977, **99**, 8145.

408. N. Aktekin, H. O. Pamuk, and C. Trindle, *Chim. Acta Turc.*, 1981, **9**, 47.

409. C. Kooistra, L. A. Sluyterman, and H. M. Buck, *Recl. J. R. Neth. Chem. Soc.*, 1982, **101**, 396.

410. Y. Hoppilliard, G. Bouchoux, and P. Jaudon, *Nouv. J. Chim.*, 1982, **6**, 43.

411. C. Glidewell and G. S. M. Mollison, *J. Mol. Struct.*, 1981, **72**, 203.

412. O. Kikuchi, *Tetrahedron Lett.*, 1980, 1055.

413. M. Teitell and J. M. Fox, *Int. J. Quant. Chem.*, 1982, **22**, 583.

414. J. Jalonen, J. Taskinen, and C. Glidewell, *Int. J. Mass Spectrom. Ion Phys.*, 1983, **46**, 243.

415. J. R. Bews and C. Glidewell, *THEOCHEM*, 1983, **8**, 353.

416. I. Tvaroska, S. Bystricky, P. Malon, and K. Blaha, *Coll. Czech. Chem. Comm.*, 1982, **47**, 17.

417. V. Z. Gabdrakipov, I. A. Slygina, and O. V. Agashkin, *Zh. Fiz. Khim.*, 1980, **54**, 491.

418. W. B. Jennings and S. D. Worley, *Tetrahedron Lett.*, 1977, 1435.

419. N. Aktekin, H. O. Pamuk, and C. Trindle, *Chim. Acta Turc.*, 1981, **9**, 127.

420. N. L. Bauld and D. Wirth, *J. Comput. Chem.*, 1981, **2**, 1.

421. J. E. Bartmess, *J. Am. Chem. Soc.*, 1982, **104**, 335.

422. H. Bock, R. Dammel, and L. Horner, *Chem. Ber.*, 1981, **114**, 220.

423. P. Livant, M. L. McKee, and S. D. Worley, *Inorg. Chem.*, 1983, **22**, 895.

424. F. Y. Hansen and H. L. McMurry, *J. Mol. Struct.*, 1979, **57**, 209.

425. M. Eckert-Maksic and Z. B. Maksic, *J. Chem. Soc. Perkin II*, 1981, 1462.

426. W. Thiel, *J. Am. Chem. Soc.*, 1981, **103**, 1420.

427. A. Schweig and W. Thiel, *J. Am. Chem. Soc.*, 1981, **103**, 1425.

428. W. Thiel, *J. Chem. Soc. Faraday II*, 1980, **76**, 302.

429. J. M. Bonicam, *Univ. Microfilms Int.*, Order no. 7818348 (*Diss. Abstr. Int. B*, 1978, **39**, 1759).

430. E. Haselbach, T. Bally, Z. Lanyiova, and P. Baertschi, *Helv. Chem. Acta*, 1979, **62**, 583.

431. J. J. C. Van Lier, M. T. Smits, and H. M. Buck, *Eur. J. Biochem.*, 1983, **132**, 55.

432. A. P. Zeif, V. D. Sokolowski, and P. I. Vadash, *React. Kin. Catal. Lett.*, 1979, **10**, 71.

433. G. Frenking, F. Marschner, and H. Goetz, *Phosphorus Sulfur*, 1980, **8**, 343.

434. J. F. Garvey and J. A. Hashmall, *J. Org. Chem.*, 1978, **43**, 2380.

435. M. J. S. Dewar, R. C. Haddon, W.-K. Li, W. Thiel, and P. K. Weiner, *J. Am. Chem. Soc.*, 1975, **97**, 4540.

436. M. J. S. Dewar and W. Thiel, *J. Am. Chem. Soc.*, 1975, **97**, 3978.

437. G. Klopman and P. Andreozzi, *Bull. Chem. Soc. Belg.*, 1977, **86**, 481.

438. A. F. Cuthbertson and C. Glidewell, *Inorg. Chim. Acta*, 1981, **49**, 91.

439. M. C. Boehm and R. Gleiter, *Z. Naturforsch. [B]*, 1981, **36**, 498.

440. C. Roberts and J. C. Walton, *J. Chem. Soc. Perkin II*, 1981, 553.

441. T. Oie, G. H. Loew, S. K. Burt, J. S. Binkley, and R. D. MacElroy, *J. Am. Chem. Soc.*, 1982, **104**, 6169.

442. G. Bertholon, C. Decoret, M. Perrin, J. Roger, A. Thozet, and B. Tinland, *Int. J. Quant. Chem.*, 1981, **19**, 1167.

443. R. Arnaud, J. Douady, and R. Subra, *Nouv. J. Chim.*, 1981, **5**, 181.

444. G. R. Pack, *Cancer Biochem. Biophys.*, 1981, **5**, 183.

445. H. Goetz, G. Frenking, and F. Marschner, *Phosphorus Sulfur*, 1978, **4**, 309.

446. J. A. Mosbo, R. K. Atkins, P. L. Bock, and B. N. Storhoff, *Phosphorus Sulfur*, 1981, **11**, 11.

447. J. T. DeSanto, J. A. Mosbo, B. N. Storhoff, P. L. Bock, and R. E. Bloss, *Inorg. Chem.*, 1980, **19**, 3086.

448. R. M. Minyaev and V. I. Minkin, *Zh. Strukt. Khim.*, 1979, **20**, 842.

449. D. E. Cabelli, *Univ. Microfilms Int.*, Order no. 8009836, (*Diss. Abstr. Int. B*, 1980, **40**, 5257).

450. P. G. Nelson, *J. Chem. Res. (S)*, 1980, **3**, 106.

451. E.-U. Wuerthwein and P. v. R. Schleyer, *Angew. Chem.*, 1979, **91**, 598.

452. D. T. Clark and M. Z. Abraham, *J. Polym. Sci.,Polym. Chem. Ed.*, 1981, **19**, 2689.

453. B. J. Orchard, S. K. Tripethy, A. J. Hopfinger, and P. J. Taylor, *J. Appl. Phys.*, 1981, **52**, 5949.

454. T. Yamabe, *Kyoto Daigaku Genshiro Jikkenso* (*Techn. Rep.*), 1982, KURRI-TR-217, 48-51.

455. D. T. Clark and H. S. Munro, *Polym. Degrad. Stab.*, 1982, **4**, 83.

456. A. A. Jones, J. F. O'Gara, P. T. Inglefield, J. T. Bendler, A. F. Yee, and K. L. Ngai, *Macromolecules*, 1983, **16**, 658.

457. I. Ohmine and K. Morokuma, *J. Chem. Phys.*, 1981, **74**, 564.

458. A. P. Zeif and V. D. Sokolovskii, *J. Mol. Catal.*, 1980, **9**, 1.

459. F. M. Menger, J. Grossman, and D. C. Liotta, *J. Org. Chem*, 1983, **48**, 905.

460. S. Olivella and J. Vilarrasa, *J. Heterocycl. Chem.*, 1981, **18**, 1189.

461. G. Orlandi and K. Schulten, *Chem. Phys. Lett.*, 1979, **64**, 370.

462. M. Eckert-Maksic, *J. Chem. Soc. Perkin II*, 1981, 62.

463. G. Frenking, H. Goetz, and F. Marschner, *Phosphorus Sulfur*, 1979, **7**, 295.

464. V. I. Minkin and R. M. Minyaev, *Zh. Org. Khim.*, 1979, **15**, 225.

465. R. Pearlman and N. Bodor, *A. C. S. Symp. Ser.*, 1979, **112** (Computer assisted drug design), 489.

466. A. Karpfen, P. Schuster, and H. Berner, *J. Org. Chem.*, 1979, **44**, 374.

467. J. Mirek and A. Sygula, *Z. Naturforsch. [A].*, 1982, **37**, 1276.

468. P. Bischof, in "*Comput. Methods Chem.*," (Proc.Int.Symp.), J. Bargon, Ed., Plenum, New York, 1980, p.133.

469. M. J. S. Dewar and S. Olivella, *J. Am. Chem. Soc.*, 1979, **101**, 4958.

470. O. Kikuchi, Y. Sato, and K. Suzuki, *Bull. Chem. Soc. Japan*, 1980, **53**, 2675.

471. T. Clark, *J. Am. Chem. Soc.*, 1979, **101**, 7746.

472. S. M. Khalil, *Iraqi J. Sci.*, 1978, **19**, 67.

473. S. M. Khalil, *Egypt. J. Chem.*, 1978, **21**, 465.

474. J. Spanget-Larsen, *J. Mol. Struct.*, 1979, **51**, 301.

475. R. V. Metzger, *J. Chem. Phys.*, 1981, **74**, 3458.

476. Y. Hoppilliard and G. Bouchoux, *Org. Mass. Spectrom.*, 1982, **17**, 534.

477. J. Giordan, H. Bock, M. Eiser, and H. W. Roesky, *Phosphorus Sulfur*, 1982, **13**, 19.

478. S. F. Nelsen, E. Haselbach, R. Gschwind, U. Klemm, and S. Lanyova, *J. Am. Chem. Soc.*, 1978, **100**, 4367.

479. D. Wilhelm, T. Clark, P. v. R. Schleyer, J. L. Courtneidge, and A. G. Davies, *J. Am. Chem. Soc.*, 1984, **106**, 361.

480. W. A. M. Castenmiller and H. M. Buck, *Tetrahedron*, 1979, **35**, 397.

481. W. S. Verwoerd, *Surf. Sci.*, 1980, **99**, 581.

482. L. A. Gibov, M. E. Elyashberg, and M. M. Raikshlat, *J. Mol. Struct.*, 1979, **53**, 81.

483. S. P. McManus, M. R. Smith, and S. G. Schafer, *J. Comput. Chem.*, 1982, **3**, 229.

484. M. Shibata, T. J. Zielinski, and R. Rein, *Int. J. Quant. Chem.*, 1980, **18**, 323.

485. A. Mondragon and A. O. Blake, *Int. J. Quant. Chem.*, 1982, **22**, 89.

486.   M. Askari and K. Karimian, *THEOCHEM*, 1983, **9**, 347.

487.   J. J. Oleksik and A. G. Turner, *Inorg. Chim. Acta*, 1982, **59**, 165.

488.   M. J. S. Dewar and G. P. Ford, *J. Am. Chem. Soc.*, 1979, **101**, 783.

489.   A. J. Bracuti and Y. P. Carignan, Report, 1980, ARLCD-YTR-80002, AD-E00435; Order No. AD-086092, Available from NTIS; (cited in *Gov. Rep. Announce. Index (US)*, 1980, **80**, 4579).

490.   S. D. Worley, J. H. Hargis, L. Chang, and W. B. Jennings, *Inorg. Chem.*, 1981, **20**, 2339.

491.   H. Bock, G. Braehler, U. Henkel, R. Schlecker, and D. Seebach, *Chem. Ber.*, 1980, **113**, 289.

492.   E. Egert, H. J. Lindner, W. Hillen, and M. C. Boehm, *J. Am. Chem. Soc.*, 1980, **102**, 3707.

493.   H.-J. Bestmann, *Pure Appl. Chem.*, 1980, **52**, 771.

494.   R. C. C. Perera and B. L. Henke, *J. Chem. Phys.*, 1979, **70**, 5398.

495.   S. F. Nelsen, C. R. Kessel, and P. J. Brien, *J. Am. Chem. Soc.*, 1980, **102**, 702.

496.   E. Kaufmann, H. Mayr, J. Chandrasekhar, and P. v. R. Schleyer, *J. Am. Chem. Soc.*, 1981, **103**, 1375.

497.   J. M. Lluch and J. Bertran, *Tetrahedron*, 1979, **35**, 2601.

498.   D. Wilhelm, T. Clark, P. v. R. Schleyer, and A. G. Davies, *J. Chem. Soc. Chem. Comm.*, 1984, 558.

499.   W. Neugebauer, A. J. Kos, and P. v. R. Schleyer, *J. Organomet. Chem.*, 1982, **228**, 107.

500.   P. v. R. Schleyer and E.-U. Wuerthwein, *J. Chem. Soc. Chem. Comm.*, 1982, 542.

501.   W. W. Schoeller and U. H. Brinker, *J. Am. Chem. Soc.*, 1978, **100**, 6012.

502.   T. Clark, A. J. Kos, P. v. R. Schleyer, W. H. DeWolf, and F. Bickelhaupt, *J. Chem. Soc. Chem. Comm.*, 1983, 685.

503.   D. Wilhelm, T. Clark, and P. v. R. Schleyer, *Tetrahedron Lett.*, 1983, 3985.

504.   T. Clark, D. Wilhelm, and P. v. R. Schleyer, *Tetrahedron Lett.*, 1982, 3547.

505.   K. Schoetz, T. Clark, H. Schaller, and P. v. R. Schleyer, *J. Org. Chem.*, 1984, **49**, 735.

506.   W. Neugebauer, T. Clark, and P. v. R. Schleyer, *Chem. Ber.*, 1983, **116**, 3283.

507.   A. F. Cuthbertson, C. Glidewell, and D. C. Liles, *THEOCHEM*, 1982, **4**, 273.

508.   S. F. Nelsen, L. A. Grezzo, and W. C. Hollinsed, *J. Org. Chem.*, 1981, **46**, 283.

509.   R. Gleiter, R. Haider, and H. Quast, *J. Chem. Res. (S)*, 1978, 138.

510.   P. v. R. Schleyer, *Pure Appl. Chem.*, 1983, **55**, 355.

511.   P. v. R. Schleyer, *Pure Appl. Chem.*, 1984, **56**, 151.

512.   M. J. S. Dewar, *Chemistry in Britain*, 1975, **11**, 97.

513.   M. J. S. Dewar, E. F. Healy, and J. J. P. Stewart, *J. Comput. Chem.*, 1984, **5**, 358.

CHAPTER 5

AB INITIO METHODS

5.1. *AB INITIO* MOLECULAR ORBITAL THEORY

The term *ab initio* implies a rigorous, nonparametrized molecular orbital treatment derived from first principles. This is not completely true. There are a number of simplifying assumptions in *ab initio* theory, but the calculations are more complete, and therefore more expensive, than those of the semiempirical methods discussed in Chapter 4. It is possible to obtain chemical accuracy *via ab initio* calculations, but the cost in computer time is enormous, and only small systems can be treated this accurately at present. In practice most calculations are performed at lower levels of theory than would be considered definitive, and the shortcomings are taken into account.

Like the semiempirical calculations, *ab initio* theory makes use of the Born–Oppenheimer approximation that the nuclei remain fixed on the time scale of electron movement, that is, that the electronic wave function is unaffected by nuclear motion. This is a very good approximation in nearly all cases. Only for an extremely flat potential surface, as, for instance, in some Jahn–Teller systems, may significant coupling exist between the vibrational and electronic wave functions.

The first stage of all *ab initio* calculations to be considered here is a single-determinant LCAO–SCF calculation, as outlined in Section 3.1. In contrast to the NDO methods, however, there are many different possible choices of atomic orbitals (the *basis set*). Almost all modern *ab initio* calculations employ *Gaussian type orbital* (GTO) basis sets. These bases, in which each atomic orbital is made up of a number of Gaussian probability functions, have considerable advantages over other types of basis set for the evaluation of one- and two-electron integrals. They are therefore much faster computationally than, for instance, equivalent Slater orbitals. The GAUSSIAN series of programs deals, as the name implies, exclusively with Gaussian-type orbitals and includes several optional GTO basis sets of varying size. This is one of the main advantages of such a widely distributed program system: The methods and basis sets used become standard and a direct comparison with literature data is often possible.

The simplest of the optional basis sets in GAUSSIAN82 are the STO-nG bases, of which STO-3G is the only one to have found wide use, although the bases STO-2G to STO-6G were originally tested.[1] STO-nG is an abbreviation for *Slater-Type-Orbitals* simulated by n *G*aussian functions each. This means that each atomic orbital consists of n Gaussian functions added together. The coefficients of the Gaussian functions are selected so as to give as good a fit as possible to the corresponding Slater-type orbitals. In early tests on the STO-nG bases it was found that basis sets with $n \geqslant 3$ gave very similar results, so the smallest of these, STO-3G, was chosen for an extensive series of calculations in the early 1970s. STO-3G is a *minimal basis set*. This means that it has only as many orbitals as are necessary to accommodate the electrons of the neutral atom. Because a complete set of p-orbitals must be added to maintain spherical symmetry, the elements boron to neon each have five atomic orbitals: $1s$, $2s$, $2p_x$, $2p_y$, and $2p_z$. For beryllium and lithium a minimal basis set actually requires only $1s$- and $2s$-orbitals. In STO-3G, however, the three $2p$-orbitals are also included for these elements in order to give a consistent description across the periodic table. Because there is only one best fit to a given type of Slater orbital ($1s$, $2p$, etc.) for each number of Gaussian functions, all STO-3G basis sets for any row of the periodic table are identical except for the *exponents* of the Gaussian functions. These are expressed as a *scale factor*, the square of which is used to multiply all exponents in the original best-fit Gaussian functions. In this way the ratios of the exponents of the individual Gaussians to each other remain constant, but the effective exponent of the entire orbital can be varied. The STO-nG basis sets were developed by optimizing the scale factors for each element until a minimum (most negative) value for the total energy was found. This process is usually performed for the ground-state atoms, but in the STO-nG bases the scale factors normally used are based on a series of calculations on small molecules. The exponents, or scale factors, can be considered to be a measure of the extent of the orbital. A low exponent indicates a diffuse (and therefore relatively high-energy) orbital; high exponents indicate compact orbitals close to the nucleus.

The STO-3G basis set is very economical, having only one *basis function* (or atomic orbital) per hydrogen atom (the $1s$), five per atom from Li to Ne ($1s$, $2s$, $2p_x$, $2p_y$, and $2p_z$) and nine per atom for the second-row elements

Na–Ar ($1s$, $2s$, $2p_x$, $2p_y$, $2p_z$, $3s$, $3p_x$, $3p_y$, and $3p_z$). Note that the fully occupied shells, or core orbitals, are included in the calculation, in contrast to the valence-electron approximation used by MINDO/3 and MNDO. The performance of STO-3G was extensively compared with that of MINDO/3 in the discussion that followed the announcement of the latter method,[2-4] and a comprehensive collection of STO-3G geometries was published in a review article.[5] Although STO-3G was found to perform better than MINDO/3, at least for the problems selected, and although it was the standard basis set for *ab initio* optimizations for several years, it was eventually replaced by small split-valence basis sets and is now hardly used. Its great weaknesses proved to be overestimation of the stability of small rings, overemphasis of the π-acceptor characteristics of electropositive elements of the first row, and total failure for the second-row electropositive elements, especially sodium. These weaknesses can be traced back to some of the simplifications used to derive the basis set, and serve to illustrate the improvements inherent in split-valence bases.

The greatest problem of any minimal basis set is its inability to expand or contract its orbitals to fit the molecular environment. Consider an oxygen *p*-orbital in water and in planar H_3O^+. In the former, the *p*-orbital perpendicular to the molecular plane is doubly occupied, with the two electrons being attracted by a total of 10 nuclear charges (8 for oxygen and 1 each for the two hydrogens) and repelled by eight other electrons.

In planar H_3O^+, on the other hand, the orbital remains unchanged, but its two electrons are now attracted by 11 nuclear charges and repelled by only eight other electrons. The orbital can therefore be stabilized by contraction to bring the electrons closer to the nuclei than in water, in which there is one less nuclear charge (contraction also increases the electron–electron repulsion). A minimal basis set, however, does not have the capability to expand or contract the orbital because the exponent is fixed. This restriction leads to compromise solutions for most molecules, and severely influences comparisons between charged and uncharged species. The problem also manifests itself for significantly anisotropic molecules. The lone-pair orbitals in water, for instance, need to be more diffuse than the OH bonding orbitals. In minimal-basis-set calculations, however, the same atomic orbitals must be used for both types of molecular orbitals.

One solution to the problem is to use *split-valence* or *double zeta* basis sets. In these bases the atomic orbitals are split into two parts, an inner, compact orbital and an outer, more diffuse one. The coefficients of these two types of orbital can be varied independently during construction of the molecular orbitals in the SCF procedure. Thus the size of the atomic orbital that contributes to the molecular orbital can be varied within the limits set by the inner and outer basis functions, as shown in Fig. 5.1.1. Split-valence basis sets split only the valence orbitals in this way, whereas double zeta bases also have split core orbitals ("double zeta" simply implies two different exponents). The split-valence basis set most widely used for the early calculations was 4-31G. This nomenclature means that the core orbitals consist of *4* and the inner and outer valence orbitals of *3* and *1* Gaussian functions, respectively. The most common procedure in the early 1970s was to perform geometry optimizations with the STO-3G basis set and then to do *single point* calculations (a single SCF calculation without geometry optimization) on the STO-3G optimized geometry using 4-31G. A common convention is to label such calculations 4-31G//STO-3G, where the basis set named before the double slash is that used for the single-point calculation at a geometry optimized using the basis set named after the double slash. The grouping of several Gaussians together in one set of functions with constant coefficients relative to each other is called *contraction*. The 4-31G basis set for a second-row element might be described as being contracted to 4431/ 431. This means that the Gaussian functions (*primitive Gaussians*) are grouped together in shells of *4*, *4*, *3*, and *1* for the $1s$, $2s$, $3s_I$, and $3s_O$ orbitals, respectively, and *4*, *3*, and *1* for $2p$, $3p_I$, and $3p_O$, respectively (the subscripts I and O indicate inner and outer orbitals). The relative weights of the primitive Gaussians within the individual orbitals remain constant, but the coefficients of the orbitals as a whole can be varied.

An important feature of the standard basis sets in GAUSSIAN82 and the earlier GAUSSIAN programs is that the exponents of the s- and p-functions

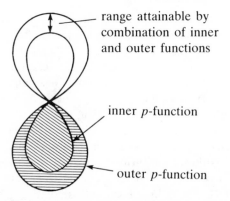

range attainable by combination of inner and outer functions

inner p-function

outer p-function

FIG. 5.1.1. Schematic representation of the effect of split valence orbitals. The size of the orbital can be varied between the limits set by the inner and outer functions.

within any given shell are equal. This restriction results in a certain loss of flexibility, but allows faster integral processing and therefore saves computer time. In practice, this constraint does not appear to be important for basis sets containing an adequate number of primitive Gaussians, although small bases may benefit from the use of separate exponents for *s*- and *p*-orbitals. The contraction scheme shown above for 4-31G does not take the common exponent for *s*- and *p*-orbitals in each shell into account, but rather describes independent *s*- and *p*-shells.

The advent of optimization procedures that use analytical gradients (see below) led to the development of split-valence basis sets with fewer primitive Gaussians than 4-31G. The basis set used most commonly for geometry optimizations is now 3-21G, which uses three primitive Gaussians for the core orbitals and a two/one split for the valence functions. Because the procedures used to calculate the atomic forces are very sensitive to the number of primitive Gaussians, a 3-21G optimization can be up to twice as fast as the same calculation with 4-31G, although the difference is not large for single-point calculations. 3-21G has now replaced STO-3G for all but the largest molecules, and is the standard basis set for initial geometry optimizations. 3-21G is available for all elements from hydrogen to chlorine, whereas the GAUSSIAN82 program contains the 4-31G basis set for only hydrogen to fluorine. The 4-31G basis for phosphorus, sulfur, and chlorine has been published,[6] but is not included as a standard option in the program. Care should be taken when using literature values for 4-31G calculations on beryllium. The original basis set[7] contained an error, which was corrected in a later publication.[8] The 4-31G basis for lithium and beryllium is actually a 5-21G basis because a problem known as "falling in" was encountered during the optimization of the basis set. "Falling in" means that the valence orbitals tend to collapse inward in order to improve the description of the core electrons, but at the cost of a very poor representation of the valence orbitals. This problem is especially acute for basis sets with few (less than five) primitive Gaussians for the core orbitals. It was avoided when constructing the 3-21G bases by optimizing the valence functions for a 6-21G basis set and then optimizing a three-Gaussian core using the fixed valence functions from 6-21G.

The next step in improving a basis set is usually the addition of *d*-orbitals for all heavy (nonhydrogen) atoms. For most organic compounds these do not function as *d*-orbitals in the normal sense of being involved in bond formation as in transition-metal compounds. Their purpose is far more to allow a shift of the center of, for instance, a *p*-orbital away from the position of the nucleus. This *polarization* is illustrated in Fig. 5.1.2. Mixing the *d*-orbital, which has lower symmetry, with the *p*-orbital results in a deformation of the resulting orbital to one side of the atom. This adjustment is particularly important for compounds containing small rings and for compounds of the second-row elements. The most commonly used *polarization basis set* (i.e., one including *d*-orbitals) in the GAUSSIAN82 program is

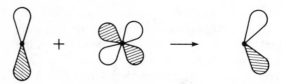

FIG. 5.1.2.   Polarization of a p-orbital by mixing with a d-function.

6-31G*. This basis uses six primitive Gaussians for the core orbitals, a three/one split for the s- and p-valence orbitals, and a single set of six d-functions (indicated by the asterisk). Six d-functions (equivalent to five d- and one s-orbital) are used for computational convenience, although the GAUSSIAN programs can also handle basis sets with five real d-orbitals. A further development is the 6-31G** basis, in which a set of p-orbitals has been added to each hydrogen in the 6-31G* basis set. p-Orbitals perform the same function for the s-valence orbital of hydrogen as the d-orbitals do for p-valence orbitals. The most common use of 6-31G* is for single-point calculations on geometries optimized with a smaller basis, usually 3-21G. Advances in computer technology and programs are, however, already making 6-31G* optimizations on fairly large systems possible, so that 3-21G may well suffer the same fate as STO-3G within a few years. Larger basis sets such as 6-311G*, which has triply split (three/one/one) valence orbitals, and basis sets with two sets of d-functions are now being used increasingly often for single-point calculations.

One further type of basis set contained in the standard GAUSSIAN82 program, the *diffuse function augmented* bases, is intended for use in calculations on anions or molecules that require very good descriptions of nonbonding electron pairs. These basis sets are obtained by adding a single set of very diffuse s- and p-orbitals (with exponents between 0.1 and 0.01) to the heavy atoms in a standard basis such as 6-31G*. This basis is then designated 6-31+G*, or 6-31++G* if diffuse s-functions are added also to hydrogen. The purpose of the diffuse functions is to improve the basis set at large distances from the nucleus and thus describe the high-energy electron pairs associated with anions better. The 3-21+G basis set,[9] which is not included as an option in GAUSSIAN82, has been used as a standard basis for anion calculations in the same way as 3-21G has for neutral or positively charged species.

The number of basis functions rises rapidly with increasing sophistication of the basis set. This is important for two reasons, first because the number of basis functions the program can handle is limited and second because the computer time required is roughly proportional to the fourth power of the number of basis functions. Table.5.1.1 shows the number of basis functions per atom for some of the basis sets available in GAUSSIAN82. The 6-31G* and 6-31G** basis sets have six d-orbitals per atom, the other polarization basis sets only five.

TABLE 5.1.1
Numbers of Basis Functions per Atom for Some Common Basis Sets.

	Basis Set						
Atom	STO-3G	3-21G	3-21G(*)	3-21+G	6-31G*	6-31G**	6-311G*
H	1	2	2	2	2	5	3
Li–Ne	5	9	9	13	15	15	18
Na–Ar	9	13	18	17	19	19	22

After selection of an appropriate basis set for the problem at hand (usually the largest practicable), the type of calculation to be performed should be chosen. The normal practice for *ab initio* calculations differs from that used with semiempirical methods, in which almost all calculations are performed with full geometry optimization. Because of the cost of *ab initio* calculations and the possibility of improving the basis set, it is often necessary to optimize the geometry with a small basis set and then perform single-point calculations with better bases or with a correction for electron correlation. This practice allows energy calculations at levels of theory that are too high for practical full geometry optimization. Usually the results are good approximations to those that would be obtained through a full geometry optimization at the higher level, although in some cases the inclusion of *d*-orbitals can lead to large changes in the structure, so that single-point calculations on a geometry obtained with a nonpolarization basis set may be misleading. A single-point UHF calculation using the 6-31G* basis set on a 3-21G optimized geometry may be conveniently labeled UHF/6-31G*//3-21G. The corresponding designation for a fully optimized 6-31G* structure is UHF/6-31G*//6-31G*.

The usual choice of basis set for a first geometry optimization is 3-21G, followed by single-point 6-31G* calculations, probably with some sort of correction for electron correlation. For some types of problems, however, more sophisticated basis sets are needed to obtain realistic results. As mentioned above, diffuse-augmented (+) basis sets are necessary for anion calculations or for problems, such as the calculation of proton affinities, in which an adequate description of lone pairs is important. For problems involving three-membered rings or the electronegative elements of the second-row of the periodic table (Si–Cl), the inclusion of *d*-orbitals is critical. Inversion barriers are also consistently calculated to be too low (and pyramidal structures too flat) if polarization functions are not used. This problem is particularly acute for small diffuse-augmented basis sets such as 3-21+G. The 3-21G(*) basis set offered as an option in GAUSSIAN82 (*via* the keyword "3-21G*") is relatively economical for use with the second-row because it includes *d*-orbitals only for Na–Cl, not for Li–F. The 3-21G(*)

basis set for the first-row elements is thus identical to 3-21G. Although it is not included as a standard option in GAUSSIAN82 the minimal MINI-1 basis set[10] has been found to give surprisingly good results, and should be considered for systems that are too large for 3-21G optimizations. Although the basis is no larger than STO-3G (three primitive Gaussians per orbital), it gains some extra flexibility by using different exponents for the *s*- and *p*-functions in a given shell. This makes calculations with MINI-1 slower than those with STO-3G, but gives a better description of many molecules. The use of nonstandard basis sets will be illustrated in the input examples for GAUSSIAN82.

An *ab initio* calculation is in many ways very similar to the corresponding MNDO or MINDO/3 job. Geometry optimizations may be performed in three different ways with GAUSSIAN82, however. The first of these methods, commonly known as "BERNY" optimization (after H. Bernhard Schlegel, who developed the method[11] and wrote the original program), is usually the fastest, and is selected automatically by GAUSSIAN82 if no other mode of optimization is specified. BERNY optimization uses analytically calculated atomic forces and a guessed force-constant matrix, which is continuously updated during the optimization, to predict the position of the minimum-energy structure. The calculation of atomic forces generally requires approximately the same amount of time as the SCF calculation that precedes it. BERNY optimization is very fast and effective for classical, acyclic molecules, but may either be very slow or fail entirely for molecules with unusual force constants or cyclic structures. These problems can often be overcome either by reading in a modified set of force constants or by the use of an appropriate strategy in writing the *Z*-matrix, as outlined in Section 3.2. If, however, the geometry optimization still fails, the second type of optimization available in GAUSSIAN82, the Murtagh–Sargent method,[12] should be used. This method, which does not rely on a guessed force-constant matrix, is usually slower but more reliable than BERNY. It too uses analytically evaluated atomic forces, but employs a different strategy to predict the minimum-energy structure. The third optimization method available in GAUSSIAN82, the Fletcher–Powell method,[13] should be used only for calculations for which no analytical atomic forces are available. It will be selected automatically by the program in these cases. The atomic forces are then evaluated by the finite-difference method, as shown in Section 2.1. This procedure can be used with any type of energy calculation, but is extremely slow in comparison with methods that use analytical gradients (forces). The strategy used to convert the atomic forces to geometry changes is similar to the Davidon–Fletcher–Powell method used by MOPAC and other MNDO and MINDO/3 programs, except that the semiempirical programs use analytical forces, rather than those calculated by finite difference. Some versions of GAUSSIAN76 also use this combination of analytical forces with a Davidon–Fletcher–Powell algorithm, which is not as fast as BERNY, but gives very reliable optimizations.

Having reached the minimum-energy structure, the program then performs a Mulliken population analysis similar to that in the MOPAC program. Overlap populations, atomic charges and so forth are, however, often misleading in large-basis-set *ab initio* calculations. The main problem is that the population analysis is based on the coefficients of the atomic orbitals in the individual molecular orbitals. But the larger, more diffuse orbitals may perform functions other than those for which they are intended. The extent of these orbitals is so great that they reach into regions of space that would normally be assigned to other atoms. They may therefore be used to improve the description of an atom other than the one on which they are centered. In that case, however, the population of the orbital involved is naturally assigned to the atom on which it is centered, not to the one that is actually using the basis function. This problem can be solved by dividing the molecule into regions of space assigned to the individual atoms. The total electron density in the region assigned to each atom can then be integrated to calculate the atomic charge, which, however, is then dependent on the way in which the molecule is divided. Such an analysis is not included as a standard feature of GAUSSIAN82, but can be added to some earlier programs in the GAUSSIAN series. This use of spare basis functions from a neighboring atom by electron-rich centers also has energetic consequences. A fluorine atom alone, for instance, gives a worse energy with an inadequate basis set than when associated with a lithium atom, from which it can "borrow" extra basis functions. These energetic consequences of this basis-set imbalance are known as *basis set superposition error* (BSSE) and can be corrected for by using the counterpoise technique, which will be illustrated in Section 5.3.

The next step in many calculations is to apply some sort of correction for electron correlation to the SCF energy. The GAUSSIAN programs have traditionally used Rayleigh–Schrödinger many-body perturbation theory (RSMBPT) as applied to molecular systems by Møller and Plesset[14] and implemented by Pople and his coworkers.[15] These methods, which can be terminated at second (MP2), third (MP3), or fourth order (MP4), are based on perturbation theory and, like many methods for calculating the correlation energy, rely on a good description of the virtual orbitals in the original SCF wave function. The calculated correlation energy is therefore dependent on the quality of the basis set, so that 3-21G, for instance, is not normally used for Møller–Plesset (MP) calculations. Møller–Plesset calculations do not give the full correlation energy (the MP2 correction is normally estimated to be about half the total correlation energy), but are very fast in comparison with classical configuration interaction calculations and appear to reproduce the energetic effects of correlation well in comparisons between molecules. To save some computer time and space, the core orbitals are normally omitted from MP calculations. This technique was not used in many earlier MP2 calculations, so one should check MP2 energies from the literature carefully before comparing them with values calculated more recently. Geometry optimization with MP2 may sometimes be necessary. This is possible with both the BERNY and the Murtagh–Sargent methods, but in these cases the core orbitals are included in the MP2 correction. A single-point calculation using the keyword "MP2" will automatically give a calculation without the core-orbital contribution to the MP2 energy, but an optimization with the same keyword will give the full MP2 calculation. The keyword "MP2=FULL" can be used to specify the complete calculation in a single-point. In general, polarization basis sets are needed to obtain good results with MP2 optimizations.[16]

One of the disadvantages of MP2 and the other correlation methods included in GAUSSIAN82 is that they deliver only a corrected total energy and no other information. In many cases a population analysis of the correlated wave function or an indication of the main electronic state being mixed into the SCF wave function would provide useful information. Electron correlation will be treated in more detail in Section 5.4.

There are several good bibliographies of *ab initio* calculations. The series by Richards et al.[17] and the recent compilation by Ohno and Morokuma[18] list the published calculations by molecular formula. The GAUSSIAN programs and the methods and basis sets available in them have recently been described by Hehre et al.,[19] who also assessed the performance at each level of theory for typical problems. Finally, the *Carnegie–Mellon Quantum Chemistry Archive*[20] is an invaluable source of data for STO-3G, 3-21G, and 6-31G* calculations.

5.2. THE GAUSSIAN PROGRAMS

The first program in the GAUSSIAN series, GAUSSIAN70,[21] was introduced in 1970 and made available *via* QCPE in 1971. It was able to perform single-point calculations or optimizations by cyclic variation of all parameters using Gaussian basis sets containing *s*- and *p*-orbitals. The most well-known standard basis sets to be built into the program were STO-3G and 4-31G. Two features of GAUSSIAN70, its speed and the simplicity of the input structure, made it the first *ab initio* program to find wide acceptance and to be used extensively outside the laboratory in which it was written. GAUSSIAN70 was limited to calculations including up to 70 basis functions, 35 atoms, and no *d*-orbitals. The optimization procedure used involved changing a geometrical parameter by a small amount and recalculating the energy. This procedure was repeated to obtain a third point, and the three total energies were then used in a parabolic extrapolation to predict an optimum value for the parameter being varied. This process had to be repeated again and again for each geometrical parameter until a constant geometry and energy were obtained. This not only required a great deal of computer time, but also demanded a certain amount of expertise on the part of the user if the optimization was to be carried out effectively.

Some structures can never be optimized in this way because their energies may sink only when two or more parameters are varied at once. Nevertheless, an extensive "STO-3G chemistry" arose from GAUSSIAN70, with geometry optimizations normally being performed with STO-3G and single-point calculations with 4-31G.

The next program in the series, GAUSSIAN76,[22] appeared 5 years later and was able to handle *d*-orbitals. The QCPE versions of the program have no post-SCF (MP2, etc.) options and are limited to cyclic optimizations as in GAUSSIAN70. Many versions of GAUSSIAN76 were, however, converted to enable them to perform MP2 calculations and other types of optimization. The first of these was a Fletcher–Powell procedure in which the atomic forces (gradients) were calculated by finite difference.[13] This was very slow by today's standards, but represented a tremendous acceleration in comparison with cyclic optimizations. It was also an "automatic" procedure that could usually be relied on to find the minimum without aid from the user. In some GAUSSIAN76 installations this was later replaced by optimization procedures that used analytically evaluated atomic forces. Many updated versions of GAUSSIAN76 are still being used.

GAUSSIAN76 was replaced in 1980 by GAUSSIAN80, the last GAUSSIAN program to be distributed by QCPE. GAUSSIAN80 was markedly changed from the GAUSSIAN76 program, notably in the input structure, geometry optimization, and post-SCF sections. The route cards used to specify the options to be used in the calculation in GAUSSIAN76 were largely replaced by keywords, and the variables in the *Z*-matrix were given symbolic names. The optimization procedures in GAUSSIAN80 were essentially those in GAUSSIAN82, except that analytical gradients with post-SCF procedures were not available. GAUSSIAN80 was the first GAUSSIAN program with post-SCF procedures to be distributed by QCPE. Most of the examples given in this chapter for GAUSSIAN82 will also run on GAUSSIAN80.

GAUSSIAN82[23] is the successor to GAUSSIAN80. GAUSSIAN82 is a considerably more precise program than GAUSSIAN80, and includes additions to the post-SCF procedures and to the range of basis sets available. Many of the tolerances (cutoffs for integrals, SCF convergence criteria, etc.) have been made smaller than those in GAUSSIAN80. A new package for calculating electrostatic properties is also included. This program, which will be available for VAX, CDC, IBM, and CRAY1 computers, is available directly from Professor J. A. Pople at Carnegie–Mellon University. GAUSSIAN82, unlike MMP2, is supplied as a source (FORTRAN) version with a very comprehensive manual. The cost of the program is reasonable, although the software agreement naturally places restrictions on the distribution of the program and its derivatives. One feature of GAUSSIAN82 that increases its usefulness enormously is the fact that it makes a record (the archive) of all successful calculations. The *Carnegie–Mellon Quantum Chemistry Archive*[20] contains all such records from Professor Pople's group, a vast amount of information (geometries, energies, dipole moments, etc.) for a large number of calculations (over 20,000). The *Archive* is available in printed form from Professor Pople or may be accessed directly by telephone on the VAX 11/780 at Carnegie–Mellon. The BROWSE program used to process the *Archive* allows individual calculations to be accessed, tables of all calculations on a given stoichiometry with a given basis set to be made, and so on. Details of this service are available from Professor Pople.

GAUSSIAN82 is made up of a series of *links*, each of which is a separate program for some computers. Each link performs its own specific task within the overall calculation and communicates with the other links *via* a series of disk files. The links are grouped together in *overlays*, of which there are 12 in the current GAUSSIAN82 program. These overlays are not what is normally understood by this term (a section of the program that is loaded all at once at run time), but are simply sets of links that are grouped together for convenience and normally use a common set of control options. The concept of overlays and links is important when writing nonstandard routes for GAUSSIAN82. GAUSSIAN82 links are identified by three- or four-digit numbers. The last two digits are the number of the link within the overlay and the first one or two give the number of the overlay. Thus link 301 is link 1 in overlay 3, and so on. The overlays in GAUSSIAN82 are arranged as follows:

Overlay 0. Overlay 0 performs two functions: It sets up some of the the disk files and machine parameters needed to run the job, and it defines the running sequence of the individual links by interpreting the keywords or route cards in terms of program options.

Overlay 1. Overlay 1 is the geometry overlay. It is responsible for reading the geometry input (usually the title, charge, multiplicity and Z-matrix) and for controlling the optimization procedures. The program returns to overlay 1 to calculate new Z-matrices during an optimization. There are five links in overlay 1: *link 101*, which reads the input; *links 102, 103*, and *105*, which control Fletcher–Powell, BERNY, and Murtagh–Sargent optimization procedures, respectively; and *link 106*, which calculates a force-constant matrix by finite-difference. This last link is seldom used.

Overlay 2. Overlay 2 is responsible for converting the Z-matrix read or produced by overlay 1 into Cartesian coordinates for the nondummy atoms, transforming these coordinates to a standard orientation with the center of mass at the origin, and determining the symmetry (point group and framework group) of the molecule. This symmetry information is then used by later overlays to save time in the calculation of integrals, to determine the symmetries of the molecular orbitals and the electronic state and so on.

Overlay 3. Overlay 3 is divided into *link 301*, which specifies the basis set, and *links 302, 303, 307, 310, 312, 314*, and *316*, which calculate the various types of integrals and their first derivatives. Link 301 may either assign basis functions to the atomic centers using one of the internally stored basis sets or construct the basis set from exponents and coefficients supplied in the input. Links 302, 311, and 314 are usually used by GAUSSIAN82 to calculate one-electron integrals, two electron-integrals for s- and p-orbitals, and two-electron integrals for s-, p-, d-, and f-orbitals, respectively. Links 312 or 310 may be used instead of links 311 and 314, but are not as sophisticated, and are used mainly for test purposes. Link 303 calculates the dipole integrals, which are required only if a population analysis is to be performed. Links 307 and 316 calculate the first derivatives of the one- and two- electron integrals for use in calculating the atomic forces. This task is normally performed by overlay 7, but these two links are used for MP2 optimizations and some other types of calculation.

Overlay 4. The purpose of overlay 4, which contains only one link, *link 401*, is to produce an initial guess (a trial set of molecular orbitals) for the SCF procedure to use as a starting point. Link 401 can produce several types of initial guess, but the normal procedure is to perform an extended Hückel or INDO calculation and to project the orbitals thus obtained to fit the basis set to be used in the *ab initio* calculation. Other alternatives include initial guesses based on an extended Hückel calculation without projection or produced by diagonalization of the core Hamiltonian. Link 401 may also alter the electronic configuration by swapping the occupancies of two or more molecular orbitals in the initial guess.

Overlay 5. Overlay 5 differs from most of the other overlays in GAUSSIAN82 in that its constituent links do not share a common set of options (the variables that control the links once they have been called). This is not important, as only one link is ever used in any one call to overlay 5, so that the options need not be transferable. The separate links in overlay 5 perform different types of SCF calculation, and can be regarded as separate programs. *Link 501* is the closed-shell (RHF) link. This link is that normally used for systems with no unpaired electrons. *Link 502* is the UHF equivalent of link 501 for open-shell systems. *Link 503* performs either UHF or RHF calculations by the steepest-descent (or direct minimization, SCFDM) technique. This method is practically certain to converge, but may be extremely slow. It can also be relied on to retain the electronic configuration of the initial guess, whereas links 501 and 502 may change it during the SCF calculation. *Link 505* is used less often than links 501 and 502 because it performs RHF calculations for open-shell systems (ROHF). This procedure is useful where spin contamination makes the results of UHF calculations unreliable, but the ROHF procedure in GAUSSIAN82 is not compatible with the post-SCF links. Link 505 may also give convergence problems for systems that converge easily using link 502.

Overlay 6. Overlay 6 is responsible for the analysis of the wave function produced by overlay 5. *Link 601* performs a population analysis, giving eigenvalue and eigenvector tables, overlap populations, orbital populations, and the dipole moment. It also calculates orbital spin densities and the

Fermi contact terms for open-shell species. Note that even if one of the post-SCF procedures is used, link 601 gives an analysis of the single-determinant wave function, not that given by the correlation method. *Link 602*, the second link in overlay 6, calculates a series of electric properties, multipole moments, and so on from the SCF wave function. It may also be used to generate a map of the electrostatic potential above the surface of the molecule.

Overlay 7. Overlay 7 is the first- and second-derivative equivalent of overlay 3, and also calculates the atomic forces using these derivatives. *Links 701, 702*, and *703* calculate the first derivatives of the integrals calculated by links 302, 311, and 312, respectively. These three links are used in SCF optimizations with analytical forces. MP2 optimizations use links 307 and 316. *Links 707* and *708* are the second derivative equivalents of *links 701*-3, and *link 716* converts the first derivatives to atomic forces.

Overlays 8 and 9. These two overlays perform the post-SCF calculations. The two links in overlay 8 set up and perform the transformation from atomic to molecular integrals. *Link 901* carries out the initial stages of all post-SCF procedures, and can calculate the MP2 energy. *Links 903, 904*, and *905* are "in-core" programs for closed-shell, open-shell, and complex MP2 calculations, respectively. In-core programs avoid the large input–output overhead of the more general programs, but can handle systems only up to a limited size. *Links 909-913* perform MP3, MP4, or CI calculations. An important feature of overlay 9 is *link 902*, which tests the stability of the Hartree–Fock wave function, allowing it to become unsymmetrical or complex and allowing an RHF wave function to become UHF. *Link 918* produces a new initial guess for a less constrained wave function if link 902 detects an instability.

Overlay 10. This overlay consists of two links, *1001* and *1002*, which calculate the derivatives of the MP2 and CI energies for optimizations with either of these methods.

Overlay 99. This is the final overlay in any GAUSSIAN82 job. It "cleans up" on machines that do not do so automatically, generates and writes an archive entry if the calculation was successful, and finishes the job with one of the many quotations stored in the program.

5.3. GAUSSIAN82 INPUT AND OUTPUT EXAMPLES

The following GAUSSIAN82 input and output examples were chosen to illustrate some techniques used in *ab initio* calculations and to show many features of the output. The techniques used, such as counterpoise and FOGO calculations, are less important; they were chosen in order to illustrate the input for jobs that require something more than simply changing the keywords. The examples are intended to complement those given in the GAUSSIAN82 manual, but some overlap is inevitable, especially for the simple calculations. Much of the discussion of the output draws comparisons with the MOPAC outputs given in Chapter 4. The program used for these examples was a CDC version derived from the original (Release A) VAX 11/780 source, and so there may be small differences in the outputs compared with those obtained with other versions of the program.

Figure 5.3.1(a) shows one possible input for a simple GAUSSIAN82 job, the optimization of CH_2 within C_{2v} symmetry using the 3-21G basis set. The job begins with a keyword or route card that specifies the options to be used by the program. A "#" rather than "#P" at the beginning of this line would give less comprehensive output printing than will be used in this example. The program options, separated by spaces, are then specified. "RHF/3-21G" requests a restricted Hartree–Fock calculation with the 3-21G basis set, and "OPT" requests a geometry optimization with the default method, which is BERNY in this case. The title for the molecule then appears after a blank card. The title is used in the archive entry for the job, and so should be as informative as possible. The point group need not be given in the title because it is determined by the program and added to the archive entry. The title may be continued on further cards, up to a maximum of five, but must be terminated by a blank card. The next line of the input gives the charge (0) and the multiplicity (1) for the molecule. The multiplicity corresponds exactly to the multiplicity of the electronic state: one for a singlet, two for a doublet, three for a triplet, and so on. GAUSSIAN82 checks that the charge and multiplicity are compatible with the total number of atoms in the molecule and stops the calculation if this is not the case. The Z-matrix that follows is exactly analogous to that given in Fig. 3.3.2 except that the bond

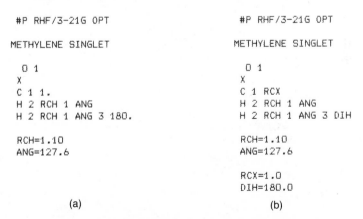

```
#P RHF/3-21G OPT                        #P RHF/3-21G OPT

METHYLENE SINGLET                       METHYLENE SINGLET

   0 1                                     0 1
X                                       X
C 1 1.                                  C 1 RCX
H 2 RCH 1 ANG                           H 2 RCH 1 ANG
H 2 RCH 1 ANG 3 180.                    H 2 RCH 1 ANG 3 DIH

RCH=1.10                                RCH=1.10
ANG=127.6                               ANG=127.6

                                        RCX=1.0
                                        DIH=180.0

        (a)                                     (b)
```

FIG. 5.3.1. Two possible GAUSSIAN82 inputs for methylene: (a) giving constant geometrical parameters in the Z-matrix; (b) using a separate block to assign valuables to the geometric constants.

lengths and angles have been replaced by the *symbolic variables* RCH and ANG. Symbolic variables, which can be assigned any name as long as it begins with a letter, are assigned values in the cards following the blank card that ends the Z-matrix. In this case the two CH bond lengths are given the name RCH, and thus held symmetrical, and RCH is assigned an initial value of 1.1 Å . Similarly the angles ANG are defined as measuring 127.6°. If some geometric variables are to be held constant during the optimization, they may be given numerical values in the Z-matrix, as for the 180° dihedral angle in Fig. 5.3.1(a), or they may be assigned values in a separate list of variables after those to be optimized, as shown for the same calculation in Fig. 5.3.1(b). In either case the input is terminated by a blank card.

The output produced by the above calculation is shown in Fig. 5.3.2. The program first identifies itself with a header giving the version used, the date, and the starting time. The keywords used are then given, and the program next converts these to a route, which is simply a series of cards that define which links are to be called, in which order, and with which options. The GAUSSIAN82 concept of overlays and links is important for understanding the route. Each line of the route corresponds to one overlay, the number of which is the first one given. All numbers between the two slashes are options to be used by the links within the overlay. The numbers after the

```
                      HELLO !!!


        ********************************************************
        GAUSSIAN 82: REVISION B: CDC VERSION:  SEPTEMBER-1982
                       84/04/22. 14.56.12.
        ********************************************************
        -----------------
        #P RHF/3-21G OPT
        -----------------
        1//1,3;
        2//2;
        3/5=5,11=1,25=11,30=1/1,2,3,11;
        4/7=1,16=1/1;
        5/6=7/1;
        6//1;
        7/27=1/1,2,16;
        1//3(1);
        99//99;
        2//2;
        3/5=5,11=1,25=11,30=1/1,2,11;
        5/6=7/1;
        7/27=1/1,2,16;
        1//3(-4);
        3/5=5,11=1,25=11,30=1/1,3;
        6//1;
        99//99;


                  ### NEXT LINK= 101  LL=   0
```

FIG. 5.3.2. GAUSSIAN82 output for the input shown in Fig. 5.3.1(a).

```
------------------
METHYLENE SINGLET
------------------
SYMBOLIC Z-MATRIX
   CHARGE = 0 MULTIPLICITY = 1
X
C    1    1.00000
H    2    RCH      1    ANG
H    2    RCH      1    ANG      3    180.00000 0
     VARIABLES
  RCH                1.10000
  ANG              127.60000
```

```
   ### NEXT LINK= 103   LL=   0
```

(ENTER 103)

GRADGRADGRADGRADGRADGRADGRADGRADGRADGRADGRADGRADGRADGRADGRADGRADGRADGRADGRAD

BERNY OPTIMIZATION

INITIALIZATION PASS

```
                       -----------------------------
                       !   INITIAL PARAMETERS     !
                       ! (ANGSTROMS AND DEGREES)  !
-----------------------                     -----------------------
!    NAME        VALUE    DERIVATIVE INFORMATION (ATOMIC UNITS)    !
-------------------------------------------------------------------
!    RCH        1.1000    ESTIMATE D2E/DX2                         !
!    ANG      127.6000    ESTIMATE D2E/DX2                         !
-------------------------------------------------------------------
```

GRADGRADGRADGRADGRADGRADGRADGRADGRADGRADGRADGRADGRADGRADGRADGRADGRADGRADGRAD

```
   ### NEXT LINK= 202   LL=   0
```

(ENTER 202)

```
--------------------------------------------------------------------
                Z-MATRIX (ANGSTROMS AND DEGREES)
CD CENT ATOM N1    LENGTH     N2    ALPHA     N3    BETA      J
--------------------------------------------------------------------
1        X
2   1    C    1  1.000000 (  1)
3   2    H    2  1.100000 (  2)  1  127.600 (  4)
4   3    H    2  1.100000 (  3)  1  127.600 (  5)  3  180.000 (  6)  0
--------------------------------------------------------------------
```

```
                   Z-MATRIX ORIENTATION:
-------------------------------------------------------------
CENTER      ATOMIC            COORDINATES (ANGSTROMS)
NUMBER      NUMBER        X           Y           Z
-------------------------------------------------------------
             -1        0.000000    0.000000    0.000000
   1          6        0.000000    0.000000    1.000000
   2          1         .871519    0.000000    1.671160
   3          1        -.871519     .000000    1.671160
-------------------------------------------------------------
```

FIG. 5.3.2. *(continued)*

```
                    DISTANCE MATRIX (ANGSTROMS)
                1           2           3
  1   C   0.000000
  2   H   1.100000    0.000000
  3   H   1.100000    1.743037    0.000000
STOICHIOMETRY     CH2
FRAMEWORK GROUP   C2V[C2(C),SGV(H2)]
DEG. OF FREEDOM    2
FULL POINT GROUP                    C2V     NOP  4
LARGEST ABELIAN SUBGROUP            C2V     NOP  4
LARGEST CONCISE ABELIAN SUBGROUP C2       NOP  2
                    STANDARD ORIENTATION:
----------------------------------------------------------------
CENTER      ATOMIC          COORDINATES (ANGSTROMS)
NUMBER      NUMBER          X           Y           Z
----------------------------------------------------------------
   1          6         0.000000    0.000000    -.167790
   2          1         0.000000     .871519     .503370
   3          1         0.000000    -.871519     .503370
----------------------------------------------------------------


    ### NEXT LINK=  301   LL=    0



                                          (ENTER 301)
STANDARD BASIS:  3-21G                 (S, S=P, 5D, 7F)
 13 BASIS FUNCTIONS        21 PRIMITIVE GAUSSIANS
  4 ALPHA ELECTRONS        4 BETA ELECTRONS
    NUCLEAR REPULSION ENERGY   6.0764354690 HARTREES
THERE ARE   7 SYMMETRY ADAPTED BASIS FUNCTIONS OF A1  SYMMETRY.
THERE ARE   0 SYMMETRY ADAPTED BASIS FUNCTIONS OF A2  SYMMETRY.
THERE ARE   2 SYMMETRY ADAPTED BASIS FUNCTIONS OF B1  SYMMETRY.
THERE ARE   4 SYMMETRY ADAPTED BASIS FUNCTIONS OF B2  SYMMETRY.
RAFFENETTI 1 INTEGRAL FORMAT.
TWO-ELECTRON INTEGRAL SYMMETRY IS TURNED ON.


    ### NEXT LINK=  302   LL=    0


                                          (ENTER 302)


    ### NEXT LINK=  303   LL=    0


                                          (ENTER 303)


    ### NEXT LINK=  311   LL=    0


                                          (ENTER 311)
STANDARD CUTOFFS SELECTED IN SHELL.
  1314 INTEGRALS PRODUCED FOR A TOTAL OF      1314


    ### NEXT LINK=  401   LL=    0
```

FIG. 5.3.2. (*continued*)

                                    (ENTER 401)

PROJECTED HUCKEL GUESS.
INITIAL GUESS ORBITAL SYMMETRIES.
        OCCUPIED: (A1) (A1) (B2) (A1)
        VIRTUAL: (B1) (B2) (A1)

   ### NEXT LINK= 501   LL=   0

                                    (ENTER 501)

RHF CLOSED SHELL SCF.
REQUESTED CONVERGENCE ON DENSITY MATRIX= .1000E-06 WITHIN 32 CYCLES.
ITER ELECTRONIC-ENERGY   CONVERGENCE   EXTRAPOLATION
---- -----------------   -----------   -------------
RHFCLO REQUIRES
  1   -.446052815583466E+02
  2   -.447154150287677E+02   .1823E-01
  3   -.447261497216077E+02   .9821E-02
  4   -.447278695204261E+02   .4502E-02
  5   -.447281983357930E+02   .2234E-02
  6   -.447282661661268E+02   .1050E-02   3-POINT.
  7   (NON-VARIATIONAL)
  8   -.447282844887363E+02   .8844E-05
  9   -.447282844913998E+02   .2192E-05
 10   -.447282844916767E+02   .1032E-05
 11   -.447282844917122E+02   .4110E-06
 12   -.447282844917177E+02   .2093E-06
 13   -.447282844917190E+02
SCF DONE: E(RHF) = **-38.6518490227**   A.U. AFTER 13 CYCLES
          CONVG =   .9880E-07   -V/T = 2.0016

   ### NEXT LINK= 601   LL=   0

                                    (ENTER 601)

ORBITAL SYMMETRIES.
        OCCUPIED: (A1) (A1) (B2) (A1)
        VIRTUAL: (B1) (A1) (B2) (B2) (A1) (B1) (A1) (B2) (A1)
THE ELECTRONIC STATE IS 1-A1.
        MOLECULAR ORBITAL COEFFICIENTS
                            1   2   3   4   5
                          (A1)   (A1)   (B2)   (A1)   (B1)
        EIGENVALUES -- -11.22336   -.89674   -.56667   -.37519   .07590
  1 1   C 1S   .98681   -.20013   0.00000   -.11033   0.00000
  2   2S (I)   .09180   .21978   0.00000   .10639   0.00000
  3   2PX (I)   0.00000   0.00000   0.00000   0.00000   .38155
  4   2PY (I)   0.00000   0.00000   -.38161   0.00000   0.00000
  5   2PZ (I)   .00489   .13254   0.00000   -.41966   0.00000
  6   2S (0)   -.04999   .56888   0.00000   .53150   0.00000
  7   2PX (0)   0.00000   0.00000   0.00000   0.00000   .74430
  8   2PY (0)   0.00000   0.00000   -.24177   0.00000   0.00000
  9   2PZ (0)   -.01195   .10945   0.00000   -.46552   0.00000
 10 2   H 1S (I)   -.00141   .17430   -.24205   -.10157   0.00000
 11   1S (0)   .01526   .05545   -.25888   -.13129   0.00000
 12 3   H 1S (I)   -.00141   .17430   .24205   -.10157   0.00000
 13   1S (0)   .01526   .05545   .25888   -.13129   0.00000

                    FIG. 5.3.2. (*continued*)

                              251

				6 (A1)	7 (B2)	8 (B2)	9 (A1)
EIGENVALUES --				.28866	.34193	.92144	.99119
1 1	C	1S		.13334	0.00000	0.00000	-.05867
2		2S	(I)	-.05525	0.00000	0.00000	.03904
3		2PX	(I)	0.00000	0.00000	0.00000	0.00000
4		2PY	(I)	0.00000	-.33259	.71359	0.00000
5		2PZ	(I)	-.22560	0.00000	0.00000	-.96294
6		2S	(0)	-1.79002	0.00000	0.00000	.52763
7		2PX	(0)	0.00000	0.00000	0.00000	0.00000
8		2PY	(0)	0.00000	-1.29252	-1.23251	0.00000
9		2PZ	(0)	-.90712	0.00000	0.00000	1.30207
10 2	H	1S	(I)	.02141	.05096	.54650	-.31013
11		1S	(0)	1.37521	1.39154	-.03949	-.24768
12 3	H	1S	(I)	.02141	-.05096	-.54650	-.31013
13		1S	(0)	1.37521	-1.39154	.03949	-.24768

FULL MULLIKEN POPULATION ANALYSIS.

				1	2	3	4	5
1 1	C	1S		2.05204				
2		2S	(I)	.01335	.13610			
3		2PX	(I)	0.00000	0.00000	0.00000		
4		2PY	(I)	0.00000	0.00000	0.00000	.29125	
5		2PZ	(I)	.00000	-.00000	0.00000	0.00000	.38742
6		2S	(0)	-.07999	.26950	0.00000	0.00000	-.00000
7		2PX	(0)	0.00000	0.00000	0.00000	0.00000	0.00000
8		2PY	(0)	0.00000	0.00000	0.00000	.09760	0.00000
9		2PZ	(0)	.00000	-.00000	0.00000	0.00000	.22196
10 2	H	1S	(I)	-.00090	.01033	0.00000	.04195	.02298
11		1S	(0)	.00294	-.00030	0.00000	.03667	.01787
12 3	H	1S	(I)	-.00090	.01033	0.00000	.04195	.02298
13		1S	(0)	.00294	-.00030	0.00000	.03667	.01787

				6	7	8	9	10
6		2S	(0)	1.21722				
7		2PX	(0)	0.00000	0.00000			
8		2PY	(0)	0.00000	0.00000	.11690		
9		2PZ	(0)	-.00000	0.00000	0.00000	.45766	
10 2	H	1S	(I)	.03026	0.00000	.04652	.04063	.19858
11		1S	(0)	-.05178	0.00000	.05854	.04826	.11063
12 3	H	1S	(I)	.03026	0.00000	.04652	.04063	-.00034
13		1S	(0)	-.05178	0.00000	.05854	.04826	-.00995

				11	12	13
11		1S	(0)	.17513		
12 3	H	1S	(I)	-.00995	.19858	
13		1S	(0)	-.03441	.11063	.17513

GROSS ORBITAL CHARGES.

				1
1 1	C	1S		1.98948
2		2S	(I)	.43900
3		2PX	(I)	0.00000
4		2PY	(I)	.54609
5		2PZ	(I)	.69109
6		2S	(0)	1.36369
7		2PX	(0)	0.00000
8		2PY	(0)	.42461
9		2PZ	(0)	.85741
10 2	H	1S	(I)	.49070
11		1S	(0)	.35361
12 3	H	1S	(I)	.49070
13		1S	(0)	.35361

FIG. 5.3.2. (*continued*)

```
          CONDENSED TO ATOMS (ALL ELECTRONS)
               1         2         3
1   C   5.703425   .303983   .303983
2   H    .303983   .594967  -.054644
3   H    .303983  -.054644   .594967
          TOTAL ATOMIC CHARGES.
               1
1   C   6.311390
2   H    .844305
3   H    .844305
DIPOLE MOMENT (DEBYE): X= 0.0000   Y= .0000   Z= 2.1990   TOTAL= 2.1990


   ### NEXT LINK= 701   LL=    0


                              (ENTER 701)


   ### NEXT LINK= 702   LL=    0


                              (ENTER 702)


   ### NEXT LINK= 716   LL=    0


                              (ENTER 716)


***** AXES RESTORED TO ORIGINAL SET *****
--------------------------------------------------------------
CENTER     ATOMIC            FORCES (HARTREES/BOHR)
NUMBER     NUMBER         X          Y          Z
--------------------------------------------------------------

   1         6      -.000000    .000000   -.001629
   2         1       .000869    .000000    .000815
   3         1      -.000869   -.000000    .000815

--------------------------------------------------------------
         MAX   .001629   RMS    .000781
--------------------------------------------------------------
                    Z-MATRIX (ANGSTROMS AND DEGREES)
CD CENT ATOM N1    LENGTH    N2    ALPHA    N3    BETA    J
--------------------------------------------------------------

1        X
2   1    C    1   1.000000 (  1)
3   2    H    2   1.100000 (  2)  1  127.600 (  4)
4   3    H    2   1.100000 (  3)  1  127.600 (  5)  3  180.000 (  6)  0
--------------------------------------------------------------


--------------------------------------------------------------
          INTERNAL COORDINATE FORCES (HARTREES/BOHR OR /RADIAN)
CENT ATOM N1    LENGTH    N2    ALPHA    N3    BETA    J
--------------------------------------------------------------

     X
1    C    1   -.000000 (  1)
2    H    2    .001186 (  2)  1   .000239 (  4)
3    H    2    .001186 (  3)  1   .000239 (  5)  3   .000000 (  6)  0
--------------------------------------------------------------
              MAX    .001186   RMS    .000698
```

FIG. 5.3.2. (*continued*)

253

```
      ### NEXT LINK= 103   LL=    0

                                  (ENTER 103)
GRADGRADGRADGRADGRADGRADGRADGRADGRADGRADGRADGRADGRADGRADGRADGRADGRADGRADGRADGRAD

BERNY OPTIMIZATION

SEARCH FOR A LOCAL MINIMUM.

STEP NUMBER   1 OUT OF A MAXIMUM OF  12
ALL QUANTITIES PRINTED IN INTERNAL UNITS (HARTREES-BOHRS-RADIANS)

SECOND DERIVATIVE MATRIX NOT UPDATED -- FIRST STEP

THE SECOND DERIVATIVE MATRIX:
                       RCH       ANG
          RCH        .67295
          ANG       0.00000    .45874
      EIGENVALUES ---   .45874    .67295

LINEAR SEARCH NOT ATTEMPTED -- FIRST POINT

VARIABLE     OLD X    -DE/DX   DELTA X   DELTA X   DELTA X    NEW X
                                (LINEAR)   (QUAD)    (TOTAL)

   RCH       2.07870   .00237  0.00000    .00352    .00352   2.08222
   ANG       2.22704   .00048  0.00000    .00104    .00104   2.22808

           ITEM               VALUE     THRESHOLD  CONVERGED?

MAXIMUM FORCE                .002371     .000450    NO
RMS     FORCE                .001711     .000300    NO
MAXIMUM DISPLACEMENT         .003524     .001800    NO
RMS     DISPLACEMENT         .002598     .001200    NO
PREDICTED CHANGE IN ENERGY  -.000004
GRADGRADGRADGRADGRADGRADGRADGRADGRADGRADGRADGRADGRADGRADGRADGRADGRADGRADGRADGRAD

    ### NEXT LINK= 202   LL=    0

                                  (ENTER 202)
-------------------------------------------------------------------------
                     Z-MATRIX (ANGSTROMS AND DEGREES)
CD CENT ATOM  N1    LENGTH    N2    ALPHA    N3    BETA      J
-------------------------------------------------------------------------

1      X
2   1  C    1  1.000000 (  1)
3   2  H    2  1.101865 (  2)  1  127.660 (  4)
4   3  H    2  1.101865 (  3)  1  127.660 (  5)  3  180.000 (  6)  0
-------------------------------------------------------------------------
                     Z-MATRIX ORIENTATION:
-------------------------------------------------------------------------
CENTER      ATOMIC           COORDINATES (ANGSTROMS)
NUMBER      NUMBER             X          Y          Z
-------------------------------------------------------------------------
             -1           0.000000   0.000000   0.000000
    1         6           0.000000   0.000000   1.000000
    2         1            .872295   0.000000   1.673207
    3         1           -.872295    .000000   1.673207
-------------------------------------------------------------------------
```

FIG. 5.3.2. (continued)

```
                   DISTANCE MATRIX (ANGSTROMS)
              1              2            3
  1   C   0.000000
  2   H   1.101865    0.000000
  3   H   1.101865    1.744589    0.000000
STOICHIOMETRY     CH2
FRAMEWORK GROUP    C2V[C2(C),SGV(H2)]
DEG. OF FREEDOM     2
FULL POINT GROUP                    C2V      NOP   4
LARGEST ABELIAN SUBGROUP            C2V      NOP   4
LARGEST CONCISE ABELIAN SUBGROUP C2          NOP   2
                   STANDARD ORIENTATION:
------------------------------------------------------------
CENTER     ATOMIC           COORDINATES (ANGSTROMS)
NUMBER     NUMBER         X           Y           Z
------------------------------------------------------------
   1          6        0.000000    0.000000    -.168302
   2          1        0.000000     .872295     .504905
   3          1        0.000000    -.872295     .504905
------------------------------------------------------------


    ### NEXT LINK=  301   LL=    0


                                    (ENTER 301)
STANDARD BASIS:  3-21G              (S, S=P, 5D, 7F)
  13 BASIS FUNCTIONS      21 PRIMITIVE GAUSSIANS
   4 ALPHA ELECTRONS       4 BETA ELECTRONS
     NUCLEAR REPULSION ENERGY    6.0663958264 HARTREES
THERE ARE    7 SYMMETRY ADAPTED BASIS FUNCTIONS OF A1  SYMMETRY.
THERE ARE    0 SYMMETRY ADAPTED BASIS FUNCTIONS OF A2  SYMMETRY.
THERE ARE    2 SYMMETRY ADAPTED BASIS FUNCTIONS OF B1  SYMMETRY.
THERE ARE    4 SYMMETRY ADAPTED BASIS FUNCTIONS OF B2  SYMMETRY.
RAFFENETTI 1 INTEGRAL FORMAT.
TWO-ELECTRON INTEGRAL SYMMETRY IS TURNED ON.



    ### NEXT LINK=  302   LL=    0


                                    (ENTER 302)


    ### NEXT LINK=  311   LL=    0


                                    (ENTER 311)


STANDARD CUTOFFS SELECTED IN SHELL.
   1314 INTEGRALS PRODUCED FOR A TOTAL OF      1314


    ### NEXT LINK=  501   LL=    0
```

FIG. 5.3.2. (*continued*)

```
RHF CLOSED SHELL SCF.
REQUESTED CONVERGENCE ON DENSITY MATRIX=   .1000E-06 WITHIN  32 CYCLES.
ITER  ELECTRONIC-ENERGY      CONVERGENCE   EXTRAPOLATION
----  -----------------      -----------   -------------
RHFCLO REQUIRES
  1   -.447166224241487E+02
  2   -.447182494751717E+02    .7478E-04
  3   -.447182496251098E+02    .3286E-04
  4   -.447182496395471E+02    .1553E-04
  5   -.447182496420644E+02    .7073E-05     4-POINT.
  6    (NON-VARIATIONAL)
  7   -.447182496427047E+02
SCF DONE: E(RHF) = -38.6518538163     A.U. AFTER   7 CYCLES
          CONVG  =    .9101E-07         -V/T =  2.0017


   ### NEXT LINK=  701   LL=   0
```

```
   ### NEXT LINK=  702   LL=   0
```

```
   ### NEXT LINK=  716   LL=   0
```

```
***** AXES RESTORED TO ORIGINAL SET *****
-----------------------------------------------------------
CENTER    ATOMIC              FORCES (HARTREES/BOHR)
NUMBER    NUMBER          X           Y           Z
-----------------------------------------------------------
   1        6        -.000000     .000000    -.000164
   2        1         .000038     .000000     .000082
   3        1        -.000038    -.000000     .000082
-----------------------------------------------------------
          MAX     .000164    RMS      .000069
-----------------------------------------------------------
                   Z-MATRIX (ANGSTROMS AND DEGREES)
CD CENT ATOM N1    LENGTH     N2    ALPHA     N3     BETA      J
-----------------------------------------------------------
1       X
2   1   C    1  1.000000 ( 1)
3   2   H    2  1.101865 ( 2) 1  127.660 ( 4)
4   3   H    2  1.101865 ( 3) 1  127.660 ( 5) 3 180.000 ( 6) 0
-----------------------------------------------------------
```

FIG. 5.3.2. (*continued*)

```
---------------------------------------------------------------
                    INTERNAL COORDINATE FORCES (HARTREES/BOHR OR /RADIAN)
   CENT ATOM N1    LENGTH      N2     ALPHA      N3      BETA       J
---------------------------------------------------------------
       X
   1  C    1   -.000000 (  1)
   2  H    2    .000080 (  2)  1   .000087 (  4)
   3  H    2    .000080 (  3)  1   .000087 (  5)  3   .000000 (  6)  0
---------------------------------------------------------------
               MAX    .000087    RMS    .000068

   ### NEXT LINK= 103   LL=    0


                          (ENTER 103)
GRADGRADGRADGRADGRADGRADGRADGRADGRADGRADGRADGRADGRADGRADGRADGRADGRADGRADGRAD

BERNY OPTIMIZATION

SEARCH FOR A LOCAL MINIMUM.

STEP NUMBER   2 OUT OF A MAXIMUM OF  12
ALL QUANTITIES PRINTED IN INTERNAL UNITS (HARTREES-BOHRS-RADIANS)

UPDATE SECOND DERIVATIVES USING INFORMATION FROM POINTS:  1   2

THE SECOND DERIVATIVE MATRIX:
                     RCH      ANG
           RCH      .62204
           ANG     -.01506   .45428
     EIGENVALUES ---  .45294   .62339

VARIABLE      OLD X    -DE/DX    DELTA X    DELTA X    DELTA X    NEW X
                                 (LINEAR)   (QUAD)     (TOTAL)

   RCH       2.08222   .00016    .00033    -.00006     .00027    2.08249
   ANG       2.22808   .00017    .00010     .00032     .00042    2.22850

         ITEM                VALUE     THRESHOLD   CONVERGED?

MAXIMUM FORCE           .000173     .000450      YES
RMS     FORCE           .000167     .000300      YES
MAXIMUM DISPLACEMENT    .000415     .001800      YES
RMS     DISPLACEMENT    .000348     .001200      YES
PREDICTED CHANGE IN ENERGY  -.000000


OPTIMIZATION COMPLETED.
   -- STATIONARY POINT FOUND.
   -- LAST STEP NOT IMPLEMENTED.

                 ---------------------------
                 !  OPTIMIZED PARAMETERS   !
                 ! (ANGSTROMS AND DEGREES) !
---------------------------------------------------------------
 !    NAME        VALUE    DERIVATIVE INFORMATION (ATOMIC UNITS)    !
---------------------------------------------------------------
 !    RCH       1.1019   -DE/DX =   .000161                        !
 !    ANG     127.6597   -DE/DX =   .000173                        !
---------------------------------------------------------------

GRADGRADGRADGRADGRADGRADGRADGRADGRADGRADGRADGRADGRADGRADGRADGRADGRADGRADGRAD
```

FIG. 5.3.2. (*continued*)

NEXT LINK= 301   LL=   0

                                    (ENTER 301)
STANDARD BASIS: 3-21G   (S, S=P, 5D, 7F)
  13 BASIS FUNCTIONS   21 PRIMITIVE GAUSSIANS
   4 ALPHA ELECTRONS   4 BETA ELECTRONS
     NUCLEAR REPULSION ENERGY   6.0663958264 HARTREES
THERE ARE   7 SYMMETRY ADAPTED BASIS FUNCTIONS OF A1 SYMMETRY.
THERE ARE   0 SYMMETRY ADAPTED BASIS FUNCTIONS OF A2 SYMMETRY.
THERE ARE   2 SYMMETRY ADAPTED BASIS FUNCTIONS OF B1 SYMMETRY.
THERE ARE   4 SYMMETRY ADAPTED BASIS FUNCTIONS OF B2 SYMMETRY.
RAFFENETTI 1 INTEGRAL FORMAT.
TWO-ELECTRON INTEGRAL SYMMETRY IS TURNED ON.

   ### NEXT LINK= 303   LL=   0

                                    (ENTER 303)

   ### NEXT LINK= 601   LL=   0

                                    (ENTER 601)
ORBITAL SYMMETRIES.
      OCCUPIED: (A1) (A1) (B2) (A1)
      VIRTUAL: (B1) (A1) (B2) (B2) (A1) (B1) (A1) (B2) (A1)
THE ELECTRONIC STATE IS 1-A1.
      MOLECULAR ORBITAL COEFFICIENTS
                          1   2   3   4   5
                        (A1)   (A1)   (B2)   (A1)   (B1)
      EIGENVALUES -- -11.22394 -.89621 -.56591 -.37546   .07580
  1 1 C 1S   .98681   -.20013 0.00000 -.11041 0.00000
  2   2S   (I)   .09178   .21972 0.00000   .10657 0.00000
  3   2PX (I)   0.00000   0.00000 0.00000 0.00000   .38167
  4   2PY (I)   0.00000   0.00000 -.38130 0.00000 0.00000
  5   2PZ (I)   .00489   .13232 0.00000 -.41942 0.00000
  6   2S   (O)   -.04991   .56939 0.00000   .53184 0.00000
  7   2PX (O)   0.00000   0.00000 0.00000 0.00000   .74421
  8   2PY (O)   0.00000   0.00000 -.24227 0.00000 0.00000
  9   2PZ (O)   -.01195   .10950 0.00000 -.46490 0.00000
 10 2 H 1S   (I)   -.00139   .17394 -.24175 -.10179 0.00000
 11   1S   (O)   .01521   .05566 -.25929 -.13197 0.00000
 12 3 H 1S   (I)   -.00139   .17394   .24175 -.10179 0.00000
 13   1S   (O)   .01521   .05566   .25929 -.13197 0.00000
                          6   7   8   9
                        (A1)   (B2)   (B2)   (A1)
      EIGENVALUES --   .28816   .34118   .92232   .99167
  1 1 C 1S   .13324   0.00000 0.00000 -.05849
  2   2S   (I)   -.05536   0.00000 0.00000   .04142
  3   2PX (I)   0.00000   0.00000 0.00000 0.00000
  4   2PY (I)   0.00000   -.33292   .71540 0.00000
  5   2PZ (I)   -.22630   0.00000 0.00000 -.96472
  6   2S   (O)   -1.78323   0.00000 0.00000   .52199
  7   2PX (O)   0.00000   0.00000 0.00000 0.00000
  8   2PY (O)   0.00000 -1.28887 -1.22997 0.00000

FIG. 5.3.2. (*continued*)

258

9			2PZ (O)	-.90704	0.00000	0.00000	1.30153
10	2	H	1S (I)	.02175	.05156	.54633	-.30768
11			1S (O)	1.37174	1.38646	-.04181	-.24745
12	3	H	1S (I)	.02175	-.05156	-.54633	-.30768
13			1S (O)	1.37174	-1.38646	.04181	-.24745

FULL MULLIKEN POPULATION ANALYSIS.

				1	2	3	4	5
1	1	C	1S	2.05209				
2			2S (I)	.01334	.13611			
3			2PX (I)	0.00000	0.00000	0.00000		
4			2PY (I)	0.00000	0.00000	0.00000	.29078	
5			2PZ (I)	.00000	-.00000	0.00000	0.00000	.38688
6			2S (O)	-.08003	.26983	0.00000	0.00000	-.00000
7			2PX (O)	0.00000	0.00000	0.00000	0.00000	0.00000
8			2PY (O)	0.00000	0.00000	0.00000	.09773	0.00000
9			2PZ (O)	.00000	-.00000	0.00000	0.00000	.22155
10	2	H	1S (I)	-.00088	.01023	0.00000	.04164	.02290
11			1S (O)	.00293	-.00035	0.00000	.03665	.01797
12	3	H	1S (I)	-.00088	.01023	0.00000	.04164	.02290
13			1S (O)	.00293	-.00035	0.00000	.03665	.01797

				6	7	8	9	10
6			2S (O)	1.21908				
7			2PX (O)	0.00000	0.00000			
8			2PY (O)	0.00000	0.00000	.11739		
9			2PZ (O)	-.00000	0.00000	0.00000	.45654	
10	2	H	1S (I)	.03001	0.00000	.04649	.04067	.19813
11			1S (O)	-.05204	0.00000	.05872	.04853	.11081
12	3	H	1S (I)	.03001	0.00000	.04649	.04067	-.00033
13			1S (O)	-.05204	0.00000	.05872	.04853	-.00990

				11	12	13
11			1S (O)	.17595		
12	3	H	1S (I)	-.00990	.19813	
13			1S (O)	-.03435	.11081	.17595

GROSS ORBITAL CHARGES.

				1
1	1	C	1S	1.98949
2			2S (I)	.43903
3			2PX (I)	0.00000
4			2PY (I)	.54510
5			2PZ (I)	.69017
6			2S (O)	1.36483
7			2PX (O)	0.00000
8			2PY (O)	.42554
9			2PZ (O)	.85648
10	2	H	1S (I)	.48976
11			1S (O)	.35492
12	3	H	1S (I)	.48976
13			1S (O)	.35492

CONDENSED TO ATOMS (ALL ELECTRONS)

		1	2	3
1	C	5.703689	.303474	.303474
2	H	.303474	.595689	-.054482
3	H	.303474	-.054482	.595689

FIG. 5.3.2. (*continued*)

```
              TOTAL ATOMIC CHARGES.
                     1
        1   C   6.310638
        2   H    .844681
        3   H    .844681
      DIPOLE MOMENT (DEBYE): X= 0.0000   Y=  .0000   Z= 2.1960   TOTAL= 2.1960


          ### NEXT LINK= 9999   LL=    0



                              (ENTER9999)
          FAU ERLG\CLARK      \84/04/22\FOPT \RHF       \3-21G     \C1H2     \1
          \\#P RHF/3-21G OPT\\METHYLENE SINGLET\\0,1\X\C,1,1.\H,2,RCH,1,ANG\H,2,
          RCH,1,ANG,3,180.,0\\RCH=1.10186\ANG=127.65973\\HF=-38.6518538\RMSD=.91
          0\RMSF=.694\DIP=2.19603\PG=C02V\\


      SIC AS THE CAWSE OF EWERY THING IS, SIC WILBE THE EFFECT.

      -- PROVERBS AND REASONS OF THE YEAR 1585
                 AS REPRINTED IN PAISLEY MAGAZINE 1828.



                     BYE,BYE !!!



      THIS JOB IS ARCHIVED AS NUMBER  1072
      LENGTH OF THE ARCHIVED RECORD    253
```

FIG. 5.3.2.   *(continued)*

second slash denote the individual links to be called within the overlay. Thus
the card reading

$$7/27 = 1/1,2,16 \ ;$$

indicates that links 701, 702, and 716 will be called with option 27 set to 1.
The options for the individual overlays and links are listed in the program
manual. The numbers in parentheses immediately before the semicolons at
the ends of some of the route cards indicate optional jumps. In this case,

$$1//3(1) \ ;$$

means that the program will skip forward one overlay (i.e., not call the next
one) if the optimization is not complete, and

$$1//3(-4) \ ;$$

instructs the program to skip four overlays backwards, again if the geometry
is not yet optimized. Each overlay definition is ended by a semicolon. Figure
5.3.3 shows the route printed in Fig. 5.3.2 in diagrammatic form.

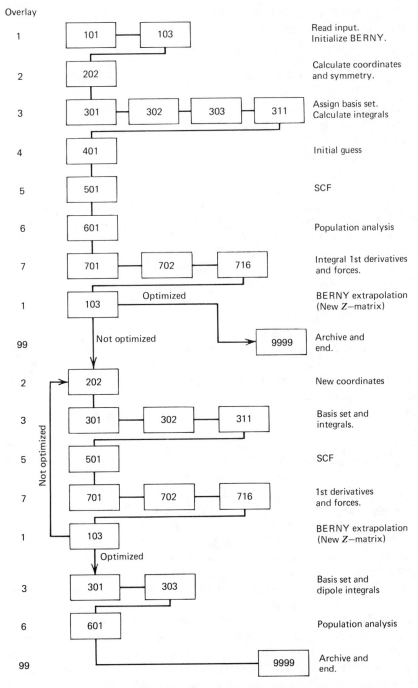

Overlay

1	101 — 103	Read input. Initialize BERNY.
2	202	Calculate coordinates and symmetry.
3	301 — 302 — 303 — 311	Assign basis set. Calculate integrals
4	401	Initial guess
5	501	SCF
6	601	Population analysis
7	701 — 702 — 716	Integral 1st derivatives and forces.
1	103 — Optimized	BERNY extrapolation (New Z–matrix)
99	Not optimized → 9999	Archive and end.
2	202	New coordinates
3	301 — 302 — 311	Basis set and integrals.
5	501	SCF
7	701 — 702 — 716	1st derivatives and forces.
1	103	BERNY extrapolation (New Z–matrix)
	Optimized	
3	301 — 303	Basis set and dipole integrals
6	601	Population analysis
99	9999	Archive and end.

Not optimized

FIG. 5.3.3. Schematic diagram of the route generated by release A of GAUSSIAN82 for a closed shell 3-21G optimization.

The geometry input is first read by link 101. Link 103 then initiates the BERNY optimization procedure. The calculation then moves on to link 202, which calculates atomic coordinates and determines the symmetry of the molecule. Link 301 next assigns the atomic orbitals of the basis set to the appropriate centers. Of the available links for the evaluation of integrals, only links 302 (one-electron integrals), 311 (two-electron integrals involving *s*- and *p*-functions), and 303 (dipole integrals) are required. Link 314, which calculates two- electron integrals involving *s*-, *p*-, *d*-, and *f*-functions, is not needed because 3-21G has only *s*- and *p*-orbitals. Link 303 is called in order that the dipole moment may be calculated in the population analysis performed for the initial geometry. The next overlay to be called consists of only one link, 401, which produces an initial guess for the SCF procedure *via* an extended Hückel calculation. The closed-shell link from overlay 5, link 501, then performs the actual SCF calculation by an iterative procedure. At this stage link 601 gives an analysis of the wave function produced for the original geometry. The procedure up to this point corresponds exactly to that for a single-point SCF calculation, except that link 103 was called to initiate the optimization procedure.

The real geometry optimization process begins with overlay 7, from which links 701, 702, and 716 are called to calculate the first derivatives of the integrals and to convert these to atomic forces. These forces are then used by link 103 to decide whether the input geometry is a stationary point (minimum or transition state). If a stationary point is found (i.e., if the input geometry was already optimized) the program moves on to the next link, 9999, which writes an archive entry and terminates the job. If, as is more likely, the geometry is not optimized the program jumps over one overlay and moves on to link 202, which calculates new coordinates from the Z-matrix predicted by link 103 to be the minimum-energy geometry. Overlay 3 then assigns the basis set to the new atomic centers and calculates new one- and two-electron integrals. Link 303 is not required in this case because there probably will be no need to perform a population analysis for this geometry. The program moves directly to link 501 for the next SCF calculation. There is no need to call link 401 to produce an initial guess, because the density matrix from the last SCF calculation can be used. If the geometry has not changed too much this will be more suitable than the result of an extended Hückel calculation as a starting point for the new SCF. Later releases of GAUSSIAN82 do call link 401 at this point, but not to produce a new initial guess.

Once the new energy has been calculated, the program moves on to overlay 7, which calculates the forces as before. These forces are used by link 103 to decide whether the geometry is optimized, and if not, to produce a new Z-matrix. In that case the calculation skips back four overlays to link 202, which then calculates coordinates for the next point. This process is repeated until link 103 detects a minimum, in which case the basis set and the dipole integrals are produced by overlay 3 so that a population analysis can be performed on the optimized geometry. The final link is once more 9999, which makes the archive entry and ends the job. These individual program steps can now be followed through the output shown in Fig. 5.3.2.

Having determined the route in overlay 0, which that therefore does not itself appear, the calculation moves on to link 101. The geometry, charge, multiplicity, and title from the input are printed in much the same form as they were read in. Link 103 next reports that it will do a BERNY optimization, and that this is the initialization step. The symbolic variables and their values are printed in a table along with some information on the second derivatives (D2E/DX2) to be used in estimating the minimum-energy geometry. Because the second derivatives are not usually calculated explicitly in a BERNY optimization the program must start with guessed values. During the optimization more and more information about the potential surface accumulates and the guessed second derivatives are constantly improved.

Link 202 then prints the Z-matrix in a form more like that given in Fig. 3.3.2. The Cartesian coordinates calculated from this Z-matrix are then given, exactly as for MNDO calculations. In this table the dummy is still present, with the atomic code number -1, and the orientation is that obtained from the Z-matrix directly (i.e., the first atom lies at the origin, the second on the z-axis, and so on). A distance matrix for the nondummy atoms is then printed, as in MOPAC. The next stage of the calculation determines the symmetry (point group and framework group) of the molecule. Because the symmetry calculations are based on the Z-matrix, and not on the coordinates, the point group calculated corresponds to the symmetry constraints for the geometry optimization, which may be lower than the real symmetry of either the initial or final geometries. GAUSSIAN82 cannot, for instance, determine a new symmetry if an unsymmetrical initial geometry optimizes to a symmetrical structure. In this example the framework group has been found to be $C_{2v}[C_2(C), \sigma_v(H_2)]$. Framework groups[24] were devised to convey more information about the molecular structure than simple point groups do by indicating atoms that lie on or in a symmetry element. In this example the point group is C_{2v} with one carbon (C) on the C_2 axis, and two hydrogens (H_2) in one of the σ_v planes. Atoms that do not lie on or in any symmetry element are assigned to the remainder of space, designated X.

The program also calculates the number of degrees of freedom (geometrical parameters to be optimized) for the molecule within the given symmetry. This information can be used to judge whether the molecule has been fully or partly optimized. "Fully optimized" in this context means optimized within the given symmetry constraints, not necessarily completely optimized. A calculation that GAUSSIAN82 indicates to be a *partial* optimization is one in which not enough parameters have been optimized to give the lowest-energy structure within the molecular point group. This means either that the geometry is not correctly optimized or that one or more parameters have been deliberately held constant, as may be the case in a reaction-path calculation. The full point group and the Abelian subgroups are used by other links either to save time by use of symmetry or to determine the symmetry of the individual molecular orbitals or the total wave function. The final task of link 202 is to shift the molecule to the so-called standard orientation, in which the center of mass lies at the origin and the principal axis along the Cartesian z-axis. This orientation facilitates the use of symmetry later in the calculation and also gives easily interpreted molecular orbitals in the population analysis. The molecular orbitals printed in link 601 therefore correspond to the standard orientation, not to the Z-matrix orientation.

Next, link 301 assigns the atomic orbitals from the 3-21G basis set to the correct atomic centers. In this example there are nine basis functions for the carbon atom ($1s$, $2s_I$, $3 \times 2p_I$, $2s_O$, and $3 \times 2p_O$) and two ($1s_I$ and $1s_O$) for each hydrogen. These orbitals are composed of 3, 2, 2, 2, 2, 1, 1, 1, and 1 primitive Gaussians for carbon and 2 and 1 Gaussians for each hydrogen, giving a total of 21. The molecule has a total of eight electrons, two more than in the equivalent MNDO calculation because the carbon $1s$ orbital is included. Because this is a closed-shell singlet calculation there are four spin-up (alpha) and four spin-down (beta) electrons paired in four orbitals.

Link 301 then makes use of the molecular point group to form symmetry-adapted basis functions, which correspond exactly to the orbital combinations used to derive the methylene orbitals in Fig. 3.6.1. The symmetries of these basis functions are denoted by Mulliken symbols, the meanings of which are summarized in Table 5.3.1. Figure 5.3.4 shows the symmetry-adapted basis functions for this calculation. The hydrogen basis functions are combined in symmetrical and antisymmetrical pairs, as in Fig. 3.6.1. The only difference is that there are more combinations in this example because of the two sets of valence functions and because the carbon $1s$ orbital is included. Finally link 301 prints two messages stating that the Raffenetti method for writing and storing integrals will be used and that symmetry will be used when calculating the two electron-integrals. These two options, which speed up the calculation, will be selected by the program whenever possible.

The remaining links in overlay 3 then calculate the various types of integrals. Link 311 prints information about the number of integrals calculated (as would link 314 if it were used), in this small calculation only 1314. Because integrals beneath a certain value (the cutoff) are ignored, the number calculated is dependent on the geometry as well as on the number of basis functions. Compact molecules require more integrals to be calculated than more extensive ones do.

The next stage of the procedure is an extended Hückel calculation to determine an initial guess for the SCF procedure. Link 401 prints the symmetries of the initial guess molecular orbitals, which correspond to those given in Fig. 3.6.3. The four occupied molecular orbitals and the lowest three virtual orbitals are given. All the necessary integrals and an initial

TABLE 5.3.1
Mulliken Symmetry Symbols

Symbol	Meaning
A	Symmetrical with respect to a 360/$n°$ rotation about the n-fold principal axis
B	Antisymmetrical with respect to a 360/$n°$ rotation about the n-fold principal axis
E	Doubly degenerate
T	Triply degenerate
	Subscripts
1	Symmetrical with respect to a 180° rotation about a C_2 axis perpendicular to the principal axis, or with respect to reflection in a σ_v plane if there are no such C_2 axes
2	Antisymmetrical with respect to a 180° rotation about a C_2 axis perpendicular to the principal axis, or with respect to reflection in a σ_v plane if there are no such C_2 axes
g	Symmetrical with respect to inversion
u	Antisymmetrical with respect to inversion
	Superscripts
'	Symmetrical with respect to reflection in a σ_h plane
''	Antisymmetrical with respect to reflection in a σ_h plane

guess are now available, so the program can proceed to link 501 for the SCF calculation. The convergence criterion, namely, that the density matrix should change by less than 10^{-7} (.1000E-06) between cycles, was met within 13 cycles (out of a maximum of 32). The electronic energy and the convergence on the density matrix are printed for every cycle because the extra printing option was used in the input. The column labeled "EXTRAPOLATION" indicates that the program extrapolated to speed convergence after cycle 6 using information from the prior three cycles. This results in a decrease in the change between consecutive density matrices from 10^{-3} to 10^{-5}. The limit of 32 cycles in the SCF procedure is often not large enough, and can be extended by using the keyword "SCFCYC=64," for instance, to give 64 cycles. In some cases the SCF procedure does not even converge slowly, but simply oscillates. Such calculations do not benefit at all from being allowed to continue for more cycles, and the direct minimization procedure (SCFDM) must be used in such cases.

The total energy is then given in atomic units (a.u., also known as Hartrees), one of which is equivalent to 627.5 kcal mol^{-1}. The total energies given by *ab initio* calculations are sums of the electronic and core repulsion energies, as for MNDO. They are, however, seldom converted to heats of formation, because these would be significantly in error due to neglect of

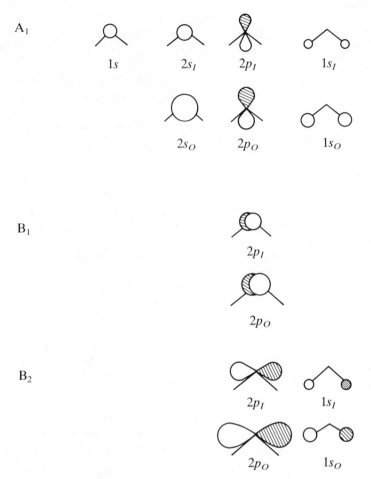

FIG. 5.3.4. Symmetry-adapted 3-21G basis functions for C_{2v} methylene.

electron correlation. This problem can be corrected by the parametrization in semiempirical calculations. *Ab initio* total energies are therefore used to calculate reaction energies or to assess the relative stabilities of isomers. The actual convergence on the density matrix (0.5737×10^{-8}) is now printed along with the virial ratio (V/T). This is the ratio between the calculated potential and kinetic energies, which should be 2 according to the virial theorem of classical mechanics.

On completion of the SCF calculation the program moves on to link 601, which performs a Mulliken population analysis. This procedure is analogous, but not identical, to the population analysis given by MOPAC. First, however, link 601 prints a table of the molecular orbitals with the Mulliken symbols for their symmetry types. GAUSSIAN82 prints only the five lowest virtual orbitals, so that in this case 9 orbitals out of a total of 13 (the number of basis functions) are printed. The eigenvalues, or energy levels, of the molecular orbitals are given in atomic units, not in electron volts as in MOPAC. The first orbital has a_1 symmetry, and is essentially the pure carbon $1s$-orbital. Note that the carbon $1s$- and $2s$-orbitals are not orthogonal, so that a small amount of $2s$ is mixed into molecular orbital 1. Because of the symmetry of the molecule the other a_1 symmetry-adapted basis functions can also contribute slightly. This core orbital has a very negative energy (-11.223 a.u. $= -304$ eV), and lies far lower in the MO manifold than the valence orbitals, which normally have energies in the range -40–0 eV.

Molecular orbitals 2, 3, and 4 are the σ_{CH}, π_{CH_2}, and n-orbitals obtained in the MNDO calculation, but the effect of the split-valence basis can now be seen. The $2a_1$-orbital (i.e. the second lowest-energy orbital of a_1 symmetry, molecular orbital 2) has a larger coefficient (0.56888) for the carbon $2s_O$-orbital than for the $2s_I$ (0.21978), but the coefficients for the inner parts of the carbon p_z- and hydrogen $1s$-orbitals (0.13254 and 0.17430, respectively) are larger than those for the corresponding outer basis functions (0.10945 and 0.05545). The entire orbital is therefore relatively compact except for the carbon $2s$-contribution. Similarly, the $1b_2$-orbital (molecular orbital 3) is composed largely of contributions from the inner orbitals. The n lone pair ($3a_1$), however, has a carbon $2s_O$ coefficient 5 times larger than that for the $2s_I$. The outer contributions for the carbon p_z- and hydrogen orbitals are also larger than the inner ones. This is a reflection of the anisotropy effect outlined in Section 5.1. The $1b_1$ LUMO is a carbon p_x-orbital in the standard orientation.

One important difference to note between the orbitals shown in Fig. 5.3.2 and those given for the MOPAC calculations is that in the former the sum of the squares of the coefficients in a given molecular orbital is not unity. Because the individual atomic orbitals of the basis set overlap with each other (i.e., they are not orthogonal), the simple formula used for the MNDO population analysis cannot be used. To take this overlap into account, a *density matrix* must first be formed and then be multiplied by the *overlap matrix*. An element $P_{i,j}$ in the density matrix is simply the sum of the products of the coefficients of the atomic orbitals i and j multiplied by the number of electrons in the orbital:

$$P_{i,j} = \Sigma \, c_{i,k} \cdot c_{j,k} \cdot Q_k \, ,$$

where N is the total number of orbitals; $c_{i,k}$ and $c_{j,k}$ are the coefficients of atomic orbitals i and j, respectively, in molecular orbital k; and Q_k is the number of electrons in this orbital. The diagonal elements of the overlap matrix, which depends only on the geometry and the basis set, are all unity. The off-diagonal elements correspond to the overlaps between the orbitals in the basis set. In MNDO, which uses orthogonal atomic orbitals and neglects differential diatomic overlap, all off-diagonal elements are zero, so that the diagonal elements of the density matrix are the only ones that do not disappear in the population analysis. These diagonal elements are therefore:

$$P_{i,i} = \Sigma\, c_{i,k} \cdot c_{i,k} \cdot Q_k = \Sigma\, c_{i,k}{}^2 \cdot Q_k\,.$$

This is the simple expression used previously in considering the charge densities in MNDO calculations.

Because, however, calculations the off-diagonal elements of the overlap matrix for an *ab initio* calculation are not zero the product of the density matrix and the overlap matrix must be used to analyze the wave function. This procedure is the Mulliken population analysis. The Mulliken population matrix is printed after the eigenvectors by link 601. Because the matrix is symmetrical, only the trianglular area below the diagonal is printed. Figure 5.3.5 shows the summations performed on such a matrix as part of the population analysis. The orbital charges are the sums of the columns (or rows, because the matrix is symmetrical) of matrix elements. The charge on the ith orbital, q_i, is given by:

$$q_i = \Sigma\, M_{i,k}\,,$$

where N is the number of basis functions and $M_{i,k}$ is the kth element in the ith column of the Mulliken population analysis matrix. The atomic charges are simply the sums of the electron densities calculated for all the atomic orbitals assigned to the atom in question, as shown in Fig. 5.3.5(a). Figure 5.3.5(b) shows the blocking of the matrix used to produce the table headed

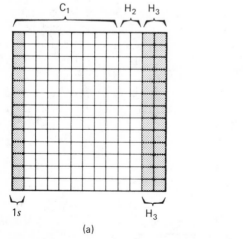
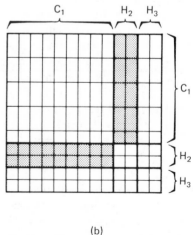

FIG. 5.3.5. The summations performed on the Mulliken matrix to produce a population analysis: (a) for the orbital and atomic charges, (b) for the overlap populations. The gray shaded area indicates the elements summed to give the C_1–H_2 overlap.

"CONDENSED TO ATOMS" in Fig. 5.3.2. The elements within a given block are summed to give the corresponding entry in the table. Of primary interest are the off-diagonal elements, which give an indication of the overlap between two atoms. Note that the overlap between atoms 1 and 2, for instance, appears twice in the matrix, once in row 1, column 2; and once in row 2, column 1. The C_1–H_2 overlap is therefore $0.303983 \times 2 = 0.607966$. The H–H overlap is -0.109288; that is, the hydrogen atoms repel each other slightly.

After the completion of link 601 the program proceeds to the optimization. Links 701 and 702 calculate the integral first derivatives and link 716 converts them to atomic forces, first in terms of x-, y-, and z-coordinates for the Z-matrix orientation (AXES RESTORED TO ORIGINAL SET), and then in terms of the internal coordinates, which are the bond lengths, angles, and dihedral angles given in the Z-matrix. The molecule must be restored to the Z-matrix orientation in order for this transformation to be made. The forces are 0.001261 Hartrees/Bohr for the CH bond lengths, and 0.0315 Hartrees/radian for the dummy–C–H angles. These forces are given in a format similar to that of the Z-matrix itself.

Link 103 next tests for a minimum after estimating the second-derivative matrix on the basis of the size of the first derivatives. Note that in this first cycle the off-diagonal elements, of which only one is printed because the matrix is symmetrical, are zeroes. These off-diagonal elements will be improved as the optimization proceeds and more information about the potential surface is obtained. The eigenvalues obtained by diagonalization of this matrix are simply the diagonal elements for the first cycle. The optimization procedure then works out the changes to be made in the Z-matrix in order to reach the predicted minimum. The bond lengths of this table are given in Bohrs, and the angles in radians. In this case the old value of RCH (OLD X) is 2.0787 Bohrs ($= 1.1$ Å) and the first derivative of the total energy with respect to changes in RCH ($-DE/DX$) was 0.00237 Hartrees/ Bohr. Because this is the first step, the program can base its geometry changes only on the estimated second derivatives, so that the only contribution to the projected change in RCH (DELTA X) comes from the quadratic extrapolation (LINEAR SEARCH NOT ATTEMPTED). The predicted value for RCH is then simply OLD X plus DELTA X. The maximum and root mean square forces and displacements are then compared with the threshold values (below which the structure is considered to be optimized), and all four are found wanting. The predicted stabilization compared with the old geometry is then given in Hartrees.

The program then moves through links 202, 301, 302, and 311 as before. Note that link 303 is not used. Link 501 then calculates the new energy using the last density matrix as an initial guess. After recalculation of the atomic forces, the second-derivative matrix is updated using the first derivatives from the two geometries that were already been calculated. The diagonal elements have changed slightly from those used in the first step, and it is now possible to assign values to the off-diagonal elements. Link 103 is then used once more to predict the minimum-energy geometry. This process is repeated until both the maximum and the root mean square forces and displacements are below the threshold values. Once this occurs the optimized values of RCH and ANG are printed with the corresponding first derivatives. After reassigning the basis set and calculating the dipole integrals, the program moves on to a population analysis of the optimized geometry. The final task is to write an archive entry for this calculation. The individual fields of the archive entry are separated by backward slashes. The entry in this case gives the site at which the job was run (FAU ERLG = Friedrich–Alexander–Universität Erlangen), the name of the user, the date, and then information about the calculation. In this example the individual fields have the following meanings:

FOPT	Fully optimized
RHF	Restricted Hartree–Fock
3-21G	The basis set
C1H2	The Stoichiometry (CH_2)

The input for any subsequent job starting from the optimized geometry is then given in compressed form. It can be reformatted from the archive entry in order to do other calculations on the optimized geometry. After the values of geometric variables RCH and ANG, the Hartree–Fock energy is given (HF=-38.6518538), as are the root mean square force and displacements, the dipole moment, and the point group (C_{2v}). GAUSSIAN82 then takes its leave by printing one of an almost inexhaustible supply of quotations.

The second input example, shown in Fig. 5.3.6, demonstrates a different type of job, an open-shell single-point calculation using the 6-31G* basis set with subsequent calculation of the MP2 correction for electron correlation. The geometry used is the UHF/6-31G* optimized structure taken from entry number 1332 in the *Carnegie–Mellon Quantum Chemistry Archive.*[20] The

```
# UMP2/6-31G*

METHYLENE TRIPLET MP2/6-31G*//6-31G*   (CMU 1332)

   0 3
   X
   C 1 1.
   H 2 RCH 1 ANG
   H 2 RCH 1 ANG 3 180.

   RCH=1.07134
   ANG=114.78339
```

FIG. 5.3.6. One possible GAUSSIAN82 input for a MP2/6-31G* single-point calculation on triplet methylene.

only keyword required is "UMP2/6-31G*." A single-point calculation is assumed if OPT is not used. The only other difference, apart from the values assigned to the geometric variables, is that the multiplicity is 3 for a triplet, instead of the "1" for the singlet carbene. The "UMP2" keyword implies an UHF calculation followed by an MP2 calculation.

The output for the job is shown in Fig. 5.3.7. The route is similar to the beginning of that shown in Fig. 5.3.3. The differences in the first six overlays are that link 103 is not needed because no geometry optimization is to be performed, link 314 is called because the basis set contains d-orbitals, and link 502 (UHF) is used rather than link 501 (RHF). The options in overlay 3 specify the 6-31G* basis set. After the population analysis (overlay 6), links 801, 802, and 901 are used for the MP2 calculation. The output is analogous to that for the singlet until link 301. The calculation now uses five alpha and three beta electrons for the triplet state, exactly as MNDO used four and two, respectively. The two extra electrons are once again attributable to the

```
                        HELLO !!!


        *****************************************************
        GAUSSIAN 82: REVISION B: CDC VERSION:  SEPTEMBER-1982
                        84/04/22. 15.18.01.
        *****************************************************
        --------------
        # UMP2/6-31G*
        --------------
        1//1;
        2//2;
        3/5=1,6=6,7=1,25=14/1,2,3,11,14;
        4/7=2,16=1/1;
        5/6=7/2;
        6//1;
        8/10=1/1,2;
        9//1;
        99/5=1/99;

           ### NEXT LINK= 101   LL=   0



        --------------------------------------------------
        METHYLENE TRIPLET MP2/6-31G*//6-31G* (CMU 1332)
        --------------------------------------------------

        SYMBOLIC Z-MATRIX
          CHARGE = 0 MULTIPLICITY = 3
        X
        C    1    1.00000
        H    2    RCH      1    ANG
        H    2    RCH      1    ANG      3    180.00000 0
             VARIABLES
          RCH              1.07134
          ANG            114.78339
```

FIG. 5.3.7. The output produced by GAUSSIAN82 from the input shown in Fig. 5.3.6.

```
                        Z-MATRIX (ANGSTROMS AND DEGREES)
CD CENT ATOM  N1     LENGTH      N2    ALPHA      N3     BETA      J
```

CD	CENT	ATOM	N1	LENGTH	N2	ALPHA	N3	BETA	J
1		X							
2	1	C	1	1.000000 (1)					
3	2	H	2	1.071340 (2)	1	114.783 (4)			
4	3	H	2	1.071340 (3)	1	114.783 (5)	3	180.000 (6)	0

```
                   Z-MATRIX ORIENTATION:
```

CENTER NUMBER	ATOMIC NUMBER	COORDINATES (ANGSTROMS) X	Y	Z
	-1	0.000000	0.000000	0.000000
1	6	0.000000	0.000000	1.000000
2	1	.972669	0.000000	1.449094
3	1	-.972669	.000000	1.449094

```
               DISTANCE MATRIX (ANGSTROMS)
            1          2          3
1  C   0.000000
2  H   1.071340   0.000000
'3 H   1.071340   1.945337   0.000000
STOICHIOMETRY    CH2(3)

FRAMEWORK GROUP   C2V[C2(C),SGV(H2)]
DEG. OF FREEDOM    2
FULL POINT GROUP                    C2V     NOP  4
LARGEST ABELIAN SUBGROUP            C2V     NOP  4
LARGEST CONCISE ABELIAN SUBGROUP C2      NOP  2
               STANDARD ORIENTATION:
```

CENTER NUMBER	ATOMIC NUMBER	COORDINATES (ANGSTROMS) X	Y	Z
1	6	0.000000	0.000000	-.112273
2	1	0.000000	.972669	.336820
3	1	0.000000	-.972669	.336820

NEXT LINK= 301 LL=   0

```
STANDARD BASIS: 6-31G(D)                (S, S=P, 6D, 7F)
 19 BASIS FUNCTIONS       36 PRIMITIVE GAUSSIANS
  5 ALPHA ELECTRONS       3 BETA ELECTRONS
     NUCLEAR REPULSION ENERGY    6.1992963799 HARTREES
THERE ARE  10 SYMMETRY ADAPTED BASIS FUNCTIONS OF A1  SYMMETRY.
THERE ARE   1 SYMMETRY ADAPTED BASIS FUNCTIONS OF A2  SYMMETRY.
THERE ARE   3 SYMMETRY ADAPTED BASIS FUNCTIONS OF B1  SYMMETRY.
THERE ARE   5 SYMMETRY ADAPTED BASIS FUNCTIONS OF B2  SYMMETRY.
REGULAR INTEGRAL FORMAT.
TWO-ELECTRON INTEGRAL SYMMETRY IS TURNED OFF.
```

FIG. 5.3.7. (*continued*)

```
### NEXT LINK=  302   LL=    0



### NEXT LINK=  303   LL=    0



### NEXT LINK=  311   LL=    0


STANDARD CUTOFFS SELECTED IN SHELL.
    2260 INTEGRALS PRODUCED FOR A TOTAL OF      2260


### NEXT LINK=  314   LL=    0


    5303 INTEGRALS PRODUCED FOR A TOTAL OF      7563


### NEXT LINK=  401   LL=    0


PROJECTED HUCKEL GUESS.
INITIAL GUESS ORBITAL SYMMETRIES.
  ALPHA ORBITALS
       OCCUPIED: (A1) (A1) (B2) (A1) (B1)
       VIRTUAL:  (B2) (A1)
  BETA ORBITALS
       OCCUPIED: (A1) (A1) (B2)
       VIRTUAL:  (A1) (B1) (B2) (A1)


### NEXT LINK=  502   LL=    0


UHF OPEN SHELL SCF.
REQUESTED CONVERGENCE ON DENSITY MATRIX =  .1000E-06 WITHIN  32 CYCLES.
SCF DONE: E(UHF) = -38.9214953474     A.U. AFTER   16 CYCLES
            CONVG  =    .5026D-08           -V/T =  2.0000
            S**2   =   2.0148
ANNIHILATION OF THE LARGEST SPIN CONTAMINANT,
S**2 BEFORE ANNIHILATION   2.0148,  AFTER    2.0001


### NEXT LINK=  601   LL=    0


ORBITAL SYMMETRIES.
  ALPHA ORBITALS
     OCCUPIED: (A1) (A1) (B2) (A1) (B1)
     VIRTUAL:  (A1) (B2) (A1) (B1) (B2) (A1) (A1) (B2) (A2) (A1)
               (B1) (B2) (A1) (A1)
  BETA ORBITALS
     OCCUPIED: (A1) (A1) (B2)
     VIRTUAL:  (A1) (B1) (A1) (B2) (B2) (A1) (B1) (A1) (A1) (B2)
               (A2) (A1) (B1) (B2) (A1) (A1)
  THE ELECTRONIC STATE IS 3-B1.
```

FIG. 5.3.7. (*continued*)

ALPHA MOLECULAR ORBITAL COEFFICIENTS

				1	2	3	4	5
				(A1)	(A1)	(B2)	(A1)	(B1)
	EIGENVALUES --			-11.25948	-.94845	-.61328	-.45322	-.40772
1 1	C	1S		-.99482	-.21022	-.00000	.05188	-.00000
2		2S	(I)	-.03106	.42033	-.00000	-.11864	-.00000
3		2PX	(I)	-.00000	.00000	-.00000	.00000	.61592
4		2PY	(I)	.00000	-.00000	.47213	.00000	.00000
5		2PZ	(I)	-.00017	.02736	-.00000	.57195	-.00000
6		2S	(0)	.00738	.45926	.00000	-.29017	.00000
7		2PX	(0)	-.00000	-.00000	.00000	.00000	.51151
8		2PY	(0)	-.00000	-.00000	.20824	.00000	.00000
9		2PZ	(0)	.00089	.01509	.00000	.41933	.00000
10		XX		.00316	-.00199	-.00000	-.00684	-.00000
11		YY		.00373	.01437	.00000	.02576	-.00000
12		ZZ		.00332	.00270	.00000	-.00163	.00000
13		XY		.00000	-.00000	.00000	-.00000	.00000
14		XZ		-.00000	-.00000	-.00000	.00000	-.00158
15		YZ		-.00000	-.00000	.02990	.00000	-.00000
16 2	H	1S	(I)	.00022	.15114	.23529	.11551	.00000
17		1S	(0)	-.00199	.03053	.15926	.11145	-.00000
18 3	H	1S	(I)	.00022	.15114	-.23529	.11551	-.00000
19		1S	(0)	-.00199	.03053	-.15926	.11145	.00000

				6	7	8	9	10
				(A1)	(B2)	(A1)	(B1)	(B2)
	EIGENVALUES --			.25848	.32724	.71868	.73001	.76723
1 1	C	1S		-.11958	.00000	.02646	.00000	-.00000
2		2S	(I)	.14943	-.00000	-.07546	.00000	.00000
3		2PX	(I)	.00000	.00000	.00000	1.04940	-.00000
4		2PY	(I)	-.00000	-.34055	-.00000	-.00000	-.73328
5		2PZ	(I)	.15824	-.00000	.94792	-.00000	-.00000
6		2S	(0)	1.94156	-.00000	-.14752	-.00000	.00000
7		2PX	(0)	-.00000	.00000	-.00000	-1.10407	.00000
8		2PY	(0)	-.00000	-1.66242	.00000	.00000	1.96267
9		2PZ	(0)	.68716	-.00000	-1.16239	.00000	-.00000
10		XX		.00551	.00000	-.03703	-.00000	-.00000
11		YY		-.02274	.00000	.05769	-.00000	-.00000
12		ZZ		-.00014	.00000	-.02629	.00000	-.00000
13		XY		-.00000	-.00000	-.00000	-.00000	-.00000
14		XZ		-.00000	.00000	.00000	-.00592	.00000
15		YZ		-.00000	.01140	-.00000	-.00000	-.11115
16 2	H	1S	(I)	-.06533	.01639	.24365	-.00000	-.42013
17		1S	(0)	-1.43432	1.85058	.00827	-.00000	-.62924
18 3	H	1S	(I)	-.06533	-.01639	.24365	.00000	.42013
19		1S	(0)	-1.43432	-1.85058	.00827	.00000	.62924

BETA MOLECULAR ORBITAL COEFFICIENTS.

				1	2	3	4	5
				(A1)	(A1)	(B2)	(A1)	(B1)
	EIGENVALUES --			-11.21150	-.77680	-.58039	.14580	.17625
1 1	C	1S		-.99635	-.17869	-.00000	-.09223	-.00000
2		2S	(I)	-.02268	.34386	.00000	.09785	.00000
3		2PX	(I)	-.00000	-.00000	.00000	.00000	-.35843
4		2PY	(I)	-.00000	-.00000	.42249	.00000	.00000
5		2PZ	(I)	-.00069	.13509	-.00000	-.31330	-.00000
6		2S	(0)	.00791	.28360	-.00000	.60729	.00000
7		2PX	(0)	-.00000	-.00000	.00000	.00000	-.75122
8		2PY	(0)	-.00000	-.00000	.15248	-.00000	.00000
9		2PZ	(0)	.00082	.03425	.00000	-.58709	-.00000

FIG. 5.3.7. (*continued*)

10	XX	.00212	−.02507	.00000	−.00319	−.00000
11	YY	.00029	.04994	.00000	.00579	.00000
12	ZZ	.00183	−.00703	.00000	−.02995	.00000
13	XY	.00000	.00000	.00000	−.00000	−.00000
14	XZ	−.00000	−.00000	.00000	.00000	−.01945
15	YZ	−.00000	−.00000	.03740	−.00000	−.00000
16 2 H 1S (I)		.00043	.21159	.26181	−.07246	.00000
17 1S (O)		−.00209	.12530	.21916	−.19966	.00000
18 3 H 1S (I)		.00043	.21159	−.26181	−.07246	−.00000
19 1S (O)		−.00209	.12530	−.21916	−.19966	−.00000

		6	7	8	9	10
		(A1)	(B2)	(B2)	(A1)	(B1)
EIGENVALUES --		.29163	.36254	.80083	.81415	.90065
1 1 C 1S		−.12748	.00000	−.00000	.06402	.00000
2	2S (I)	.13382	−.00000	−.00000	.58574	−.00000
3	2PX (I)	.00000	−.00000	.00000	.00000	−1.16258
4	2PY (I)	−.00000	−.34423	.71919	.00000	.00000
5	2PZ (I)	.17027	−.00000	−.00000	.56011	.00000
6	2S (O)	2.02171	−.00000	.00000	−1.49171	.00000
7	2PX (O)	.00000	.00000	−.00000	−.00000	.95724
8	2PY (O)	−.00000	−1.68266	−2.00916	−.00000	.00000
9	2PZ (O)	.74638	−.00000	.00000	−.91038	−.00000
10	XX	−.01918	.00000	.00000	−.00824	.00000
11	YY	.00075	.00000	−.00000	.18534	−.00000
12	ZZ	.00323	.00000	.00000	.02303	−.00000
13	XY	.00000	−.00000	.00000	.00000	.00000
14	XZ	−.00000	−.00000	.00000	.00000	−.01466
15	YZ	−.00000	−.01639	.09735	.00000	−.00000
16 2 H 1S (I)		−.02398	−.03123	.42989	.55042	.00000
17 1S (O)		−1.46530	1.88000	.67219	.13887	.00000
18 3 H 1S (I)		−.02398	.03123	−.42989	.55042	−.00000
19 1S (O)		−1.46530	−1.88000	−.67219	.13887	−.00000

FULL MULLIKEN POPULATION ANALYSIS.

		1	2	3	4	5
1 1 C 1S		2.06118				
2	2S (I)	−.02245	.31047			
3	2PX (I)	0.00000	0.00000	.37936		
4	2PY (I)	0.00000	0.00000	0.00000	.40141	
5	2PZ (I)	.00000	−.00000	0.00000	0.00000	.34613
6	2S (O)	−.03271	.26365	0.00000	0.00000	−.00000
7	2PX (O)	0.00000	0.00000	.17950	0.00000	0.00000
8	2PY (O)	0.00000	0.00000	0.00000	.09272	0.00000
9	2PZ (O)	.00000	.00000	0.00000	0.00000	.13952
10	XX	−.00006	−.00624	0.00000	0.00000	−.00000
11	YY	−.00116	.01423	0.00000	0.00000	.00000
12	ZZ	−.00036	−.00087	0.00000	0.00000	−.00000
13	XY	0.00000	0.00000	0.00000	0.00000	0.00000
14	XZ	0.00000	0.00000	−.00000	0.00000	0.00000
15	YZ	0.00000	0.00000	0.00000	.00000	0.00000
16 2 H 1S (I)		−.00228	.03500	0.00000	.07673	.01579
17 1S (O)		−.00179	.02074	0.00000	.04053	.00909
18 3 H 1S (I)		−.00228	.03500	0.00000	.07673	.01579
19 1S (O)		−.00179	.02074	0.00000	.04053	.00909

		6	7	8	9	10
6	2S (O)	.37567				
7	2PX (O)	0.00000	.26164			
8	2PY (O)	0.00000	0.00000	.06661		
9	2PZ (O)	−.00000	0.00000	0.00000	.17724	

FIG. 5.3.7. (*continued*)

10			XX		-.00378	0.00000	0.00000	0.00000	.00069
11			YY		.00839	0.00000	0.00000	0.00000	-.00048
12			ZZ		-.00015	0.00000	0.00000	0.00000	.00007
13			XY		0.00000	0.00000	0.00000	0.00000	0.00000
14			XZ		0.00000	0.00000	0.00000	0.00000	0.00000
15			YZ		0.00000	0.00000	0.00000	0.00000	0.00000
16	2	H	1S	(I)	.03688	0.00000	.04168	.01254	-.00079
17			1S	(O)	.01225	0.00000	.03503	.01251	-.00142
18	3	H	1S	(I)	.03688	0.00000	.04168	.01254	-.00079
19			1S	(O)	.01225	0.00000	.03503	.01251	-.00142

					11	12	13	14	15
11			YY		.00338				
12			ZZ		-.00011	.00007			
13			XY		0.00000	0.00000	.00000		
14			XZ		0.00000	0.00000	0.00000	.00000	
15			YZ		0.00000	0.00000	0.00000	0.00000	.00229
16	2	H	1S	(I)	.00644	-.00023	0.00000	0.00000	.00384
17			1S	(O)	.00405	-.00037	0.00000	0.00000	.00068
18	3	H	1S	(I)	.00644	-.00023	0.00000	0.00000	.00384
19			1S	(O)	.00405	-.00037	0.00000	0.00000	.00068

					16	17	18	19
16	2	H	1S	(I)	.20486			
17			1S	(O)	.09140	.10246		
18	3	H	1S	(I)	-.00039	-.00569	.20486	
19			1S	(O)	-.00569	-.01491	.09140	.10246

GROSS ORBITAL CHARGES.

					TOTAL	ALPHA	BETA	SPIN
1	1	C	1S		1.99632	.99812	.99821	-.00009
2			2S	(I)	.67028	.39332	.27695	.11637
3			2PX	(I)	.55886	.55886	.00000	.55886
4			2PY	(I)	.72863	.39214	.33650	.05564
5			2PZ	(I)	.53540	.50160	.03380	.46781
6			2S	(O)	.70933	.46055	.24877	.21178
7			2PX	(O)	.44114	.44114	.00000	.44114
8			2PY	(O)	.31274	.18020	.13254	.04767
9			2PZ	(O)	.36686	.35783	.00903	.34880
10			XX		-.01424	-.00051	-.01373	.01322
11			YY		.04522	.00891	.03631	-.02740
12			ZZ		-.00256	.00169	-.00425	.00593
13			XY		.00000	.00000	.00000	.00000
14			XZ		.00000	.00000	.00000	.00000
15			YZ		.01134	.00461	.00673	-.00212
16	2	H	1S	(I)	.51578	.23965	.27613	-.03647
17			1S	(O)	.30456	.11111	.19344	-.08233
18	3	H	1S	(I)	.51578	.23965	.27613	-.03647
19			1S	(O)	.30456	.11111	.19344	-.08233

CONDENSED TO ATOMS (ALL ELECTRONS)

		1	2	3
1	C	5.645554	.356885	.356885
2	H	.356885	.490124	-.026670
3	H	.356885	-.026670	.490124

TOTAL ATOMIC CHARGES.

		1
1	C	6.359324
2	H	.820338
3	H	.820338

FIG. 5.3.7. (*continued*)

```
        ATOMIC SPIN DENSITIES.
               1          2          3
 1   C   2.324103   -.043246   -.043246
 2   H    -.043246   -.084869    .009310
 3   H    -.043246    .009310   -.084869
DIPOLE MOMENT (DEBYE): X=  .0000   Y= -.0000   Z=  .5826   TOTAL=  .5826
FERMI CONTACT ANALYSIS (ATOMIC UNITS).
               1
 1   C    .613330
 2   H   -.027445
 3   H   -.027445


  ### NEXT LINK=  801   LL=    0

RANGE OF M.O."S USED FOR CORRELATION:    2  19

  ### NEXT LINK=  802   LL=    0

  ### NEXT LINK=  901   LL=    0

EXPSYM   FULL IN-CORE ALGORITHM.

EXPSYM   FULL IN-CORE ALGORITHM.
NORM(A1)=  .10131D+01
E(MP2)=      -.81808942E-01       E(CORR)=    -.39003304289E+02
(S**2,0)=  .20148D+01          (S**2,1)=  .20050D+01

  ### NEXT LINK= 9999   LL=    0

  FAU ERLG\CLARK    \84/04/22\SP   \UMP2-FC   \6-31G*    \C1H2(3)   \0
  \\# UMP2/6-31G*\\METHYLENE TRIPLET MP2/6-31G*//6-31G* (CMU 1332)\\0,3\
  X\C,1,1.\H,2,1.07134,1,114.78339\H,2,1.07134,1,114.78339,3,180.,0\\HF=
  -38.9214953\MP2=-39.0033043\S2=2.015\RMSD=.502\DIP=.58259\PG=C02V\\
```

I WOULD TAKE COUNSEL OF MYSELF.
I WOULD STOP AND LOOK WITHIN
AND LOOKING WITHIN, LOOK BACK, ALSO
THAT I MAY LOOK AHEAD WITH CLEARER UNDERSTANDING
OF THE WAY I HAVE BEEN MOVING, AND IN WHAT DIRECTION.
I NEED TO KNOW IF I AM GOING FORWARD OR RETREATING,
WHETHER I HAVE BEEN WASTING, OR ENJOYING
THE PRECIOUS MOMENTS OF LIFE. THERE HAVE BEEN FRICTIONS,
ANNOYANCES AND SOMETIMES WRATH,
BUT WERE THEY BECAUSE I WAS RIGHT AND OTHERS WRONG.....
HAVE I HAD MY THOUGHTS TOO SHARPLY FOCUSED
ON THAT WHICH PLEASED ME, SERVED MY SELF-ESTEEM,
UNDERGIRDED MY SECURITY,
OF WHICH I DID NOT INQUIRE WHETHER IT SERVED OR HAMPERED OTHERS.....
HAVE I BEEN TRYING TO STOP THE CLOCK
TO HOLD THE WORLD
IN PERPETUATION OF WHAT WAS AN IS ALREADY SLIPPING AWAY.....
HAVE I BEEN CRITICAL OF OTHERS FOR WHAT REALLY NEEDED CHANGING IN ME...
LET ME INDEED TAKE COUNSEL OF MYSELF
AND SET MY DIRECTIONS STRAIGHT.

                    R.T. WESTON AS ADAPTED BY D. OSBORN 1967

              BYE,BYE !!!

 THIS JOB IS ARCHIVED AS NUMBER 1074
 LENGTH OF THE ARCHIVED RECORD   287

FIG. 5.3.7. (*continued*)

277

carbon $1s$-orbital. There are now 19 basis functions (13 as before, and 6 d-functions) with 36 primitive Gaussians. In this calculation the Raffenetti integral format and symmetry cannot be used because they are not compatible with the MP2 procedure.

The program then calculates the integrals as before. Note that link 311 produces 2260 integrals involving s- and p-orbitals. The increase from 1314 for 3-21G to 2260 for 6-31G* is due to the increase in the number of primitive Gaussians in the s- and p-part of the basis. Link 314 is then called to calculate the integrals involving d-orbitals, an additional 5303 in this case, giving a total of 7563. Link 401 produces alpha and beta initial guesses, which are identical except for the occupancies of the orbitals. The open-shell SCF calculation is similar to the closed-shell version, the only diference in the output being that the calculated $\langle S^2 \rangle$ value is printed. In this case the calculation gives $\langle S^2 \rangle = 2.0148$, compared with the expectation value of 2 for a pure triplet. Link 502 then tries to eliminate the contribution from the state contributing most to the spin contamination. $\langle S^2 \rangle$ values before and after this *spin annihilation* are printed, and the density matrix is corrected. The UHF energy, however, is not changed.

The alpha and beta orbitals are now printed exactly as demonstrated for triplet CH_2 in Fig. 4.4.7. The electronic state is calculated to be $^3b_1$. The superscript "3" indicates a triplet and the Mulliken symbol for the state is obtained by multiplying the symmetries of the two singly occupied orbitals ($a_1 \times b_1 = b_1$). For closed-shell molecules, the wave function always corresponds to the totally symmetrical species (a_1 or σ_g) because the square of each symmetry type always gives the totally symmetrical species. The orbital symmetries must be squared because the orbitals are all doubly occupied. Radical electronic states are given by the symmetry of the singly occupied orbital. The d-orbitals (10–15) are denoted as x^2, y^2, z^2, xy, xz, and yz. Their coefficients are small, indicating that their role is to polarize the p-functions.

The full Mulliken population analysis given represents the sums of the alpha and beta contributions, which are calculated as for the closed-shell case, but with singly occupied molecular orbitals. The gross orbital charges are given next, the total charges being the sums of the alpha and beta contributions and the spin densities being their differences, as for UHF/MNDO. The Fermi contact analysis is a measure of the spin–spin coupling between the unpaired electron and the nuclei.

The calculation then moves on to the MP2 section, which produces very little output. Molecular orbitals 2–19 are used for the MP2 calculation, that is all except the carbon $1s$, which could be included using the keyword "MP2=FULL." The MP2 correction is -0.08180942 Hartrees, giving a corrected total energy of -39.003304289 Hartrees. The $\langle S^2 \rangle$ parameter is reduced from 2.0148 at the UHF level to 2.005 by the MP2 treatment. The archive entry differs from that for the singlet in that "SP" indicates a single-point, "UMP2-FC" a frozen core MP2 calculation, and the triplet state is indicated by the "(3)" in the stoichiometry field. The $\langle S^2 \rangle$ value is archived for UHF calculations, and both the Hartree–Fock and the MP2 energies are included in the archive entry.

The next example also involves methylene, but is a little more unusual, and also illustrates some of the shortcomings of single-determinant calculations. It is possible to use GAUSSIAN82 to calculate open-shell singlet states, i.e., those in which there are two unpaired electrons of opposite spin. The input for such a job is shown in Fig. 5.3.8. The keyword "UHF," which could have been replaced by "HF" in the last two examples, is essential in this case. The program would otherwise automatically use RHF calculations for a singlet. "UHF" alone, however, is not enough to give the required open-shell singlet state. Specifying UHF for a singlet usually yields exactly the same results as an RHF calculation, but at a cost of more computer time. The electronic states for the alpha and beta electrons normally given by the initial guess for a singlet are identical. In this case, however, the highest-lying beta electron must be shifted to the next-highest orbital to give the open-shell carbene, as shown in Fig. 5.3.9. The electronic state can be changed by use of the keyword "GUESS=ALTER." The remainder of the input is similar to Fig. 5.3.1 until the end of the geometry specification. It is important to note that the input to the various overlays must correspond to the order in which they are called. In the two previous examples the entire input was read by link 101, so there was no problem in determining the correct order. In this example the specification of the orbitals to be swapped in the initial guess is read by link 401, and therefore comes after the geometry specification, as shown in Fig. 5.3.8.

The geometry specification ends with a blank card after the last assignment of a value to the symbolic variables. The input for link 401 then begins with a blank card to indicate that no alpha orbitals are to be swapped. For closed-shell systems the altered configuration is specified by one set of cards, each consisting of two numbers giving the orbitals to be swapped, and

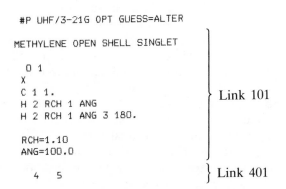

```
#P UHF/3-21G OPT GUESS=ALTER

METHYLENE OPEN SHELL SINGLET

  0 1
  X
  C 1 1.                          Link 101
  H 2 RCH 1 ANG
  H 2 RCH 1 ANG 3 180.

  RCH=1.10
  ANG=100.0

    4    5                        Link 401
```

FIG. 5.3.8. One possible GAUSSIAN82 input for the open-shell singlet configuration of methylene.

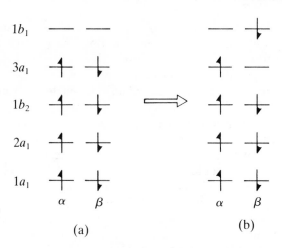

FIG. 5.3.9. The molecular orbital occupancies (a) before and (b) after changing the configuration of the initial guess for the calculation shown in Fig. 5.3.8.

terminated by a blank card. For open-shell molecules there are two such sections, one for the alpha molecular orbitals and one for the beta. The alpha section in this example therefore consists simply of the terminating blank card. The second card of the input to link 401 indicates that beta orbitals 4 and 5 in the initial guess [which corresponds to Fig. 5.3.9(a)] should be swapped to give the occupancy shown in Fig. 5.3.9(b). This is the only change required to give the correct state, and so the input to link 401 is terminated by a blank card to indicate that the beta-occupancy changes are complete. The normal procedure to use if you are not sure which electronic state will be given by the initial guess is to run the calculation with the keyword "GUESS=ONLY," in which case the job will proceed as far as link 401 and print the initial-guess molecular orbitals. This requires little computer time and allows the correct electronic state to be selected. In some cases links 501 or 502 may change the electronic state to one other than that specified in the initial guess. This is not as common in GAUSSIAN82 as it was in the earlier programs, but may still happen. In such a case, link 503 (SCFDM) should be used. In this example, however, link 502 gives the correct state, as shown in the output, which is summarized in Fig. 5.3.10.

Once again the initial stages of the output are very similar to those given in Fig. 5.3.1. In link 401, however, the orbitals to be swapped are indicated and the symmetry of the resulting initial guess is printed. What was the highest occupied beta orbital (molecular orbital 4, the $3a_1$) is now the lowest beta virtual orbital, and the former molecular orbital 5 (the $1b_1$) is now occupied, as specified in the input. The UHF SCF calculation proceeds as usual and converges after 15 cycles. The spin contamination, however, is particularly high ($\langle S^2 \rangle$ should be zero for a singlet). This is an indication that the triplet state of the same symmetry is being mixed into the desired

```
                    HELLO !!!


********************************************************
GAUSSIAN 82: REVISION B: CDC VERSION:  SEPTEMBER-1982
                84/04/22. 16.41.56.
********************************************************
----------------------------
#P UHF/3-21G OPT GUESS=ALTER
----------------------------
1//1,3;
2//2;
3/5=5,11=2,25=11,30=1/1,2,3,11;
4/7=2,8=1,16=1/1;
5/6=7/2;
6//1;
7/27=1/1,2,16;
1//3(1);
99//99;
2//2;
3/5=5,11=2,25=11,30=1/1,2,11;
5/6=7/2;
7/27=1/1,2,16;
1//3(-4);
3/5=5,11=2,25=11,30=1/1,3;
6//1;
99//99;

    ### NEXT LINK= 101   LL=    0


                                    (ENTER 101)
    ----------------------------
    METHYLENE OPEN SHELL SINGLET
    ----------------------------
SYMBOLIC Z-MATRIX
   CHARGE = 0 MULTIPLICITY = 1
X
C    1    1.00000
H    2    RCH      1    ANG
H    2    RCH      1    ANG      3    180.00000 0
    VARIABLES
  RCH              1.10000
  ANG            100.00000


    ### NEXT LINK= 103   LL=    0


                                   (ENTER 103)
GRADGRADGRADGRADGRADGRADGRADGRADGRADGRADGRADGRADGRADGRADGRADGRADGRADGRAD

BERNY OPTIMIZATION

INITIALIZATION PASS
```

FIG. 5.3.10. An extract from the GAUSSIAN82 output produced from the input shown in Fig. 5.3.8. The missing portions of the output are indicated by "⟨⟨⟨BREAK HERE⟩⟩⟩".

```
                          ---------------------------
                          !   INITIAL PARAMETERS    !
                          ! (ANGSTROMS AND DEGREES) !
------------------------------------                    ------------------------
!     NAME        VALUE   DERIVATIVE INFORMATION (ATOMIC UNITS)              !
------------------------------------------------------------------------------
!     RCH        1.1000   ESTIMATE D2E/DX2                                   !
!     ANG      100.0000   ESTIMATE D2E/DX2                                   !
------------------------------------------------------------------------------
```

GRADGRADGRADGRADGRADGRADGRADGRADGRADGRADGRADGRADGRADGRADGRADGRADGRADGRADGRADGRADGRAD

NEXT LINK= 202   LL=   0

                                        (ENTER 202)

                        Z-MATRIX (ANGSTROMS AND DEGREES)
CD CENT ATOM N1   LENGTH   N2   ALPHA   N3   BETA   J
----

1   X
2   1   C   1   1.000000 (1)
3   2   H   2   1.100000 (2)   1 100.000 (4)
4   3   H   2   1.100000 (3)   1 100.000 (5)   3 180.000 (6) 0

                              Z-MATRIX ORIENTATION:

CENTER   ATOMIC   COORDINATES (ANGSTROMS)
NUMBER   NUMBER   X   Y   Z

              -1   0.000000   0.000000   0.000000
   1   6   0.000000   0.000000   1.000000
   2   1   1.083289   0.000000   1.191013
   3   1   -1.083289   .000000   1.191013

                         DISTANCE MATRIX (ANGSTROMS)
                    1   2   3
1   C   0.000000
2   H   1.100000   0.000000
3   H   1.100000   2.166577   0.000000
STOICHIOMETRY   CH2
FRAMEWORK GROUP   C2V[C2(C),SGV(H2)]
DEG. OF FREEDOM   2
FULL POINT GROUP   C2V   NOP 4
LARGEST ABELIAN SUBGROUP   C2V   NOP 4
LARGEST CONCISE ABELIAN SUBGROUP C2   NOP 2
                         STANDARD ORIENTATION:

CENTER   ATOMIC   COORDINATES (ANGSTROMS)
NUMBER   NUMBER   X   Y   Z
----
   1   6   0.000000   0.000000   -.047753
   2   1   0.000000   1.083289   .143260
   3   1   0.000000   -1.083289   .143260

```

**FIG. 5.3.10.**  *(continued)*
```

NEXT LINK= 301   LL=   0

                                    (ENTER 301)
STANDARD BASIS: 3-21G   (S, S=P, 5D, 7F)
  13 BASIS FUNCTIONS   21 PRIMITIVE GAUSSIANS
   4 ALPHA ELECTRONS   4 BETA ELECTRONS
      NUCLEAR REPULSION ENERGY   6.0170863223 HARTREES
THERE ARE   7 SYMMETRY ADAPTED BASIS FUNCTIONS OF A1 SYMMETRY.
THERE ARE   0 SYMMETRY ADAPTED BASIS FUNCTIONS OF A2 SYMMETRY.
THERE ARE   2 SYMMETRY ADAPTED BASIS FUNCTIONS OF B1 SYMMETRY.
THERE ARE   4 SYMMETRY ADAPTED BASIS FUNCTIONS OF B2 SYMMETRY.
RAFFENETTI 2 INTEGRAL FORMAT.
TWO-ELECTRON INTEGRAL SYMMETRY IS TURNED ON.

  ### NEXT LINK= 302   LL=   0

                                    (ENTER 302)

  ### NEXT LINK= 303   LL=   0

                                    (ENTER 303)

  ### NEXT LINK= 311   LL=   0

                                    (ENTER 311)

STANDARD CUTOFFS SELECTED IN SHELL.
   1314 INTEGRALS PRODUCED FOR A TOTAL OF   1314

  ### NEXT LINK= 401   LL=   0

                                    (ENTER 401)

PROJECTED HUCKEL GUESS.
NO ALPHA ORBITALS SWITCHED.
   0 0
PAIRS OF BETA ORBITALS SWITCHED:
   4 5
INITIAL GUESS ORBITAL SYMMETRIES.
  ALPHA ORBITALS
     OCCUPIED: (A1) (A1) (B2) (A1)
     VIRTUAL: (B1) (B2) (A1)
  BETA ORBITALS
     OCCUPIED: (A1) (A1) (B2) (B1)
     VIRTUAL: (A1) (B2) (A1)

  ### NEXT LINK= 502   LL=   0

FIG. 5.3.10. (*continued*)

```
UHF OPEN SHELL SCF.
REQUESTED CONVERGENCE ON DENSITY MATRIX = .1000E-06 WITHIN  32 CYCLES.
ITER  ELECTRONIC-ENERGY        CONVERGENCE    EXTRAPOLATION
----  ------------------       -----------    -------------
  1   -.445686046375033D+02
  2   -.446628968544580D+02     .6386D-02
  3   -.446735494519201D+02     .3211D-02
  4   -.446754083313665D+02     .1672D-02
  5   -.446758242163958D+02     .1000D-02
  6   -.446759414129676D+02     .6180D-03      4-POINT.
  7   (NON-VARIATIONAL)
  8   -.446760059436763D+02     .2675D-04
  9   -.446760061199216D+02     .1472D-04
 10   -.446760061559141D+02     .8366D-05
 11   -.446760061668624D+02     .5449D-05
 12   -.446760061709574D+02     .3558D-05
 13   -.446760061726309D+02     .2349D-05      4-POINT.
 14   (NON-VARIATIONAL)
 15   -.446760061738744D+02
SCF DONE: E(UHF) = -38.6589198515   A.U. AFTER  15 CYCLES
          CONVG =    .2627D-07          -V/T = 2.0053
          S**2  =   1.0005
ANNIHILATION OF THE LARGEST SPIN CONTAMINANT.
S**2 BEFORE ANNIHILATION   1.0005,  AFTER     .0036
```

$\langle\langle\langle\langle\langle\langle\langle\langle\langle\langle\langle$ BREAK HERE $\rangle\rangle\rangle\rangle\rangle\rangle\rangle\rangle\rangle\rangle$

```
------------------------------------------------------------------
                      Z-MATRIX (ANGSTROMS AND DEGREES)
CD CENT ATOM  N1    LENGTH      N2    ALPHA      N3    BETA      J
------------------------------------------------------------------

 1       X
 2   1   C    1  1.000000 ( 1)
 3   2   H    2  1.069329 ( 2)  1  112.074 ( 4)
 4   3   H    2  1.069329 ( 3)  1  112.074 ( 5)  3  180.000 ( 6)  0
------------------------------------------------------------------
```

```
------------------------------------------------------------------
              INTERNAL COORDINATE FORCES (HARTREES/BOHR OR /RADIAN)
CENT ATOM N1     LENGTH     N2     ALPHA     N3      BETA      J
------------------------------------------------------------------

      X
 1  C    1   -.000000 ( 1)
 2  H    2    .000065 ( 2)  1  -.000019 ( 4)
 3  H    2    .000065 ( 3)  1  -.000019 ( 5)  3  -.000000 ( 6)  0
------------------------------------------------------------------
              MAX    .000065    RMS    .000039
```

NEXT LINK= 103   LL=   0

GRADGRADGRADGRADGRADGRADGRADGRADGRADGRADGRADGRADGRADGRADGRADGRADGRADGRAD

BERNY OPTIMIZATION

FIG. 5.3.10. (*continued*)

SEARCH FOR A LOCAL MINIMUM.

STEP NUMBER   5 OUT OF A MAXIMUM OF 12
ALL QUANTITIES PRINTED IN INTERNAL UNITS (HARTREES-BOHRS-RADIANS)

UPDATE SECOND DERIVATIVES USING INFORMATION FROM POINTS: 3 4 5

THE SECOND DERIVATIVE MATRIX:
```
                     RCH       ANG
        RCH         .76409
        ANG        -.09212    .36313
    EIGENVALUES ---  .34298    .78424
```

VARIABLE	OLD X	-DE/DX	DELTA X (LINEAR)	DELTA X (QUAD)	DELTA X (TOTAL)	NEW X
RCH	2.02074	.00013	.00008	.00008	.00016	2.02090
ANG	1.95606	-.00004	-.00016	.00010	-.00006	1.95600

ITEM	VALUE	THRESHOLD	CONVERGED?
MAXIMUM FORCE	.000129	.000450	YES
RMS FORCE	.000095	.000300	YES
MAXIMUM DISPLACEMENT	.000162	.001800	YES
RMS DISPLACEMENT	.000123	.001200	YES

PREDICTED CHANGE IN ENERGY -.000000

OPTIMIZATION COMPLETED.
   -- STATIONARY POINT FOUND.
   -- LAST STEP NOT IMPLEMENTED.

```
              ---------------------------
              !  OPTIMIZED PARAMETERS   !
              !  (ANGSTROMS AND DEGREES) !
```

!	NAME	VALUE	DERIVATIVE INFORMATION (ATOMIC UNITS)	!
!	RCH	1.0693	-DE/DX = .000129	!
!	ANG	112.0743	-DE/DX = -.000037	!

GRADGRADGRADGRADGRADGRADGRADGRADGRADGRADGRADGRADGRADGRADGRADGRADGRADGRADGRAD

   ### NEXT LINK= 301   LL=   0

                              (ENTER 301)
STANDARD BASIS: 3-21G   (S, S=P, 5D, 7F)
 13 BASIS FUNCTIONS   21 PRIMITIVE GAUSSIANS
  4 ALPHA ELECTRONS   4 BETA ELECTRONS

   NUCLEAR REPULSION ENERGY   6.2054259777 HARTREES
THERE ARE   7 SYMMETRY ADAPTED BASIS FUNCTIONS OF A1 SYMMETRY.
THERE ARE   0 SYMMETRY ADAPTED BASIS FUNCTIONS OF A2 SYMMETRY.
THERE ARE   2 SYMMETRY ADAPTED BASIS FUNCTIONS OF B1 SYMMETRY.
THERE ARE   4 SYMMETRY ADAPTED BASIS FUNCTIONS OF B2 SYMMETRY.
RAFFENETTI 2 INTEGRAL FORMAT.
TWO-ELECTRON INTEGRAL SYMMETRY IS TURNED ON.

FIG. 5.3.10. (*continued*)

NEXT LINK= 303   LL=   0
(ENTER 303)

NEXT LINK= 601   LL=   0
(ENTER 601)

ORBITAL SYMMETRIES.
 ALPHA ORBITALS
      OCCUPIED: (A1) (A1) (B2) (A1)
      VIRTUAL: (B1) (A1) (B2) (A1) (B2) (B1) (A1) (B2) (A1)
 BETA ORBITALS
      OCCUPIED: (A1) (A1) (B2) (B1)
      VIRTUAL: (A1) (A1) (B2) (B1) (A1) (B2) (A1) (B2) (A1)
 THE ELECTRONIC STATE IS 1-B1.
   ALPHA MOLECULAR ORBITAL COEFFICIENTS

			1	2	3	4	5
			(A1)	(A1)	(B2)	(A1)	(B1)
EIGENVALUES --			-11.18315	-.85146	-.61227	-.40512	.14392
1 1	C	1S	-.98651	.20703	.00000	.06511	-.00000
2		2S (I)	-.09099	-.22856	.00000	-.06591	.00000
3		2PX(I)	.00000	.00000	.00000	-.00000	-.35180
4		2PY(I)	.00000	.00000	.38171	-.00000	-.00000
5		2PZ(I)	-.00148	-.03978	-.00000	.47443	.00000
6		2S (0)	.04614	-.55233	-.00000	-.35245	-.00000
7		2PX(0)	-.00000	.00000	-.00000	-.00000	-.76830
8		2PY(0)	-.00000	.00000	.25663	-.00000	-.00000
9		2PZ(0)	.00560	-.02228	.00000	.54585	-.00000
10 2	H	1S (I)	.00256	-.18305	.24172	.09428	-.00000
11		1S (0)	-.01339	-.08447	.20013	.11051	.00000
12 3	H	1S (I)	.00256	-.18305	-.24172	.09428	.00000
13		1S (0)	-.01339	-.08447	-.20013	.11051	-.00000

			6	7	8	9
			(A1)	(B2)	(A1)	(B2)
EIGENVALUES --			.29596	.36880	.92371	1.00910
1 1	C	1S	.14874	.00000	.03585	.00000
2		2S (I)	-.06056	-.00000	.04366	.00000
3		2PX(I)	.00000	-.00000	.00000	-.00000
4		2PY(I)	.00000	.31284	.00000	.75399
5		2PZ(I)	-.14024	.00000	.91194	-.00000
6		2S (0)	-1.84970	.00000	-.22891	-.00000
7		2PX(0)	.00000	-.00000	-.00000	.00000
8		2PY(0)	.00000	1.66424	-.00000	-1.92444
9		2PZ(0)	-.61010	-.00000	-1.01652	.00000
10 2	H	1S (I)	.06301	.00178	.38279	.53214
11		1S (0)	1.29884	-1.71987	-.09904	.55039
12 3	H	1S (I)	.06301	-.00178	.38279	-.53214
13		1S (0)	1.29884	1.71987	-.09904	-.55039

   BETA MOLECULAR ORBITAL COEFFICIENTS.

			1	2	3	4	5
			(A1)	(A1)	(B2)	(B1)	(A1)
EIGENVALUES --			-11.18247	-.86389	-.60044	-.36532	.11781
1 1	C	1S	-.98636	.20825	.00000	.00000	.08930
2		2S (I)	-.09144	-.22467	.00000	.00000	-.04159
3		2PX(I)	.00000	.00000	.00000	.50780	.00000
4		2PY(I)	.00000	.00000	-.39046	.00000	-.00000
5		2PZ(I)	-.00189	-.10149	-.00000	-.00000	.31039

FIG. 5.3.10.   (*continued*)

6		2S	(0)	.04584	-.58868	-.00000	-.00000	-.59466
7		2PX	(0)	-.00000	.00000	.00000	.63377	.00000
8		2PY	(0)	-.00000	.00000	-.26943	-.00000	-.00000
9		2PZ	(0)	.00533	-.07613	-.00000	-.00000	.61238
10 2	H	1S	(I)	.00276	-.16916	-.23291	.00000	.07815
11		1S	(0)	-.01331	-.05838	-.19235	.00000	.22347
12 3	H	1S	(I)	.00276	-.16916	.23291	-.00000	.07815
13		1S	(0)	-.01331	-.05838	.19235	-.00000	.22347

				6	7	8	9
				(A1)	(B2)	(B1)	(A1)
	EIGENVALUES --			.28679	.38758	.94335	.98674
1 1	C	1S		.13389	-.00000	.00000	.04587
2		2S	(I)	-.05247	-.00000	.00000	.28383
3		2PX	(I)	-.00000	.00000	-1.06331	-.00000
4		2PY	(I)	.00000	-.31803	.00000	.00000
5		2PZ	(I)	-.15940	.00000	-.00000	.65791
6		2S	(0)	-1.77288	.00000	-.00000	-.47332
7		2PX	(0)	-.00000	-.00000	.99339	-.00000
8		2PY	(0)	.00000	-1.62565	-.00000	.00000
9		2PZ	(0)	-.67615	-.00000	.00000	-.76118
10 2	H	1S	(I)	.07187	-.01977	.00000	.65893
11		1S	(0)	1.27915	1.71916	.00000	-.26770
12 3	H	1S	(I)	.07187	.01977	-.00000	.65893
13		1S	(0)	1.27915	-1.71916	-.00000	-.26770

FULL MULLIKEN POPULATION ANALYSIS.

				1	2	3	4	5
1 1	C	1S		2.03659				
2		2S	(I)	.01561	.12370			
3		2PX	(I)	0.00000	0.00000	.25786		
4		2PY	(I)	0.00000	0.00000	0.00000	.29816	
5		2PZ	(I)	.00000	.00000	0.00000	0.00000	.23698
6		2S	(0)	-.06322	.20811	0.00000	0.00000	-.00000
7		2PX	(0)	0.00000	0.00000	.17023	0.00000	0.00000
8		2PY	(0)	0.00000	0.00000	0.00000	.10746	0.00000
9		2PZ	(0)	.00000	-.00000	0.00000	0.00000	.14153
10 2	H	1S	(I)	-.00154	.01513	0.00000	.05229	.00801
11		1S	(0)	.00032	.01132	0.00000	.03313	.00548
12 3	H	1S	(I)	-.00154	.01513	0.00000	.05229	.00801
13		1S	(0)	.00032	.01132	0.00000	.03313	.00548

				6	7	8	9	10
6		2S	(0)	.78006				
7		2PX	(0)	0.00000	.40167			
8		2PY	(0)	0.00000	0.00000	.13845		
9		2PZ	(0)	-.00000	0.00000	0.00000	.30430	
10 2	H	1S	(I)	.05826	0.00000	.05858	.01303	.18370
11		1S	(0)	.02773	0.00000	.05611	.01466	.08323
12 3	H	1S	(I)	.05826	0.00000	.05858	.01303	-.00011
13		1S	(0)	.02773	0.00000	.05611	.01466	-.00446

				11	12	13
11		1S	(0)	.10016		
12 3	H	1S	(I)	-.00446	.18370	
13		1S	(0)	-.01493	.08323	.10016

GROSS ORBITAL CHARGES.

				TOTAL	ALPHA	BETA	SPIN
1 1	C	1S		1.98656	.99368	.99288	.00080
2		2S	(I)	.40034	.20829	.19205	.01624
3		2PX	(I)	.42809	.00000	.42809	-.42809

FIG. 5.3.10. (*continued*)

```
 4      2PY (I)    .57646    .28359    .29286   -.00927
 5      2PZ (I)    .40547    .38606    .01942    .36664
 6      2S  (O)   1.09691    .56591    .53099    .03492
 7      2PX (O)    .57191    .00000    .57191   -.57191
 8      2PY (O)    .47530    .23178    .24352   -.01175
 9      2PZ (O)    .50123    .48448    .01674    .46774
10 2 H  1S  (I)    .46612    .24924    .21688    .03237
11      1S  (O)    .31275    .17386    .13889    .03497
12 3 H  1S  (I)    .46612    .24924    .21688    .03237
13      1S  (O)    .31275    .17386    .13889    .03497
           CONDENSED TO ATOMS (ALL ELECTRONS)
                 1         2         3

 1  C   5.737235   .352511   .352511
 2  H    .352511   .450315  -.023955
 3  H    .352511  -.023955   .450315
           TOTAL ATOMIC CHARGES.
                 1
 1  C   6.442257
 2  H    .778872
 3  H    .778872
           ATOMIC SPIN DENSITIES.
                 1         2         3
 1  C   -.133797  -.000436  -.000436
 2  H    -.000436   .062251   .005521
 3  H    -.000436   .005521   .062251
DIPOLE MOMENT (DEBYE): X=  .0000   Y=  .0000   Z=  .8090   TOTAL=  .8090
FERMI CONTACT ANALYSIS (ATOMIC UNITS).
                 1
 1  C    .156275
 2  H    .026050
 3  H    .026050



   ### NEXT LINK= 9999   LL=    0



                          (ENTER9999)
   FAU ERLG\CLARK    \84/04/22\FOPT \UHF       \3-21G    \C1H2    \5
\\#P UHF/3-21G OPT GUESS=ALTER\\METHYLENE OPEN SHELL SINGLET\\0,1\X\C,
1,1.\H,2,RCH,1,ANG\H,2,RCH,1,ANG,3,180.,0\\RCH=1.06933\ANG=112.07425\\
\4,5\\HF=-38.6662514\S2=1.002\RMSD=.921\RMSF=.325\DIP=.80896\PG=C02V\\



THE MOST SERIOUS THREAT TO THE SURVIVAL OF MANKIND IS NOT NOW
IGNORANCE IN THE TRADITIONAL SENSE,
BUT A MORALLY NEUTRAL, AN INSENSITIVE OR INHIBITED HUMAN INTELLIGENCE.


             -- MARGO JEFFERSON IN NEWSWEEK, SEPTEMBER 2, 1974



             BYE,BYE !!!



   THIS JOB IS ARCHIVED AS NUMBER  1076
   LENGTH OF THE ARCHIVED RECORD    290
```

FIG. 5.3.10. (*continued*)

wave function. The results should therefore be treated with caution; energetic comparisons with RHF calculations on the closed-shell singlet may be misleading. Spin annihilation reduces $\langle S^2 \rangle$ to 0.0036. The problem of spin contamination is often, but not always, encountered with open-shell singlet calculations. The calculation leads to a successful optimization, however, giving a structure with a 135.9° H–C–H angle, compared with 104.6° for the closed-shell singlet and 131.0° for the triplet. Note that the population analysis is carried out on the spin-annihilated wave function, so that the contamination by other states is not seen. The table of gross orbital charges, for instance, shows one excess beta electron (negative spin) shared between the carbon p_{xI} (-0.42809) and p_{xO} (-0.57191) orbitals, and one excess alpha electron mainly in the carbon p_z-orbital ($+0.36664$ and $+0.46774$ for inner and outer components, respectively).

It is, however, important to note that only the population analysis is affected by spin annihilation, not the other results of the calculation. The correct procedure to follow in order to obtain more reliable results for this problem would be first to use a larger basis set to obtain a better UHF geometry and then to carry out the optimization of both the open- and closed-shell singlets with a correlation correction, in this case a CI procedure would be better than a perturbational method such as MP2. The effect of correlation will be discussed in more detail in Section 5.4, but the purpose of this example is to illustrate one of the pitfalls of single-determinant theory. The spin contamination is indicated only by the $\langle S^2 \rangle$ value, not anywhere else in the output, and could easily be missed in a superficial examination of the output. In all other respects the results look exactly like those for a successful calculation. The results of high-level calculations on CH_2 are discussed by Salem.[25] The H–C–H angle of the open-shell singlet is on the order of 144°, and it lies about 1.5 eV higher in energy than the closed-shell singlet. The smaller angle calculated here and the fact that the open-shell singlet is found to be more stable than the closed-shell configuration are direct results of the triplet contamination.

The next example is more successful, and demonstrates the use of ghost atoms and the input of a general basis set. One technique that has been used to overcome the basis set superposition error (BSSE) is known as the *counterpoise* method.[26] Consider the complex formed between ammonia and borane(3). The 3-21G basis set for ammonia consists of 15 basis functions, which must be occupied by 10 electrons. Borane(3), BH_3, is also described with 15 basis functions at 3-21G, but has only eight electrons. The basis set for BH_3, with almost two basis functions per electron, is therefore proportionally better than that for NH_3, which has only 1.5 basis functions per electron. Furthermore, the outer valence functions for the electropositive element boron are considerably more diffuse than those for nitrogen, whose lone-pair orbital is constrained to be more compact than it would "like" to be. If NH_3 is calculated alone, the electrons are strictly confined to the orbitals of the basis set for NH_3. Nothing outside this space is included. If, however, the $BH_3:NH_3$ complex is calculated, the most diffuse boron orbitals extend into the vicinity of nitrogen. BH_3 does not, however, use these orbitals fully, especially the empty p-orbital that points towards nitrogen, as shown in Fig. 5.3.11. Two effects are now important. The first is the formation of the donor–acceptor bond between BH_3 and NH_3—the interaction that the calculation is intended to investigate. The second, however, is that the nitrogen basis set is improved by being able to "borrow" excess basis functions from boron. The lone-pair electrons are no longer confined strictly to the bounds set by the basis functions for NH_3. They can expand into the space made available by inclusion of the relatively sparsely occupied BH_3 basis functions in the calculation. This leads to an extra stabilization of the complex over and above that attributable to the donor–acceptor bond. The counterpoise method attempts to correct for this BSSE by calculating the energy of the ammonia molecule in the presence of the BH_3 basis functions, but without the nuclei and electrons attributable to the borane. This is shown diagrammatically for the boron p-orbital colinear with the BN bond in Fig. 5.3.11, although the entire BH_3 basis set is actually used. The counterpoise method slightly overestimates the BSSE correction, because it makes all the BH_3 orbitals available to the ammonia molecule, whereas in the calculation for $BH_3:NH_3$ these orbitals contain eight electrons.

There are two ways to carry out such calculations in GAUSSIAN82; one is easy and one is much more instructive. Figure 5.3.12 shows two possible inputs for a single-point counterpoise calculation for ammonia at the 3-21G optimized geometry for $BH_3:NH_3$ (from the *Carnegie–Mellon Quantum Chemistry Archive*[20]) in the presence of the basis functions for BH_3. The first input, shown in Fig. 5.3.12(a), uses the MASSAGE option to change the specification of the basis set. The input is written as usual, except for the keywords "MASSAGE" and "GUESS=CORE." After the geometry specification for link 101, however, an extra section is added to make changes to the basis set or to the nuclear charges (atomic numbers). The options used for MASSAGE are given in the GAUSSIAN82 manual, but the four cards here indicate that the nuclear charges (NUC) on centers 1, 3, 4, and 5 are to

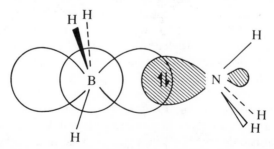

FIG. 5.3.11. The basis set superposition error for $BH_3:NH_3$. The remaining boron and nitrogen p-orbitals have been omitted for clarity.

be changed to zero. Note that the nuclear charges must be given as floating-point numbers (i.e., with decimal points), although they are converted to integers by the program. Atoms with the atomic number zero are known as *ghost atoms*. A ghost atom can be regarded as a point in space upon which orbitals can be centered. In contrast to dummy atoms, ghost atoms count as atomic centers. Atomic centers are numbered in the order in which they appear in the Z-matrix when the dummy atoms are omitted, as shown in the Z-matrices given in the outputs shown above. The input shown in Fig. 5.3.12(a) will produce exactly the same results as that shown in Fig. 5.3.12(b), in which the basis set has been specified completely. Once again the option GUESS=CORE is used. The projected Hückel and NDO initial guesses in GAUSSIAN82 cannot deal with ghost atoms, and so counterpoise calculations must use one of the alternative types of initial guess: that derived from diagonalization of the core Hamiltonian. This is not usually as good an initial guess as that given by the other options, but is applicable to all types of calculation, whatever the basis set. The other keywords in Fig. 5.3.12(b) are "RHF/GEN," specifying an RHF calculation using a general basis set included in the input, and "GFPRINT" to request printing of the basis set. The geometry input is the same as in Fig. 5.3.12(a) except that the boron atom and the three hydrogens attached to it have been given the atomic symbol BQ. (GAUSSIAN82, which has already demonstrated its literary leanings with a range of quotations, names its ghost atoms after Banquo's ghost in *MacBeth*; hence the atomic symbol BQ).

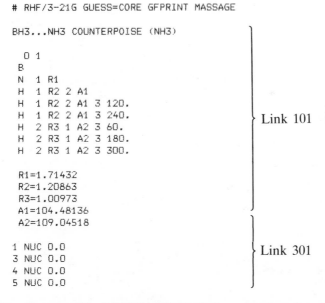

```
# RHF/3-21G GUESS=CORE GFPRINT MASSAGE

BH3...NH3 COUNTERPOISE (NH3)

  0 1
 B
 N   1 R1
 H   1 R2 2 A1
 H   1 R2 2 A1 3 120.
 H   1 R2 2 A1 3 240.            Link 101
 H   2 R3 1 A2 3 60.
 H   2 R3 1 A2 3 180.
 H   2 R3 1 A2 3 300.

 R1=1.71432
 R2=1.20863
 R3=1.00973
 A1=104.48136
 A2=109.04518

 1 NUC 0.0
 3 NUC 0.0                        Link 301
 4 NUC 0.0
 5 NUC 0.0
```

FIG. 5.3.12. Two possible GAUSSIAN82 inputs for a counterpoise calculation on NH_3: (a) using the MASSAGE option;

```
# RHF/GEN GUESS=CORE GFPRINT

BH3...NH3 COUNTERPOISE (NH3)

  0  1
BQ
N   1 R1
BQ  1 R2 2 A1
BQ  1 R2 2 A1 3 120.
BQ  1 R2 2 A1 3 240.
H   2 R3 1 A2 3 60.
H   2 R3 1 A2 3 180.
H   2 R3 1 A2 3 300.

R1=1.71432
R2=1.20863
R3=1.00973
A1=104.48136
A2=109.04518

32103210210210210210210219
  1
        1S      S 3 1.00
116.434             0.0629605
17.4314             0.363304
3.68016             0.697255
        2SP    SP 2 1.00
2.28187            -0.368662           0.231152
0.465248            1.19944            0.866764
        2SP    SP 1 1.00
0.124328            1.0                1.0
****
  2
        1S      S 3 1.00
242.766             0.0598657
36.4851             0.352955
7.81449             0.706513
        2SP    SP 2 1.00
5.42522            -0.413301           0.237972
1.14915             1.22442            0.858953
        2SP    SP 1 1.00
0.283205            1.0                1.0
****
  3 4 5 6 7 8
        1S      S 2 1.10
4.50180             0.156285
0.681444            0.904691
        1S      S 1 1.10
0.151398            1.0
****
```

Link 101

Link 301

FIG. 5.3.12. (b) specifying the complete basis set.

After the section of the input to be read by link 101 comes the specification of the basis set, which is read by link 301. The first card of this section defines the *shells* to be specified. A shell may consist of s-, p-, d-, or f-orbitals, or of a combination of s and p with common exponents, as for most of the internal basis sets in GAUSSIAN82. Thus the 3-21G basis set for carbon, for instance, consists of three shells, the $1s$, the $2s$ and p inner, and the $2s$ and p outer. In the following description a shell consisting of s-, p_x-, p_y-, and p_z-orbitals with common exponents will be denoted by sp. The three shells for the carbon 3-21G basis are therefore $1s$, $2sp_I$, and $2sp_O$. The first card of the basis set section of the input gives the number of primitive Gaussian functions for each shell. Using the above example this is three for the $1s$, two for the $2sp_I$, and one for the $2sp_O$. The shell definition card simply consists of these numbers of primitive Gaussians given atom for atom in the Z-matrix order. The atomic centers are separated by zeroes, and the entire list ends with a "9." Thus the first two centers, the ghost atom with the boron basis set and the nitrogen, are denoted by 3210, the first five hydrogens and ghost atoms with hydrogen orbitals by 210, and the last hydrogen by 219. The basis is thus

Center	Atomic No.	Shells
1	0	$1s(3)$, $2sp_I(2)$, $2sp_O(1)$
2	7	$1s(3)$, $2sp_I(2)$, $2sp_O(1)$
3–5	0	$1s_I(2)$, $1s_O(1)$
6–8	1	$1s_I(2)$, $1s_O(1)$

The individual basis functions must now be defined. This definition uses blocks of cards to assign basis functions to different atomic centers. The first line of any block gives the numbers of the centers to which the basis functions are to be assigned. In Fig. 5.3.12(b) there are three such blocks, one for center 1, one for center 2, and one for centers 3, 4, 5, 6, 7, and 8. The basis set definitions for each type of center are separated by four asterisks. An individual shell consisting of n primitive Gaussians is defined by one card that gives the name assigned to the shell ($1s$, $2sp$, etc.), the number of primitive Gaussian functions, the shell type (s, p, sp, d, or f), and the scale factor (usually 1.0, but 1.10 for hydrogen). This card is then followed by n cards giving the exponents of the individual primitive Gaussian functions with and the coefficients for s-, p-, d-, and f-orbitals within the shell. The $2sp$-shell assigned to center 1 (the boron ghost), for instance, is therefore an sp-shell consisting of two primitive Gaussians with a scale factor of 1.00. The first primitive Gaussian has an exponent of 2.28187, an s-coefficient of -0.368662, and a p-coefficient of 0.231152. The d- and f-coefficients, which are zero, need not be typed. The second primitive Gaussian of this shell has an exponent of 0.465248 and s- and p-coefficients of 1.19944 and 0.866764, respectively. Each type of center is defined in this way and the input ends with a blank card after the last four asterisks. This blank card is essential. Note that the three hydrogen ghosts and the three real hydrogen atoms can be defined together, even though they have different nuclear charges. This section of the input differs from most other types of input for GAUSSIAN82 in that the format is fixed (i.e., the data must be typed in the correct columns, as for MMP2). This format is given in the program manual.

The output produced by GAUSSIAN82 from the input given in Fig. 5.3.12(b) is shown in Fig. 5.3.13. The BQ atoms are assigned the atomic code number 0, as opposed to -1 for dummies, and are not eliminated from the Z-matrix. The symmetry package in link 202 is essentially switched off because of the ghost atoms. To determine the symmetry in this calculation it would be necessary to consider the basis set, which has not yet been assigned, for the ghost atoms. Link 301 reads the basis set and confirms that it has been correctly typed in by its silence. An error in the basis set would result in the message "SIGNIFICANT RENORMALIZATION REQUIRED" being given along with a numerical value corresponding to the

```
                        HELLO !!!


        *****************************************************
        GAUSSIAN 82: REVISION B: CDC VERSION:  SEPTEMBER-1982
                     84/05/31. 10.17.36.
        *****************************************************
        -----------------------------------
        # RHF/GEN GUESS=CORE GFPRINT
        -----------------------------------
        1//1;
        2//2;
        3/5=7,11=1,24=1,25=14,30=1/1,2,3,11,14;
        4/5=2,7=1,16=1/1;
        5//1;
        6//1;
        99/5=1/99;


            --------------------------------
            BH3...NH3 COUNTERPOISE (NH3)
            --------------------------------
        SYMBOLIC Z-MATRIX
            CHARGE = 0 MULTIPLICITY = 1
        BQ
        N    1    R1
        BQ   1    R2    2    A1
        BQ   1    R2    2    A1    3    120.00000 0
        BQ   1    R2    2    A1    3    240.00000 0
        H    2    R3    1    A2    3     60.00000 0
        H    2    R3    1    A2    3    180.00000 0
        H    2    R3    1    A2    3    300.00000 0
```

FIG. 5.3.13. The output produced by GAUSSIAN82 from the input shown in Fig. 5.3.12(b).

```
        VARIABLES
     R1              1.71432
     R2              1.20863
     R3              1.00973
     A1            104.48136
     A2            109.04518
```

                         Z-MATRIX (ANGSTROMS AND DEGREES)
     CD CENT ATOM N1   LENGTH   N2   ALPHA   N3   BETA   J

CD	CENT	ATOM	N1	LENGTH		N2	ALPHA		N3	BETA		J
1	1	BQ										
2	2	N	1	1.714320	(1)							
3	3	BQ	1	1.208630	(2)	2	104.481	(8)				
4	4	BQ	1	1.208630	(3)	2	104.481	(9)	3	120.000	(14)	0
5	5	BQ	1	1.208630	(4)	2	104.481	(10)	3	240.000	(15)	0
6	6	H	2	1.009730	(5)	1	109.045	(11)	3	60.000	(16)	0
7	7	H	2	1.009730	(6)	1	109.045	(12)	3	180.000	(17)	0
8	8	H	2	1.009730	(7)	1	109.045	(13)	3	300.000	(18)	0

                      Z-MATRIX ORIENTATION:

CENTER	ATOMIC		COORDINATES (ANGSTROMS)	
NUMBER	NUMBER	X	Y	Z
1	0	0.000000	0.000000	0.000000
2	7	0.000000	0.000000	1.714320
3	0	1.170231	0.000000	-.302236
4	0	-.585115	-1.013449	-.302236
5	0	-.585115	1.013449	-.302236
6	1	.477229	.826586	2.043809
7	1	-.954459	.000000	2.043809
8	1	.477229	-.826586	2.043809

                   DISTANCE MATRIX (ANGSTROMS)
```
              1          2          3          4          5
 1 BQ  0.000000
 2 N   1.714320   0.000000
 3 BQ  1.208630   2.331510   0.000000
 4 BQ  1.208630   2.331510   2.026899   0.000000
 5 BQ  1.208630   2.331510   2.026899   2.026899   0.000000
 6 H   2.255692   1.009730   2.582135   3.165159   2.582135
 7 H   2.255692   1.009730   3.165159   2.582135   2.582135
 8 H   2.255692   1.009730   2.582135   2.582135   3.165159
              6          7          8
 6 H   0.000000
 7 H   1.653171   0.000000
 8 H   1.653171   1.653171   0.000000
SYMM--  CANNOT COPE WITH DUMMY OR (SHIVER ) GHOST ATOMS.
SYMM--  SYMMETRY TURNED OFF.
STOICHIOMETRY    H3N
FRAMEWORK GROUP
PRSFWG-- UNABLE TO FIND LEFT SQUARE BRACKET.
NUMDOF-- UNRECOGNIZED SYMMETRIC SUBSPACE, ICHAR= E   "
DEG. OF FREEDOM   -1
FULL POINT GROUP                    NOP  1
```

FIG. 5.3.13. *(continued)*

GENERAL BASIS READ FROM CARDS: (S, S=P, 5D, 7F)

```
***********************************************************************************************************************
*            ATOMIC CENTER            *      ATOMIC ORBITAL      *                  GAUSSIAN FUNCTIONS                *
***********************************************************************************************************************
*                                     * FUNCTION   SHELL   SCALE *                                                   *
*  ATOM    X-COORD  Y-COORD  Z-COORD   * NUMBER    TYPE   FACTOR * EXPONENT    S-COEF       P-COEF     D-COEF   F-COEF *
***********************************************************************************************************************
*  BANQUO  0.00000  0.00000  0.00000        1             1.00    .116434D+03 .629605D-010.          0.       0.      *
*                                                                 .174314D+02 .363304D+000.          0.       0.      *
*                                                                 .368016D+01 .697255D+000.          0.       0.      *
*                                          2- 5           1.00    .228187D+01-.368662D+00 .231152D+000.        0.      *
*                                                                 .465248D+00 .119944D+01 .866764D+000.        0.      *
*                                          6- 9           1.00    .124328D+00 .100000D+01 .100000D+010.        0.      *
+---------------------------------------------------------------------------------------------------------------------+
* NITROGEN 0.00000  0.00000  3.23960        10            1.00    .242766D+03 .598657D-010.          0.       0.      *
*                                                                 .364851D+02 .352955D+000.          0.       0.      *
*                                                                 .781449D+01 .706513D+000.          0.       0.      *
*                                          11- 14         1.00    .545222D+01-.413301D+00 .237972D+000.        0.      *
*                                                                 .114915D+01 .122442D+01 .858953D+000.        0.      *
*                                          15- 18         1.00    .283205D+00 .100000D+01 .100000D+010.        0.      *
+---------------------------------------------------------------------------------------------------------------------+
*  BANQUO  2.21142  0.00000  -.57114        19            1.10    .450180D+01 .156285D+000.          0.       0.      *
*                                                                 .681444D+00 .904691D+000.          0.       0.      *
*                                          20            1.10    .151398D+00 .100000D+010.          0.       0.      *
+---------------------------------------------------------------------------------------------------------------------+
*  BANQUO -1.10571 -1.91514 -.57114         21            1.10    .450180D+01 .156285D+000.          0.       0.      *
*                                                                 .681444D+00 .904691D+000.          0.       0.      *
*                                          22            1.10    .151398D+00 .100000D+010.          0.       0.      *
+---------------------------------------------------------------------------------------------------------------------+
*  BANQUO -1.10571  1.91514 -.57114         23            1.10    .450180D+01 .156285D+000.          0.       0.      *
*                                                                 .681444D+00 .904691D+000.          0.       0.      *
*                                          24            1.10    .151398D+00 .100000D+010.          0.       0.      *
+---------------------------------------------------------------------------------------------------------------------+
* HYDROGEN  .90183  1.56202  3.86224        25            1.10    .450180D+01 .156285D+000.          0.       0.      *
*                                                                 .681444D+00 .904691D+000.          0.       0.      *
*                                          26            1.10    .151398D+00 .100000D+010.          0.       0.      *
+---------------------------------------------------------------------------------------------------------------------+
* HYDROGEN -1.80367  0.00000  3.86224        27            1.10    .450180D+01 .156285D+000.          0.       0.      *
*                                                                 .681444D+00 .904691D+000.          0.       0.      *
*                                          28            1.10    .151398D+00 .100000D+010.          0.       0.      *
+---------------------------------------------------------------------------------------------------------------------+
* HYDROGEN  .90183 -1.56202  3.86224        29            1.10    .450180D+01 .156285D+000.          0.       0.      *
*                                                                 .681444D+00 .904691D+000.          0.       0.      *
*                                          30            1.10    .151398D+00 .100000D+010.          0.       0.      *
***********************************************************************************************************************
```

30 BASIS FUNCTIONS   48 PRIMITIVE GAUSSIANS
5 ALPHA ELECTRONS   5 BETA ELECTRONS
NUCLEAR REPULSION ENERGY   11.9659278113 HARTREES
RAFFENETTI 1 INTEGRAL FORMAT.
TWO-ELECTRON INTEGRAL SYMMETRY IS TURNED OFF.

STANDARD CUTOFFS SELECTED IN SHELL.
    90984 INTEGRALS PRODUCED FOR A TOTAL OF   90984
SHELL: FMTGEN WAS CALLED   12754 TIMES.

      0 INTEGRALS PRODUCED FOR A TOTAL OF   90984

RHF CLOSED SHELL SCF.
REQUESTED CONVERGENCE ON DENSITY MATRIX=   .5000E-04 WITHIN 32 CYCLES.
N**3 SYMMETRY TURNED OFF IN RHFCLO.
SCF DONE: E(RHF) = -55.8832894476   A.U. AFTER 15 CYCLES
          CONVG =   .1327E-05   -V/T = 2.0038

FIG. 5.3.13.   (continued)

MOLECULAR ORBITAL COEFFICIENTS

			1	2	3	4	5
	EIGENVALUES --		-15.45915	-1.13671	-.62772	-.62772	-.40562
1 1	BQ	1S	.00040	-.00027	-.00000	-.00000	-.00313
2		2S	.00100	.00043	-.00000	-.00000	-.03511
3		1PX	-.00000	.00000	-.00033	-.00342	.00000
4		1PY	-.00000	-.00000	-.00342	.00033	.00000
5		1PZ	-.00274	.00554	-.00000	-.00000	-.02581
6		3S	-.01574	.01313	.00000	.00000	.10414
7		2PX	-.00000	-.00000	-.00119	-.01233	-.00000
8		2PY	-.00000	.00000	-.01233	.00119	-.00000
9		2PZ	-.00342	-.00259	.00000	.00000	.06755
10 2	N	1S	-.98578	.21716	.00000	.00000	-.06384
11		2S	-.10006	-.19883	.00000	.00000	.06570
12		1PX	-.00000	.00000	.03691	.38309	.00000
13		1PY	-.00000	.00000	.38309	-.03691	.00000
14		1PZ	-.00147	-.07495	-.00000	-.00000	-.45736
15		3S	.05467	-.65953	-.00000	-.00000	.26798
16		2PX	.00000	.00000	.03316	.34411	.00000
17		2PY	.00000	.00000	.34411	-.03316	-.00000
18		2PZ	-.00046	-.07725	.00000	.00000	-.55618
19 3	BQ	1S	.00023	.00001	.00010	.00102	-.00105
20		2S	.00406	-.00888	.00115	.01191	.02093
21 4	BQ	1S	.00023	.00001	-.00093	-.00042	-.00105
22		2S	.00406	-.00888	-.01089	-.00496	.02093
23 5	BQ	1S	.00023	.00001	.00083	-.00059	-.00105
24		2S	.00406	-.00888	.00974	-.00695	.02093
25 6	H	1S	-.00199	-.12031	.24209	.11031	-.04361
26		2S	-.00679	-.02205	.19980	.09104	-.06693
27 7	H	1S	-.00199	-.12031	-.02552	-.26481	-.04361
28		2S	-.00679	-.02205	-.02106	-.21855	-.06693
29 8	H	1S	-.00199	-.12031	-.21658	.15450	-.04361
30		2S	-.00679	-.02205	-.17874	.12751	-.06693

			6	7	8	9	10
	EIGENVALUES --		.22320	.28941	.34710	.34710	.36583
1 1	BQ	1S	-.00698	-.00708	.00000	-.00000	.00000
2		2S	-.07362	-.07115	.00000	-.00000	.00000
3		1PX	.00000	.00000	-.00022	.09544	.01728
4		1PY	-.00000	-.00000	-.09544	-.00022	-.00003
5		1PZ	.02910	.04682	.00000	-.00000	.00000
6		3S	.03139	.42310	-.00000	.00000	-.00000
7		2PX	-.00000	-.00000	.00229	-.97746	-.07337
8		2PY	.00000	-.00000	.97746	.00229	.00013
9		2PZ	-.26450	-.39254	-.00000	.00000	-.00000
10 2	N	1S	.09935	-.10882	.00000	.00000	.00000
11		2S	-.05022	.00914	-.00000	.00000	-.00000
12		1PX	.00000	.00000	-.00022	.09368	-.28217
13		1PY	-.00000	.00000	-.09368	-.00022	.00048
14		1PZ	.01140	.16554	-.00000	-.00000	.00000
15		3S	-.88840	1.26821	-.00000	-.00000	.00000
16		2PX	.00000	.00000	-.00069	.29467	-.97163
17		2PY	-.00000	.00000	-.29467	-.00069	.00167
18		2PZ	-.15601	.32873	-.00000	.00000	-.00000
19 3	BQ	1S	-.03226	-.01736	-.00011	.04901	.01673
20		2S	.33783	.07738	.00028	-.11911	-.19005
21 4	BQ	1S	-.03226	-.01736	.04251	-.02441	-.00834
22		2S	.33783	.07738	-.10330	.05932	.09474
23 5	BQ	1S	-.03226	-.01736	-.04239	-.02461	-.00839
24		2S	.33783	.07738	.10302	.05980	.09531
25 6	H	1S	.01076	-.03733	-.04645	.02667	.03157
26		2S	.48038	-.61026	.17943	-.10303	.71897
27 7	H	1S	.01076	-.03733	.00013	-.05356	-.06333

FIG. 5.3.13. (*continued*)

```
28        2S        .48038    -.61026   -.00048    .20691  -1.44222
29 8   H  1S        .01076    -.03733    .04632    .02689    .03176
30        2S        .48038    -.61026   -.17894   -.10387    .72326
FULL MULLIKEN POPULATION ANALYSIS.
                       1         2         3         4         5
 1 1   BQ 1S        .00002
 2        2S        .00004    .00247
 3        1PX     0.00000   0.00000    .00002
 4        1PY     0.00000   0.00000   0.00000    .00002
 5        1PZ     0.00000   0.00000   0.00000   0.00000    .00141
 6        3S       -.00012   -.00569   0.00000   0.00000   0.00000
 7        2PX     0.00000   0.00000    .00005   0.00000   0.00000
 8        2PY     0.00000   0.00000   0.00000    .00005   0.00000
 9        2PZ     0.00000   0.00000   0.00000   0.00000   -.00193
10 2   N  1S       -.00000    .00001   0.00000   0.00000    .00010
11        2S       -.00000   -.00018   0.00000   0.00000   -.00042
12        1PX     0.00000   0.00000   -.00005   0.00000   0.00000
13        1PY     0.00000   0.00000   0.00000   -.00005   0.00000
14        1PZ      -.00000   -.00180   0.00000   0.00000   -.00255
15        3S       -.00002   -.00323   0.00000   0.00000   -.00487
16        2PX     0.00000   0.00000   -.00031   0.00000   0.00000
17        2PY     0.00000   0.00000   0.00000   -.00031   0.00000
18        2PZ      -.00021   -.01354   0.00000   0.00000   -.01012
19 3   BQ 1S        .00000    .00002   0.00000   0.00000   -.00000
20        2S       -.00001   -.00067   -.00003   0.00000    .00010
21 4   BQ 1S        .00000    .00002   -.00000   -.00000   -.00000
22        2S       -.00001   -.00067   -.00001   -.00002    .00010
23 5   BQ 1S        .00000    .00002   -.00000   -.00000   -.00000
24        2S       -.00001   -.00067   -.00000   -.00002    .00010
25 6   H  1S        .00000    .00001   -.00000   -.00001    .00001
26        2S        .00000    .00041   -.00002   -.00006    .00034
27 7   H  1S        .00000    .00001   -.00001   0.00000    .00001
28        2S        .00000    .00041   -.00007   0.00000    .00034
29 8   H  1S        .00000    .00001   -.00000   -.00001    .00001
30        2S        .00000    .00041   -.00002   -.00006    .00034
                       6         7         8         9        10
 6        3S        .02253
 7        2PX     0.00000    .00031
 8        2PY     0.00000   0.00000    .00031
 9        2PZ     0.00000   0.00000   0.00000    .00916
10 2   N  1S        .00066   0.00000   0.00000   -.00019   2.04598
11        2S        .00179   0.00000   0.00000    .00333    .01998
12        1PX     0.00000   -.00072   0.00000   0.00000   0.00000
13        1PY     0.00000   0.00000   -.00072   0.00000   0.00000
14        1PZ       .00833   0.00000   0.00000    .00628   -.00000
15        3S        .01312   0.00000   0.00000    .02224   -.07893
16        2PX     0.00000   -.00282   0.00000   0.00000   0.00000
17        2PY     0.00000   0.00000   -.00282   0.00000   0.00000
18        2PZ       .04424   0.00000   0.00000    .01998   -.00000
19 3   BQ 1S       -.00007   -.00001   -.00001    .00002   -.00000
20        2S        .00264   -.00018   0.00000   -.00045   -.00006
21 4   BQ 1S       -.00007   -.00000   -.00001    .00002   -.00000
22        2S        .00264   -.00005   -.00014   -.00045   -.00006
23 5   BQ 1S       -.00007   -.00000   -.00001    .00002   -.00000
24        2S        .00264   -.00005   -.00014   -.00045   -.00006
25 6   H  1S       -.00089   -.00013   -.00040   -.00091   -.00093
26        2S       -.00362   -.00026   -.00078   -.00365    .00087
27 7   H  1S       -.00089   -.00054   0.00000   -.00091   -.00093
28        2S       -.00362   -.00104   0.00000   -.00365    .00087
29 8   H  1S       -.00089   -.00013   -.00040   -.00091   -.00093
30        2S       -.00362   -.00026   -.00078   -.00365    .00087
```

FIG. 5.3.13. (*continued*)

			11	12	13	14	15
11		2S	.10773				
12		1PX	0.00000	.29624			
13		1PY	0.00000	0.00000	.29624		
14		1PZ	.00000	0.00000	0.00000	.42960	
15		3S	.21554	0.00000	0.00000	.00000	1.01957
16		2PX	0.00000	.13731	0.00000	0.00000	0.00000
17		2PY	0.00000	0.00000	.13731	0.00000	0.00000
18		2PZ	.00000	0.00000	0.00000	.26849	.00000
19 3	BQ	1S	-.00000	.00000	0.00000	-.00000	-.00001
20		2S	.00016	.00014	0.00000	.00047	.00260
21 4	BQ	1S	-.00000	.00000	.00000	-.00000	-.00001
22		2S	.00016	.00004	.00011	.00047	.00260
23 5	BQ	1S	-.00000	.00000	.00000	-.00000	-.00001
24		2S	.00016	.00004	.00011	.00047	.00260
25 6	H	1S	.00755	.01192	.03577	.00466	.04850
26		2S	.00046	.00651	.01953	.00343	-.00484
27 7	H	1S	.00755	.04770	0.00000	.00466	.04850
28		2S	.00046	.02605	0.00000	.00343	-.00484
29 8	H	1S	.00755	.01192	.03577	.00466	.04850
30		2S	.00046	.00651	.01953	.00343	-.00484

			16	17	18	19	20
16		2PX	.23902				
17		2PY	0.00000	.23902			
18		2PZ	0.00000	0.00000	.63061		
19 3	BQ	1S	.00002	0.00000	-.00004	.00000	
20		2S	.00085	0.00000	.00389	-.00001	.00135
21 4	BQ	1S	.00000	.00001	-.00004	.00000	-.00000
22		2S	.00021	.00064	.00389	-.00000	.00024
23 5	BQ	1S	.00000	.00001	-.00004	.00000	-.00000
24		2S	.00021	.00064	.00389	-.00000	.00024
25 6	H	1S	.02389	.07167	.01203	.00000	.00006
26		2S	.01842	.05527	.01305	.00001	.00002
27 7	H	1S	.09556	0.00000	.01203	-.00000	-.00002
28		2S	.07370	0.00000	.01305	-.00000	-.00029
29 8	H	1S	.02389	.07167	.01203	.00000	.00006
30		2S	.01842	.05527	.01305	.00001	.00002

			21	22	23	24	25
21 4	BQ	1S	.00000				
22		2S	-.00001	.00135			
23 5	BQ	1S	.00000	-.00000	.00000		
24		2S	-.00000	.00024	-.00001	.00135	
25 6	H	1S	-.00000	-.00002	.00000	.00006	.17432
26		2S	-.00000	-.00029	.00001	.00002	.08267
27 7	H	1S	.00000	.00006	.00000	.00006	-.00056
28		2S	.00001	.00002	.00001	.00002	-.00698
29 8	H	1S	.00000	.00006	-.00000	-.00002	-.00056
30		2S	.00001	.00002	-.00000	-.00029	-.00698

			26	27	28	29	30
26		2S	.10644				
27 7	H	1S	-.00698	.17432			
28		2S	-.01562	.08267	.10644		
29 8	H	1S	-.00698	-.00056	-.00698	.17432	
30		2S	-.01562	-.00698	-.01562	.08267	.10644

GROSS ORBITAL CHARGES.

			1
1 1	BQ	1S	-.00032
2		2S	-.02262
3		1PX	-.00046
4		1PY	-.00046
5		1PZ	-.01702

FIG. 5.3.13. (*continued*)

```
    6       3S        .07905
    7       2PX      -.00583
    8       2PY      -.00583
    9       2PZ       .04387
   10 2  N  1S       1.98723
   11       2S        .37229
   12       1PX       .54362
   13       1PY       .54362
   14       1PZ       .73405
   15       3S       1.32217
   16       2PX       .62839
   17       2PY       .62839
   18       2PZ      1.02625
   19 3  BQ 1S       -.00009
   20       2S        .01113
   21 4  BQ 1S       -.00009
   22       2S        .01113
   23 5  BQ 1S       -.00009
   24       2S        .01113
   25 6  H  1S        .45475
   26       2S        .24875
   27 7  H  1S        .45475
   28       2S        .24875
   29 8  H  1S        .45475
   30       2S        .24875
```

CONDENSED TO ATOMS (ALL ELECTRONS)

		1	2	3	4	5	6
1	BQ	.021059	.075159	.001347	.001347	.001347	-.009961
2	N	.075159	6.703414	.008029	.008029	.008029	.327782
3	BQ	.001347	.008029	.001335	.000233	.000233	.000086
4	BQ	.001347	.008029	.000233	.001335	.000233	-.000310
5	BQ	.001347	.008029	.000233	.000233	.001335	.000086
6	H	-.009961	.327782	.000086	-.000310	.000086	.446102
7	H	-.009961	.327782	-.000310	.000086	.000086	-.030143
8	H	-.009961	.327782	.000086	.000086	-.000310	-.030143

		7	8
1	BQ	-.009961	-.009961
2	N	.327782	.327782
3	BQ	-.000310	.000086
4	BQ	.000086	.000086
5	BQ	.000086	-.000310
6	H	-.030143	-.030143
7	H	.446102	-.030143
8	H	-.030143	.446102

TOTAL ATOMIC CHARGES.

		1
1	BQ	.070376
2	N	7.786006
3	BQ	.011040
4	BQ	.011040
5	BQ	.011040
6	H	.703500
7	H	.703500
8	H	.703500

DIPOLE MOMENT (DEBYE): X= -.0000   Y= -.0000   Z= 2.2938   TOTAL= 2.2938

"I COULD HAVE DONE IT IN A MUCH MORE COMPICLATED WAY"
SAID THE RED QUEEN, IMMENSELY PROUD.

                    -- LEWIS CARROLL

        BYE,BYE !!!

FIG. 5.3.13. (continued)

largest change in an orbital coefficient needed to normalize the basis set (values of up to 10^{-4} may be given for some basis sets for second-row molecules: This is because the data published are not precise enough to fulfil the normalization criterion within the limits set in the program). GFPRINT then gives a table of the basis set, atom for atom, more or less as defined in the input. The remaining links of overlay 3 produce the integrals as before, except that link 314 is called because a general basis set may or may not contain d-orbitals. In this case it does not, so the program returns without doing anything. The closed-shell SCF calculation proceeds as usual except that the convergence criterion is relaxed for general-basis single-point calculations.

The molecular orbitals are then printed in the usual format. Note that in molecular orbital 5 (the HOMO) the most diffuse s- and p_z-orbitals of the boron ghost have significant coefficients (0.10414 and 0.06755, respectively), even though there are no electrons formally situated on the ghost BH_3 fragment. This is a direct indication of the magnitude of the BSSE. Similarly the "GROSS ORBITAL CHARGES" table shows populations of 0.07905 and 0.04387 electrons for these two orbitals, giving a total charge on the ghost BH_3 fragment of $0.070376 + 3 \cdot (0.011040) = 0.103496$ electrons. This is an indication of the extent to which the NH_3 molecule has "borrowed" basis functions from the ghost BH_3. The small negative orbital populations of some of the ghost orbitals is an artifact of the Mulliken population analysis. A phenomenon often observed with the 3-21G basis set is that the nitrogen $1s$-orbital (molecular orbital 1) has a small contribution from the outer ghost $2s$-orbital (coefficient -0.01574). This is a consequence of the smallness of the basis set used for the core orbitals in 3-21G. Contributions from neighboring diffuse s-functions to core orbitals may be very large in 3-21+G calculations, especially for second-row atoms. This is an extension of the BSSE to the nonvalence orbitals. Note that the order of stability of the atomic orbitals for a ghost atom is the reverse of that for a real atom because the electrostatic attraction from the nucleus no longer exists. For a ghost atom the innermost orbitals are the least stable and the outermost the most stable. Figure 5.3.13 shows that only the outer ghost orbitals have significant populations.

The energy given by this single-point calculation (-55.88329 a.u.) is 7 kcal mol^{-1} lower than the 3-21G total energy for the optimized structure of ammonia (-55.87220 a.u.).[20] A reasonable procedure for calculating a corrected binding energy for $BH_3 : NH_3$ would be to optimize the geometry of NH_3 at the correct distance from the ghost BH_3 fragment and to do the same for BH_3 with a ghost ammonia molecule, although the BSSE would be small in this case. The corrected binding energy could then be calculated from these two counterpoise energies and the total energy for the complex. In view of the input structure used for this job, the quotation chosen by link 9999 seems particularly appropriate.

The final example illustrates the use of a nonstandard route with

GAUSSIAN82 and some of the pitfalls that may be encountered in more complicated calculations. The technique used in this calculation is known as *floating orbital geometry optimization* (FOGO),[27] and was designed to add some polarization to the basis set without the use of *d*-orbitals. Normally the basis functions in a GAUSSIAN82 calculation are centered exactly on the nuclei. In the FOGO method, however, the outermost valence orbitals in an *sp*-basis set are allowed to wander away from the nucleus and their positions are optimized in addition to those of the nuclei. They are thus called *floating orbitals*. FOGO calculations are not as effective as addition of *d*-orbitals to the basis set because they can assign only an average polarization to the molecule, not an individual polarization for each molecular orbital. The lone-pair orbital in methylene, for instance, would prefer to have the diffuse valence functions on its side of the molecule, but the CH bonding orbitals pull in the opposite direction. The final result is a compromise. *d*-Functions allow each individual orbital to be optimally polarized. Another consequence of the FOGO method is that the virtual orbitals, which require polarizations opposite to those of the occupied molecular orbitals, are actually represented less well in a FOGO calculation than in a conventional one. This means that the MP2 procedure, for instance, gives a lower correlation contribution for the FOGO calculation because it relies heavily on a good description of the virtual orbitals. Nevertheless, the FOGO technique, which requires a more complicated geometry optimization but reduces the number of basis functions appreciably, has some advantages in terms of computer time used and allows polarization to be introduced into calculations for large molecules. GAUSSIAN82, although very flexible, is not written with such calculations in mind, so that some effort is required to run the job successfully.

In principle the FOGO technique requires only that an extra ghost atom be defined for each real atom. The outer valence orbitals are then situated on the ghost atoms, which are placed near the real nuclei. The rest of the basis set is then assigned to the real atomic centers. Figure 5.3.14 shows one way to define the input for a FOGO calculation on CH_2. The two dummies

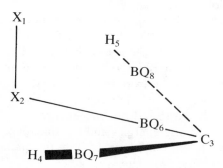

FIG. 5.3.14. The geometry used to perform a FOGO calculation on CH_2.

X_1 and X_2 facilitate the definition of the molecule without $0°$ or $180°$ bond angles. The ghost atom BQ_6 carries the outer valence functions for C_3, and BQ_7 and BQ_8 those for H_4 and H_5, respectively.

If, however, a corresponding input is written using general-basis-set specification, the calculation stops in link 202 because the ghost atoms are situated less than 0.5 Å away from the corresponding real atoms, so the program assumes that a geometry error has been made. This geometry test can be disabled by setting option 12 equal to 1 in link 202, but not *via* a keyword in Release A (the first version of GAUSSIAN82 to be published). In later versions of the program the the keyword "NOCROWD" can be used. A nonstandard route must therefore be used for the version of the program used for this example. This is not a real problem, because a dummy job with the keywords " # RHF/GEN OPT GUESS=CORE SCFCYC=99" automatically generates a route in which one only has to insert "12=1" into each overlay 2 card. In this example the unsuccessful first calculation also generated the route needed. After correction of the route, however, another problem arises. The initial guess produced by diagonalizing the core Hamiltonian does not give the ground state as reliably as the projected Hückel and NDO procedures do. In this example the carbene π orbital is doubly occupied and the *n*-orbital unoccupied in the initial guess. This gives the situation shown in Fig. 5.3.15. Structure (1) which is given by the initial

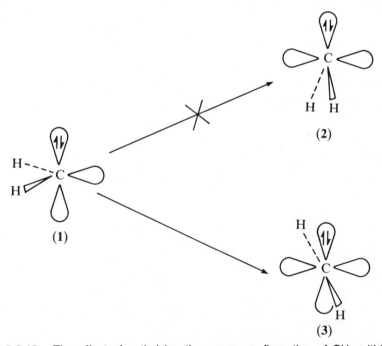

FIG. 5.3.15. The effect of optimizing the wrong configuration of CH_2 within C_{2v} symmetry.

guess, should optimize to (2), which is the lowest closed-shell singlet structure. This, however, cannot happen because the carbene is to be optimized within C_{2v} symmetry, which means that the molecular plane remains fixed. The best that the molecule can do within the given symmetry constraints is to make the linear structure (3), in which the doubly occupied p-orbital is at least no worse than the LUMO. This is exactly what happens during the optimization until BERNY notices that (3) is a transition state and not a minimum, and stops the optimization. This situation is surprisingly frequent for AX_2 molecules with one, two, or three electrons in the π and n-orbitals, and leads to many spurious linear structures. The electronic state should always be checked in such cases.

A third attempt is therefore necessary, this time setting option 8 in link 401 to 1 in order to change the configuration of the initial guess. The final input, which would have pleased the Red Queen immensely, is shown in Fig. 5.3.16. The nonstandard route is introduced by the keyword

```
# NONSTD
1//1,3;
2/12=1/2;
3/5=7,11=1,25=14,30=1/1,2,3,11,14;
4/5=2,7=1,8=1,16=1/1;
5/6=7,7=99/1;
6//1;
7/27=1/1,2,3,16;
1//3(1);
99//99;
2/12=1/2;
3/5=7,6=1,11=1,25=14,30=1/1,2,11,14;
5/6=7,7=99/1;
7/27=1/1,2,3,16;
1//3(-4);
3/5=7,6=1,11=1,25=14,30=1/1,3;
6//1;
```
} Link 0

```
CH2 SINGLET FOGO(3-21G)

 0  1
X
X   1 1.
C   2 1. 1 90.
H   3 R1 2 A1  1  90.
H   3 R1 2 A1  1 -90.
BQ  2 R2 1 90. 3 0.
BQ  3 R3 2 A2  1  90.
BQ  3 R3 2 A2  1 -90.

R1=1.1
R2=0.95
R3=1.08
A1=53.
A2=53.
```
} Link 101

FIG. 5.3.16. One possible GAUSSIAN82 input for a FOGO calculation on CH_2 using the 3-21G basis set.

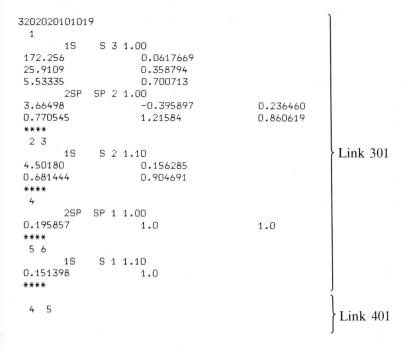

```
3202020101019
  1
       1S     S 3 1.00
 172.256           0.0617669
 25.9109           0.358794
 5.53335           0.700713
       2SP    SP 2 1.00
 3.66498          -0.395897          0.236460
 0.770545          1.21584           0.860619
 ****
  2 3
       1S     S 2 1.10
 4.50180           0.156285
 0.681444          0.904691
 ****
  4
       2SP    SP 1 1.00
 0.195857          1.0               1.0
 ****
  5 6
       1S     S 1 1.10
 0.151398          1.0
 ****

  4  5
```

> Link 301

> Link 401

FIG. 5.3.16. (*continued*)

"NONSTD." The individual overlays are typed on separate lines for clarity, but this need not be the case as long as they are separated by semicolons. The route was obtained as outlined above with the addition of "12=1" for each link 202 call and "8=1" in the overlay 4 card. This section of the input is read by overlay 0. The geometry definition that then follows corresponds to the structure shown in Fig. 5.3.14. This section, which ends with a blank card, is read by link 101. There are two remaining sections, the basis-set definition, which is read by link 301, and the change of configuration required by link 401. Because link 301 is called before link 401, the basis-set definition comes first. Center 1 is C_3, which has the 1s- and $2sp_I$-shells of the carbon basis set (320). Centers 2 and 3 are the two real hydrogens, that each have the inner 1s-orbitals (2020). The carbon ghost BQ_6 now follows with the $2sp_O$-shell (10), and finally come the two hydrogen ghosts with $1s_O$-orbitals (1019). A blank card ends this section. The configuration change is defined in two lines, the first of which indicates the orbitals (4 and 5) to be swapped. The final blank card finishes the input.

An extract from the output for this job is shown in Fig. 5.3.17. The initial section and the optimization are fairly normal except that more cycles are needed than for a normal calculation. The final structure obtained differs from the normal 3-21G geometry, however. The CH bond length, 1.1029 Å, is very similar to the 3-21G value, 1.1021 Å, but the H–C–H angle has

HELLO !!!

```
********************************************************
GAUSSIAN 82: REVISION B: CDC VERSION: SEPTEMBER-1982
              84/05/13. 11.32.42.
********************************************************
----------------------------------------------------------------------
# NONSTD 1//1,3; 2/12=1/2; 3/5=7,11=1,25=14,30=1/1,2,3,11,14; 4/5=2,7=1,8=1,16=1
/1; 5/6=7,7=99/1; 6//1; 7/27=1/1,2,3,16; 1//3(1)
; 99//99; 2/12=1/2; 3/5=7,6=1,11=1,25=14,30=1/1,2,11,14; 5/6=7,7=99/1; 7/27=1/1,
2,3,16; 1//3(-4); 3/5=7,6=1,11=1,25=14,30=1/1,3;
 6//1;
----------------------------------------------------------------------
1//1,3;
2/12=1/2;
3/5=7,11=1,25=14,30=1/1,2,3,11,14;
4/5=2,7=1,8=1,16=1/1;
5/6=7,7=99/1;
6//1;
7/27=1/1,2,3,16;
1//3(1);
99//99;
2/12=1/2;
3/5=7,6=1,11=1,25=14,30=1/1,2,11,14;
5/6=7,7=99/1;
7/27=1/1,2,3,16;
1//3(-4);
3/5=7,6=1,11=1,25=14,30=1/1,3;
6//1;
99//99;

    ### NEXT LINK= 101   LL=    0

    -----------------------
    CH2 SINGLET FOGO(3-21G)
    -----------------------
SYMBOLIC Z-MATRIX
    CHARGE = 0 MULTIPLICITY = 1
X
X    1    1.00000
C    2    1.00000    1    90.00000
H    3    R1         2    A1       1    90.00000  0
H    3    R1         2    A1       1    -90.00000 0
BQ   2    R2         1    90.00000 3    0.00000   0
BQ   3    R3         2    A2       1    90.00000  0
BQ   3    R3         2    A2       1    -90.00000 0
     VARIABLES
    R1              1.10000
    R2               .95000
    R3              1.08000
    A1             53.00000
    A2             53.00000

    ### NEXT LINK= 103   LL=    0
```

GRADGRADGRADGRADGRADGRADGRADGRADGRADGRADGRADGRADGRADGRADGRADGRADGRADGRADGRAD

FIG. 5.3.17. An extract from the output produced by GAUSSIAN82 from the input shown in Fig. 5.3.16.

BERNY OPTIMIZATION

INITIALIZATION PASS

```
                    ------------------------------
                    !    INITIAL PARAMETERS     !
                    !   (ANGSTROMS AND DEGREES) !
----------------------                      ----------------------------
!     NAME      VALUE   DERIVATIVE INFORMATION (ATOMIC UNITS)       !
------------------------------------------------------------------------
!      R1      1.1000   ESTIMATE D2E/DX2                            !
!      R2       .9500   ESTIMATE D2E/DX2                            !
!      R3      1.0800   ESTIMATE D2E/DX2                            !
!      A1     53.0000   ESTIMATE D2E/DX2                            !
!      A2     53.0000   ESTIMATE D2E/DX2                            !
------------------------------------------------------------------------
```

GRADGRADGRADGRADGRADGRADGRADGRADGRADGRADGRADGRADGRADGRADGRADGRADGRADGRADGRAD

NEXT LINK= 202   LL=   0


```
---------------------------------------------------------------------------
                    Z-MATRIX (ANGSTROMS AND DEGREES)
CD CENT ATOM N1   LENGTH     N2    ALPHA     N3     BETA      J

 1       X
 2       X    1  1.000000 (  1)
 3   1   C    2  1.000000 (  2)  1  90.000 (  8)
 4   2   H    3  1.100000 (  3)  2  53.000 (  9)  1  90.000 ( 14)  0
 5   3   H    3  1.100000 (  4)  2  53.000 ( 10)  1 -90.000 ( 15)  0
 6   4   BQ   2   .950000 (  5)  1  90.000 ( 11)  3   0.000 ( 16)  0
 7   5   BQ   3  1.080000 (  6)  2  53.000 ( 12)  1  90.000 ( 17)  0
 8   6   BQ   3  1.080000 (  7)  2  53.000 ( 13)  1 -90.000 ( 18)  0
---------------------------------------------------------------------------
                    Z-MATRIX ORIENTATION:
---------------------------------------------------------------------------
CENTER    ATOMIC              COORDINATES (ANGSTROMS)
NUMBER    NUMBER         X            Y            Z
---------------------------------------------------------------------------
           -1         0.000000     0.000000     0.000000
           -1         0.000000     0.000000     1.000000
   1        6         1.000000     0.000000     1.000000
   2        1          .338003      .878499     1.000000
   3        1          .338003     -.878499     1.000000
   4        0          .950000     0.000000     1.000000
   5        0          .350040      .862526     1.000000
   6        0          .350040     -.862526     1.000000
---------------------------------------------------------------------------
                    DISTANCE MATRIX (ANGSTROMS)
              1          2          3          4          5
 1  C   0.000000
 2  H   1.100000   0.000000
 3  H   1.100000   1.756998   0.000000
 4  BQ   .050000   1.070654   1.070654   0.000000
 5  BQ  1.080000    .020000   1.741067   1.050668   0.000000
 6  BQ  1.080000   1.741067    .020000   1.050668   1.725053
              6
```

FIG. 5.3.17. (*continued*)

```
     6  BQ    0.000000
SYMM--  CANNOT COPE WITH DUMMY OR (SHIVER ) GHOST ATOMS.
SYMM--  SYMMETRY TURNED OFF.
STOICHIOMETRY    CH2
FRAMEWORK GROUP
PRSFWG-- UNABLE TO FIND LEFT SQUARE BRACKET.
NUMDOF-- UNRECOGNIZED SYMMETRIC SUBSPACE, ICHAR= E    "
DEG. OF FREEDOM   -1
FULL POINT GROUP                              NOP  1


    ### NEXT LINK=  301   LL=    0




GENERAL BASIS READ FROM CARDS:           (S, S=P, 5D, 7F)
 13 BASIS FUNCTIONS      21 PRIMITIVE GAUSSIANS


   4 ALPHA ELECTRONS        4 BETA ELECTRONS
     NUCLEAR REPULSION ENERGY    6.0740231392 HARTREES
RAFFENETTI 1 INTEGRAL FORMAT.
TWO-ELECTRON INTEGRAL SYMMETRY IS TURNED OFF.


    ### NEXT LINK=  302   LL=    0


    ### NEXT LINK=  303   LL=    0


    ### NEXT LINK=  311   LL=    0



STANDARD CUTOFFS SELECTED IN SHELL.
   2376 INTEGRALS PRODUCED FOR A TOTAL OF      2376

    ### NEXT LINK=  314   LL=    0


     0 INTEGRALS PRODUCED FOR A TOTAL OF      2376


    ### NEXT LINK=  401   LL=    0



PAIRS OF ALPHA ORBITALS SWITCHED:
  4  5



    ### NEXT LINK=  501   LL=    0



RHF CLOSED SHELL SCF.
REQUESTED CONVERGENCE ON DENSITY MATRIX=   .1000E-06 WITHIN  99 CYCLES.
N**3 SYMMETRY TURNED OFF IN RHFCLO.
SCF DONE:  E(RHF) =  -38.6552596811    A.U. AFTER   15 CYCLES
         CONVG  =    .6407E-07            -V/T =  2.0019

      <<<<<<<<<<   BREAK HERE   >>>>>>>>>>>>>

GRADGRADGRADGRADGRADGRADGRADGRADGRADGRADGRADGRADGRADGRADGRADGRADGRADGRADGRADGRAD
```

FIG. 5.3.17.   (*continued*)

```
BERNY OPTIMIZATION


SEARCH FOR A LOCAL MINIMUM.


STEP NUMBER  14 OUT OF A MAXIMUM OF  15
ALL QUANTITIES PRINTED IN INTERNAL UNITS (HARTREES-BOHRS-RADIANS)


UPDATE SECOND DERIVATIVES USING INFORMATION FROM POINTS:  1 12 13 14


THE SECOND DERIVATIVE MATRIX:
                          R1        R2        R3        A1        A2
          R1           .58260
          R2          -.01112    .24409
          R3          -.00515    .00644    .00308
          A1           .14479   -.16556   -.00443    .83453
          A2          -.01088    .08262   -.00468   -.10910    .09671
   EIGENVALUES ---      .00212    .05886    .21688    .53023    .95293


VARIABLE     OLD X     -DE/DX    DELTA X   DELTA X   DELTA X    NEW X

                                 (LINEAR)   (QUAD)    (TOTAL)


   R1       2.08416   -.00000   -.00001   -.00001   -.00002   2.08414
   R2       1.70415    .00001    .00006    .00005    .00011   1.70426
   R3       1.56395   -.00000   -.00058   -.00024   -.00083   1.56312
   A1        .90003    .00000    .00001   -.00002   -.00001    .90002
   A2        .84443   -.00001   -.00004   -.00016   -.00020    .84423


           ITEM             VALUE     THRESHOLD  CONVERGED?

MAXIMUM FORCE             .000006      .000450      YES
RMS     FORCE             .000004      .000300      YES
MAXIMUM DISPLACEMENT      .000825      .001800      YES
RMS     DISPLACEMENT      .000383      .001200      YES
PREDICTED CHANGE IN ENERGY  -.000000


OPTIMIZATION COMPLETED.
   -- STATIONARY POINT FOUND.
   -- LAST STEP NOT IMPLEMENTED.
                    ----------------------------
                    !  OPTIMIZED PARAMETERS    !
                    !  (ANGSTROMS AND DEGREES) !
----------------------                       ----------------------
 !    NAME       VALUE    DERIVATIVE INFORMATION (ATOMIC UNITS)      !
----------------------------------------------------------------------
 !    R1        1.1029   -DE/DX =   -.000005                         !
 !    R2         .9018   -DE/DX =    .000006                         !
 !    R3         .8276   -DE/DX =   -.000001                         !
 !    A1       51.5677   -DE/DX =    .000001                         !
 !    A2       48.3825   -DE/DX =   -.000006                         !
----------------------------------------------------------------------

GRADGRADGRADGRADGRADGRADGRADGRADGRADGRADGRADGRADGRADGRADGRADGRADGRADGRAD
```

FIG. 5.3.17.   (*continued*)

BASIS READ FROM RWF: (S, S=P, 5D, 7F)
  13 BASIS FUNCTIONS   21 PRIMITIVE GAUSSIANS
   4 ALPHA ELECTRONS   4 BETA ELECTRONS
      NUCLEAR REPULSION ENERGY   6.0639732924 HARTREES
RAFFENETTI 1 INTEGRAL FORMAT.
TWO-ELECTRON INTEGRAL SYMMETRY IS TURNED OFF.

NEXT LINK= 303   LL=   0

NEXT LINK= 601   LL=   0

MOLECULAR ORBITAL COEFFICIENTS

			1	2	3	4	5
		EIGENVALUES --	-11.21722	-.88697	-.55677	-.37914	.07725
1 1	C	1S	-.98705	.19977	-.00000	-.11305	-.00000
2		2S	-.09360	-.22332	-.00000	.08885	-.00000
3		1PX	.00482	.12738	.00000	.41799	.00000
4		1PY	.00000	.00000	-.37900	-.00000	-.00000
5		1PZ	-.00000	.00000	-.00000	-.00000	.37636
6 2	H	1S	.00161	-.17361	-.23082	-.10518	-.00000
7 3	H	1S	.00161	-.17361	.23082	-.10518	-.00000
8 4	BQ	1S	.07773	-.53638	.00000	.53682	.00000
9		1PX	-.01430	.05272	-.00000	.51439	-.00000
10		1PY	.00000	.00000	-.20346	-.00000	.00000
11		1PZ	.00000	.00000	.00000	-.00000	.74940
12 5	BQ	1S	-.02854	-.07145	-.36826	-.13333	-.00000
13 6	BQ	1S	-.02854	-.07145	.36826	-.13333	-.00000

			6	7	8	9
		EIGENVALUES --	.30969	.35811	.96548	.99036
1 1	C	1S	-.14817	-.00000	.00000	-.05516
2		2S	.03354	.00000	.00000	.05089
3		1PX	-.21473	-.00000	-.00000	.96120
4		1PY	.00000	-.32321	-.78282	-.00000
5		1PZ	.00000	.00000	-.00000	-.00000
6 2	H	1S	-.02173	.06732	-.53200	-.31934
7 3	H	1S	-.02173	-.06732	.53200	-.31934
8 4	BQ	1S	3.70582	.00000	-.00000	1.19070
9		1PX	-1.22834	-.00000	.00000	-1.42559
10		1PY	.00000	-2.00970	1.44218	.00000
11		1PZ	-.00000	-.00000	.00000	.00000
12 5	BQ	1S	-2.28032	2.50336	-.18490	-.55073
13 6	BQ	1S	-2.28032	-2.50336	.18490	-.55073

FULL MULLIKEN POPULATION ANALYSIS.

			1	2	3	4	5
1 1	C	1S	2.05392				
2		2S	.01445	.13305			
3		1PX	.00000	-.00000	.38193		
4		1PY	0.00000	0.00000	0.00000	.28729	
5		1PZ	-.00000	.00000	-.00000	0.00000	.00000
6 2	H	1S	-.00086	.01095	.02338	.03900	-.00000
7 3	H	1S	-.00086	.01095	.02338	.03900	.00000
8 4	BQ	1S	-.08762	.24268	-.01352	0.00000	.00000

FIG. 5.3.17. (*continued*)

```
  9         1PX      -.00191    .00680    .23069   0.00000    .00000
 10         1PY      0.00000   0.00000   0.00000    .08114   0.00000
 11         1PZ      -.00000    .00000   -.00000   0.00000   -.00000
 12 5  BQ  1S         .00645    .00705    .02013    .04867   -.00000
 13 6  BQ  1S         .00645    .00705    .02013    .04867    .00000
                         6         7         8         9        10
  6 2  H   1S         .18896
  7 3  H   1S        -.00025    .18896
  8 4  BQ  1S         .02633    .02633   1.16385
  9         1PX       .03628    .03628   -.00000    .53515
 10         1PY       .03960    .03960   0.00000   0.00000    .08279
 11         1PZ       .00000   -.00000    .00000   -.00000   0.00000
 12 5  BQ  1S         .13790   -.02289   -.05813    .04303    .06142
 13 6  BQ  1S        -.02289    .13790   -.05813    .04303    .06142
                        11        12        13
 11         1PZ       .00000
 12 5  BQ  1S         .00000    .31863
 13 6  BQ  1S        -.00000   -.13565    .31863
        GROSS ORBITAL CHARGES.
                         1
  1 1  C   1S        1.99001
  2         2S         .43299
  3         1PX       .68611
  4         1PY       .54378
  5         1PZ       .00000
  6 2  H   1S         .47842
  7 3  H   1S         .47842
  8 4  BQ  1S        1.24177
  9         1PX       .92933
 10         1PY       .36596
 11         1PZ       .00000
 12 5  BQ  1S         .42661
 13 6  BQ  1S         .42661
        CONDENSED TO ATOMS (ALL ELECTRONS)
            1         2         3         4         5         6
 1   C   2.885089   .072477   .072477   .458242   .082303   .082303
 2   H    .072477   .188964  -.000245   .102204   .137902  -.022886
 3   H    .072477  -.000245   .188964   .102204  -.022886   .137902
 4   BQ   .458242   .102204   .102204  1.781791   .046311   .046311
 5   BQ   .082303   .137902  -.022886   .046311   .318630  -.135652
 6   BQ   .082303  -.022886   .137902   .046311  -.135652   .318630
        TOTAL ATOMIC CHARGES.
             1
 1   C    3.652889
 2   H     .478415
 3   H     .478415
 4   BQ   2.537065
 5   BQ    .426608
 6   BQ    .426608
DIPOLE MOMENT (DEBYE): X=-1.7367   Y= -.0000   Z=  .0000   TOTAL= 1.7367

   ### NEXT LINK= 9999   LL=   0

THE RED LIGHT IS ALWAYS LONGER THAN THE GREEN LIGHT.

                -- PETERS THEORY OF RELATIVITY

            BYE,BYE !!!
```

FIG. 5.3.17. (continued)

closed from 104.6° to 103.1°. The FOGO structure is shown in Fig. 5.3.18. The hydrogen floating orbitals lie 0.280 Å away from their cores (25.4% of the bond length) and the carbon outer valence functions 0.098 Å away from the corresponding nucleus. The total energy is 3.4 kcal mol^{-1} lower than the 3-21G value. The major differences are that the hydrogen valence functions have higher electron densities in the FOGO calculation, and that the dipole moment is decreased from 2.2 to 1.7 Debyes.

Errors that arise with GAUSSIAN82 are generally explained by the program or in the manual, but a few problems may remain. The following paragraphs describe some common problems and suggests solutions for them.

Initial Guess Symmetry Errors. Link 401 checks the symmetry of the initial guess to avoid partly occupied degenerate sets of orbitals. If an initial guess with a lower symmetry than the molecular point group is detected, the program aborts with the recommendation to try the keyword "NOSYMM." If this keyword is used, the calculation will run, but may give a wave function with lower symmetry than the molecular point group, as outlined for MOPAC. The correct solution to the problem is usually either to optimize the geometry with less symmetry, or to use "GUESS=ALTER" to change the electronic configuration (and hence the symmetry) of the initial guess. This situation may, however, also arise with a very poor starting geometry in which a central bond length is either much too long or much too short. In this case σ and π orbitals can exchange places in the initial guess and thus give rise to spurious partly occupied degenerate π sets.

Number of d-Orbitals. The 6-31G* and 6-31G** basis sets use six Cartesian d-functions, and all other polarization basis sets in the program five real ones. When using general-basis-set input with six d-functions the keyword "6D" must be used, or option 8 in overlay 3 must be set to 2 to force the program to use six d-functions. The job will not abort if this is not

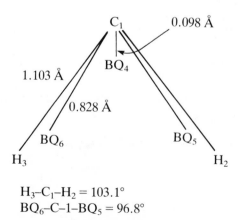

$$H_3\text{–}C_1\text{–}H_2 = 103.1°$$
$$BQ_6\text{–}C\text{–}1\text{–}BQ_5 = 96.8°$$

FIG. 5.3.18. The FOGO/3-21G structure of methylene.

done, but will perform the calculation with five real *d*-orbitals, rather than with six Cartesian *d*-functions. This gives a slightly worse energy because the six functions are equivalent to five real *d*-orbitals and one extra *s*-function.

Small Diffuse-Augmented Basis Sets. The 3-21+G and 4-31+G basis-sets are not included as options in the standard version of GAUSSIAN82. They can, however, easily be obtained by using the keyword "MASSAGE" to add diffuse *sp*-shells to the standard 3-21G and 4-31G basis sets.

Wrong Number of Negative Eigenvalues. The BERNY optimization procedure normally checks on the number of negative frequencies during an optimization, and aborts the job if, for instance, one negative value is found during a search for a minimum. Unfortunately, this test can lead to unnecessary job termination because it is based on the guessed second-derivative matrix, which may be considerably in error in the early stages of the calculation. In most cases the calculation will continue normally if it is restarted from where it broke off, and will successfully find the minimum. If it is not important that the program check the nature of the stationary point, as is usually the case in simple optimizations, the keyword "OPT=NOEIGENTEST" can be used to switch off testing. BERNY optimizations usually proceed smoothly with this option.

Cyclic Structures. As in MOPAC, cyclic structures should never be defined by simply giving bond lengths and angles consecutively around the ring. This leads to a lever effect that often causes the atomic forces to be outside the range that the optimization procedures can handle. The definitions given for cyclic molecules in Chapter 3 give good results.

Transition State Optimizations. GAUSSIAN82 can optimize transition states using BERNY. This procedure, however, is not simple to use, and transition-state searches can be very expensive. In any case some sort of preliminary reaction-path calculation should be performed to obtain an approximate geometry and to calculate a reasonable force constant matrix. Simply reading in a geometry with OPT=TS usually results in the optimization procedure not finding a negative eigenvalue and stopping the job. This problem can be overcome by using the option to turn off the eigenvalue test, but this may result in the program finding a completely different transition state from the one required, and using large amounts of computer time. If a preliminary search has been performed, the reaction coordinate is usually well suited for a transition-state optimization because it corresponds to one of the geometric variables in the Z-matrix. There are two possibile ways to obtain a better second-derivative matrix at the beginning of the calculation. The geometry closest to the transition state can be read in with the force constants from the previous run, or, if these are not available, some or all of the second derivatives may be calculated explicitly at the beginning of the optimization. Normally a transition-state search begins with a series of calculations using one parameter as a reaction coordinate. The geometry of the point nearest to the transition state is then read in along with the diagonal elements of the second-derivative matrix for all parameters except the one used previously as the reaction coordinate. The second derivative for this parameter can be calculated by finite-difference to give a complete set of second derivatives on which to base the optimization of the transition state.

In general the same principles apply to GAUSSIAN82 calculations as to MOPAC jobs. The same problems can arise with forbidden reactions or degenerate orbitals, for instance, although the symmetry package in GAUSSIAN82 often detects errors that MOPAC does not. The program is more conservative than MOPAC, and may sometimes seem fussy. It does, however, save a lot of computer time every time it finds an error. A few false alarms are a small price to pay for such thrift.

5.4. ELECTRON CORRELATION

Figures 1.1.1 and 4.4.16 illustrate the major weakness of single-determinant MO theory—it is restricted to only one electronic configuration. In reality any electronic state can be considered to be a combination of many such single configurations, and more realistic calculations should take this into account. That single-determinant theory usually works so well is a consequence of the happy accident that one configuration usually makes up 99% of the ground electronic state, so that omission of the others is not a serious limitation. It is not immediately obvious that this mixing of electronic configurations is equivalent to allowing the electrons to avoid each other, the electron correlation outlined in Section 3.1. Consider, however, σ and σ^* orbitals, as shown in Fig. 5.4.1. If the σ orbital is doubly occupied in the ground state (S^0), the two electrons will repel each other, and this repulsion will be overestimated by the SCF method, as explained in Section 3.1. If, however, one electron is promoted to the σ^* orbital to give a singly excited (S^1) state this repulsion is reduced, one electron being more concentrated on atom A and the other on atom B. If we now allow S^0 to mix with S^1 the resulting state will reflect the reduced electron-electron repulsion gained by partially unpairing the two electrons. The resulting *instantaneous* alpha and beta molecular orbitals are also shown in Fig. 5.4.1. This mixing does not lead to any static spin polarization because an equivalent S^1 state could be obtained equally well by promoting an alpha rather than a beta electron. This process of mixing electronic states is known as *configuration interaction* (CI), and is usually performed after a normal SCF calculation.

Configuration interaction works as follows: A single determinant calculation is performed as usual. The molecular orbitals thus obtained are then assumed to be constant for all configurations to be mixed in the CI. These states are therefore simply those obtained by swapping electrons between the orbitals obtained for the *reference configuration*. Because, however, there may be millions of such states it is usually impossible to consider them

FIG. 5.4.1.   Schematic diagram of the effect of mixing two electronic configurations.

all in a CI calciulation. Many schemes have been designed to select the correct states to use in the CI, including the so-called π-only CI, but the smallest one, and the one most commonly used for closed-shell molecules is a 3×3 *CI*, in which the ground state and the two singlet states obtained by promoting one and two electrons from the HOMO to the LUMO are used:

The corresponding CI for radicals is a *2 × 2 CI*, in which the ground state and one singly excited state (either HOMO→SOMO or SOMO→LUMO excitation) are used. Although such small-scale CI treatments do not give a very large fraction of the total correlation energy, they are very useful for the study of forbidden reactions, for which the mixing of only two configurations is often adequate. A 3×3 CI is included in MOPAC and MNDO/C for singlets and triplets, and a 2×2 CI for radicals in MNDO/C. The CISD (CI including single and double excitations) procedure included in GAUSSIAN82 is on a much larger scale. As the name implies, it includes all states in which one or two electrons are promoted from the ground-state configuration. The range of states used can, however, be restricted by using the WINDOW option to remove some molecular orbitals from the CI. More sophisticated CI procedures use more than one reference configuration in order to obtain better results.

The mixing of electronic states can also be achieved by considering more than one configuration at the SCF level. Such *multiconfiguration self consis-*

tent field (MCSCF) methods are widespread in high-quality calculations, but no easily available program package includes them at the moment. A similar procedure based on the valence-bond principle of mixing resonance structures (which can be regarded as electronic configurations) is the *generalized valence bond* (GVB) method, which was used extensively with the PRDDO semiempirical program, among others.

For calculations in which the energy correction for electron correlation, rather than determination of the extent to which electronic configurations are being mixed, is of prime importance, a number of other calculational methods are available. The best known are based on Rayleigh–Schrödinger many-body perturbation theory (RSMBPT), and include the Møller–Plesset techniques used by GAUSSIAN82 and a variety of perturbational treatments available in MNDO/C. These calculations are dependent on the quality of the basis-set, as they require a good description of the virtual orbitals.

Other techniques for the calculation of electron correlation include the correlated electron pairs (CEPA) and self-consistent electron pairs (SCEP) approximations, the random-phase approximation (RPA), and density functional methods, which are based on electron gas theory. Random-phase approximation and SCEP programs are available from QCPE. The reasoning behind these methods is beyond the scope of this book, but Szabo and Ostlund[28] treat RSMBPT in some detail and Salem[25] gives a good qualitative description of the effects of electron correlation.

Correlation is particularly important in all cases where two or more electronic states approach each other in energy. These include forbidden reactions, homolytic bond fissions, Jahn–Teller molecules, and many carbenes, such as the methylene example given in Section 5.3. Most radical reactions involve several electronic states, because radical excited states lie lower in energy than do those for closed-shell molecules. Any reactions in which the transition state or intermediates may have appreciable biradical character require consideration of electron correlation. Such reactions may give bogus zwitterionic states at the single-determinant level, similarly to the dissociation of H_2 to $H^+ + H^-$ in RHF calculations. It is perhaps appropriate to end by noting that the expressions "biradical" and "zwitterion" lose their absolute character as soon as electronic configurations are allowed to mix, and become simply the two extremes of a range of possible states.

REFERENCES

1. W. J. Hehre, R. F. Stewart, and J. A. Pople, *J. Chem. Phys.*, 1969, **51**, 2657; W. J. Hehre, R. Ditchfield, R. F. Stewart, and J. A. Pople, *ibid.*, 1970, **52**, 2769.
2. J. A. Pople, *J. Am. Chem. Soc.*, 1975, **97**, 5306.
3. W. J. Hehre, *J. Am. Chem. Soc.*, 1975, **97**, 5308.
4. M. J. S. Dewar, *J. Am. Chem. Soc.*, 1975, **97**, 6591.

5. W. A. Lathan, L. A. Curtiss, W. J. Hehre, J. B. Lisle, and J. A. Pople, *Prog. Phys. Org. Chem.*, 1974, **11**, 1.

6. W. J. Hehre and W. A. Lathan, *J. Chem. Phys.*, 1972, **56**, 5255.

7. R. Ditchfield, W. J. Hehre, and J. A. Pople, *J. Chem. Phys.*, 1971, **54**, 724; W. J. Hehre and J. A. Pople, *ibid.*, 1972, **56**, 4233; J. D. Dill and J. A. Pople, *ibid.*, 1975, **62**, 2921.

8. J. D. Dill and J. A. Pople, *J. Chem. Phys.*, 1975, **62**, 2921.

9. T. Clark, J. Chandrasekhar, G. W. Spitznagel, and P. v. R. Schleyer, *J. Comput. Chem.*, 1983, **4**, 294.

10. H. Tatewaki and S. Huzinaga, *J. Comput. Chem*, 1980, **1**, 205; *J. Chem. Phys.*, 1979, **71**, 4339.

11. H. B. Schlegel, *J. Chem. Phys.*, 1982, **77**, 3676.

12. B. A. Murtagh and R. W. H. Sargent, *Comput. J.*, 1970, **13**, 185.

13. See, for instance D. Poppinger, *Chem. Phys. Lett.*, 1975, **34**, 332.

14. C. Møller and M. S. Plesset, *Phys. Rev.*, 1934, **46**, 618.

15. J. A. Pople, J. S. Binkley, and R. Seeger, *Int. J. Quant. Chem.*, 1976, **S10**, 1; J. A. Pople, R. Seeger, and R. Krishnan, *ibid.*, 1977, **S11**, 165; R. Krishnan and J. A. Pople, *ibid.*, 1978, **14**, 91.

16. D. J. DeFrees, B. A. Levi, S. K. Pollack, W. J. Hehre, J. S. Binkley, and J. A. Pople, *J. Am. Chem. Soc.*, 1979, **101**, 4085.

17. W. G. Richards, T. E. H. Walker, and R. K. Hinkley, "A Bibliography of *Ab Initio* Molecular Wave Functions," Oxford University Press, Oxford, 1971; W. G. Richards, T. E. H. Walker, L. Farnell, and P. R. Scott, *ibid.*, supplement for 1970–3, 1974; W. G. Richards, P. R. Scott, E. A. Colbourn, and A. F. Marchington, *ibid.*, supplement for 1974–7, 1978; W. G. Richards, P. R. Scott, V. Sackwild, and S. A. Robins, *ibid.*, supplement for 1978–80, 1981.

18. K. Ohno and K. Morokuma, "Quantum Chemistry Literature Data Base," Elsevier, Amsterdam, 1982; *THEOCHEM*, 1982, **8**; 1983, **15**, special issues.

19. W. J. Hehre, L. Radom, P. v. R. Schleyer, and J. A. Pople, "*Ab Initio* Molecular Orbital Theory," Wiley-Interscience, New York, 1985.

20. R. A. Whiteside, M. J. Frisch, J. S. Binkley, D. J. DeFrees, H. B. Schlegel, K. Raghavachari, and J. A. Pople, "Carnegie–Mellon Quantum Chemistry Archive," 2nd Ed. Carnegie–Mellon University, Pittsburgh, 1981.

21. W. J. Hehre, W. A. Lathan, R. Ditchfield, M. D. Newton, and J. A. Pople, QCPE Program No. 236.

22. J. S. Binkley, R. A. Whiteside, P. C. Hariharan, R. Seeger, J. A. Pople, W. J. Hehre, and M. D. Newton, QCPE Program No. 368.

23. J. S. Binkley, R. A. Whiteside, K. Raghavachari, R. Seeger, D. J. DeFrees, H. B. Schlegel, M. J. Frisch, J. A. Pople, and L. R. Kahn, "GAUSSIAN82 Release A," Carnegie–Mellon University, Pittsburgh, 1982.

24. J. A. Pople, *J. Am. Chem. Soc.*, 1980, **102**, 4615.

25. L. Salem, "Electrons in Chemical Reactions," Wiley-Interscience, New York, 1982.

26. S. F. Boys and F. Bernardi, *Mol. Phys.*, 1970, **19**, 553.

27. H. Huber, *Theor. Chim. Acta*, 1980, **55**, 117.

28. A. Szabo and N. L. Ostlund, "Modern Quantum Chemistry," Macmillan, New York, 1982.

APPENDIX A

MOPAC Z-MATRICES FOR CHAPTER 3

The following Z-matrices correspond to those given in Chapter 3, but are given here in MOPAC format with the necessary symmetry cards to give the optimizations outlined in Section 3.4. The examples are numbered as in chapter 3. The definitions of the first three atoms have been filled out with zeroes, although this is only necessary for the third atom.

METHYLENE (FIG. 3.3.2)

XX	0.0	0	0.0	0	0.0	0	0	0	0
C	1.0	0	0.0	0	0.0	0	1	0	0
H	1.09	1	122.0	1	0.0	0	2	1	0
H	1.09	0	122.0	0	180.0	0	2	1	3
0									
3,1,4									
3,2,4									

ETHYLENE (FIG. 3.3.3)

C	0.0	0	0.0	0	0.0	0	0	0	0
C	1.34	1	0.0	0	0.0	0	1	0	0
H	1.09	1	122.0	1	0.0	0	2	1	0

H	1.09	0	122.0	0	180.0	0	2	1	3
H	1.09	0	122.0	0	0.0	0	1	2	3
H	1.09	0	122.0	0	180.0	0	1	2	3

0
3,1,4,5,6,
3,2,4,5,6,

ETHANE (FIG. 3.3.4)

C	0.0	0	0.0	0	0.0	0	0	0	0
C	1.54	1	0.0	0	0.0	0	1	0	0
H	1.09	1	110.0	0	0.0	0	2	1	0
H	1.09	0	110.0	0	120.0	0	2	1	3
H	1.09	0	110.0	0	240.0	0	2	1	3
H	1.09	0	110.0	0	60.0	0	1	2	3
H	1.09	0	110.0	0	180.0	0	1	2	3
H	1.09	0	110.0	0	300.0	0	1	2	3

0
3,1,4,5,6,7,8,
3,2,4,5,6,7,8,

PROPYNE (FIG. 3.3.5)

C	0.0	0	0.0	0	0.0	0	0	0	0
C	1.46	1	0.0	0	0.0	0	0	0	0
H	1.09	1	110.0	0	0.0	0	2	1	0
H	1.09	0	110.0	0	120.0	0	2	1	3
H	1.09	0	110.0	0	240.0	0	2	1	3
XX	1.0	0	90.0	0	0.0	0	1	2	3
C	1.2	1	90.0	0	180.0	0	1	6	2
XX	1.0	0	90.0	0	0.0	0	7	1	6
H	1.08	1	90.0	0	180.0	0	7	8	1

0
3,1,4,5,
3,2,4,5,

METHANOL (FIG. 3.3.6)

C	0.0	0	0.0	0	0.0	0	0	0	0
O	1.44	1	0.0	0	0.0	0	1	0	0
H	0.96	1	110.0	1	0.0	0	2	1	0
H	1.09	1	110.0	1	180.0	0	1	2	3
XX	1.0	0	125.0	1	0.0	0	1	2	3
H	1.09	1	55.0	1	90.0	0	1	5	2
H	1.09	0	55.0	1	−90.0	0	1	5	2

0
6,1,7,
6,2,7,

CYCLOPROPANE (FIG. 3.3.7)

XX	0.0	0	0.0	0	0.0	0	0	0	0
XX	1.0	0	0.0	0	0.0	0	1	0	0
C	0.9	1	90.0	0	0.0	0	2	1	0
C	0.9	0	90.0	0	120.0	0	2	1	3
C	0.9	0	90.0	0	240.0	0	2	1	3
H	1.08	1	122.0	1	0.0	0	3	2	1
H	1.08	0	122.0	0	180.0	0	3	2	1
H	1.08	0	122.0	0	0.0	0	4	2	1
H	1.08	0	122.0	0	180.0	0	4	2	1
H	1.08	0	122.0	0	0.0	0	5	2	1
H	1.08	0	122.0	0	180.0	0	5	2	1

0
3,1,4,5,
6,1,7,8,9,10,11,
6,2,7,8,9,10,11,

TWIST CYCLOPENTANE (FIG. 3.3.9)

XX	0.0	0	0.0	0	0.0	0	0	0	0
C	0.91	1	0.0	0	0.0	0	1	0	0
H	1.08	1	125.0	1	0.0	0	2	1	0

H	1.08	0	125.0	0	180.0	0	2	1	3
C	1.25	1	90.0	0	90.0	1	1	2	3
C	1.25	0	90.0	0	90.0	0	1	2	4
XX	1.0	0	90.0	0	0.0	0	1	2	3
XX	1.41	1	90.0	0	180.0	0	1	7	2
C	0.77	1	90.0	0	29.6	1	8	1	6
C	0.77	0	90.0	0	29.6	0	8	1	5
H	1.08	1	110.0	1	120.0	1	5	2	6
H	1.08	0	110.0	0	120.0	0	6	2	5
H	1.08	1	110.0	1	−120.0	1	5	2	6
H	1.08	0	110.0	0	−120.0	0	6	2	5
H	1.08	1	110.0	1	120.0	1	9	5	10
H	1.08	0	110.0	0	120.0	0	10	6	9
H	1.08	1	110.0	1	−120.0	1	9	5	10
H	1.08	0	110.0	0	−120.0	0	10	6	9

0
3,1,4,
3,2,4,
5,1,6,
9,1,10,
9,3,10,
11,1,12,
11,2,12,
11,3,12,
13,1,14,
13,2,14,
13,3,14,
15,1,16,
15,2,16,
15,3,16,
17,1,18,
17,2,18,
17,3,18,

CHAIR CYCLOHEXANE (FIG. 3.3.10)

XX	0.0	0	0.0	0	0.0	0	0	0	0
XX	0.51	1	0.0	0	0.0	0	1	0	0

C	1.45	1	90.0	0	0.0	0	2	1	0
C	1.45	0	90.0	0	120.0	0	2	1	3
C	1.45	0	90.0	0	240.0	0	2	1	3
C	1.45	0	90.0	0	180.0	0	1	2	3
C	1.45	0	90.0	0	60.0	0	1	2	3
C	1.45	0	90.0	0	−60.0	0	1	2	3
H	1.08	1	95.0	1	180.0	0	3	2	1
H	1.08	0	95.0	0	180.0	0	4	2	1
H	1.08	0	95.0	0	180.0	0	5	2	1
H	1.08	0	95.0	0	180.0	0	6	2	1
H	1.08	0	95.0	0	180.0	0	7	2	1
H	1.08	0	95.0	0	180.0	0	8	2	1
H	1.08	1	155.0	1	0.0	0	3	2	1
H	1.08	0	155.0	0	0.0	0	4	2	1
H	1.08	0	155.0	0	0.0	0	5	2	1
H	1.08	0	155.0	0	0.0	0	6	2	1
H	1.08	0	155.0	0	0.0	0	7	2	1
H	1.08	0	155.0	0	0.0	0	8	2	1

0
3,1,4,5,6,7,8,
9,1,10,11,12,13,14,
9,2,10,11,12,13,14,
15,1,16,17,18,19,20,
15,2,16,17,18,19,20,

APPENDIX **B**

OTHER USEFUL PROGRAMS

Several other programs have been mentioned or used in this book. Three of the most useful are described here.

EUCLID (QCPE 452)

EUCLID is an interactive program that can calculate bond lengths, angles, and dihedral angles from Cartesian coordinates. It can also use combinations of bond lengths, angles, and dihedral angles to calculate Cartesian coordinates. The most useful feature of EUCLID is that it can use any combination of three such internal parameters to determine the position of an atom. The program is easy to use, but the documentation is not very clear. QCPE has recently announced a new program called MMHELP (QCPE 476), which is designed to create input files for MM2.

NAMOD (QCPE 370)

The ball-and-stick plots given in Chapter 2 were produced with a program based on NAMOD. The program is supplied as a set of subroutines to be used with a main program that can be written to suit the user's own requirements. NAMOD is very economical in comparison with most other molecule-plotting programs, which are primarily intended to be used for displaying the results of X-ray analyses.

PSI77 (QCPE 340)

The molecular orbital plots shown in Fig. 4.4.4 were produced with PSI77. The program can handle either extended Hückel or STO-3G orbitals, and can be used either for pseudo-three-dimensional plots or electron-density contour diagrams. Newer programs, not yet generally available, are considerably faster than PSI77.

APPENDIX C

BOND LENGTH
TABLES

The following bond lengths, which are partly based on 6-31G*structures and partly on covalent radii, should be good starting points for geometry optimizations. The values given are for single bonds, but distances for donor–acceptor bonds (such as BN) are those for the partial multiple bonds.

	H	Li	Be	B	C	N	O	F
H	0.75	1.64	1.33	1.19	1.09	1.00	0.96	0.91
Li		2.81	2.47	2.05	2.01	1.75	1.59	1.56
Be			2.12	1.90	1.70	1.50	1.40	1.37
B				1.68	1.57	1.39	1.34	1.31
C					1.54	1.45	1.40	1.36
N						1.41	1.40	1.39
O							1.40	1.38
F								1.34

	H	Na	Mg	Al	Si	P	S	Cl
H	0.75	1.91	1.71	1.58	1.48	1.41	1.33	1.27
Na		3.13	3.00	2.84	2.72	2.64	2.56	2.36
Mg			2.90	2.72	2.60	2.52	2.44	2.23
Al				2.60	2.48	2.40	2.32	2.12
Si					2.35	2.28	2.21	2.07
P						2.21	2.14	2.10
S							2.06	2.03
Cl								1.99

	Li	Be	B	C	N	O	F
Na	2.88	2.62	2.35	2.31	2.29	2.27	2.25
Mg	2.76	2.50	2.23	2.19	2.17	2.15	2.13
Al	2.64	2.38	2.11	2.07	2.05	2.03	2.01
Si	2.52	2.26	1.99	1.95	1.93	1.91	1.89
P	2.44	2.17	1.91	1.87	1.85	1.83	1.81
S	2.36	2.09	1.83	1.79	1.77	1.75	1.73
Cl	2.33	2.06	1.80	1.76	1.74	1.72	1.70

INDEX